De Gruyter Graduate
Jelinek • Nanoparticles

Also of Interest

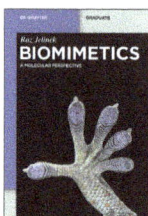

Biomimetics: A Molecular Perspective
Raz Jelinek, 2013
ISBN 978-3-11-028117-0, e-ISBN 978-3-11-028119-4

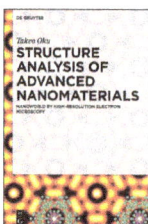

Structure Analysis of Advanced Nanomaterials: Nanoworld by High Resolution Electron Microscopy
Takeo Oku, 2014
ISBN 978-3-11-030472-5, e-ISBN 978-3-11-030501-2

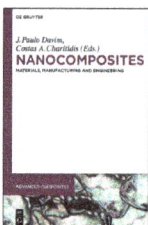

Nanocomposites: Materials, Manufacturing and Engineering
J. Paulo Davim, Constantinos A. Charitidis (Eds.) 2013
ISBN 978-3-11-026644-3, e-ISBN 978-3-11-026742-6

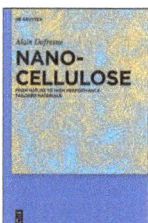

Nanocellulose: From Nature to High Performance Tailored Materials
Alain Dufresne, 2012
ISBN 978-3-11-025456-3, e-ISBN 978-3-11-025460-0

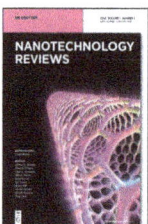

Nanotechnology Reviews
Challa Kumar (Editor-in-Chief)
ISSN 2191-9089, e-ISSN 2191-9097

www.degruyter.com

Raz Jelinek

Nanoparticles

—

DE GRUYTER

Author
Prof. Raz Jelinek
Ben Gurion University
Department of Chemistry
1 Ben Gurion Ave
Beer Sheva 84105
Israel
razj@bgu.ac.il

ISBN 978-3-11-033002-1
e-ISBN (PDF) 978-3-11-033003-8
e-ISBN (EPUB) 978-3-11-038991-3

Library of Congress Cataloging-in-Publication Data
A CIP catalog record for this book has been applied for at the Library of Congress.

Bibliographic information published by the Deutsche Nationalbibliothek
The Deutsche Nationalbibliothek lists this publication in the Deutsche Nationalbibliografie;
detailed bibliographic data are available on the Internet at http://www.dnb.de.

© 2015 Walter de Gruyter GmbH, Berlin/Boston
Typesetting: PTP-Berlin, Protago T$_E$X-Production GmbH
Printing and binding: CPI books GmbH, Leck
Cover image: The figure is the total electron charge density isosurface plot of a Si quantum dot
(Kindly provided by Lin-Wang Wang, CMSN, Berkley Lab, USA.)
♾ Printed on acid-free paper
Printed in Germany

MIX
Papier aus verantwor-
tungsvollen Quellen
FSC® C083411

www.degruyter.com

Preface

Like many researchers who "came of age" (scientifically speaking) during the early days of the nanotechnology "revolution" in the late 1980s and 1990s, I have been fascinated by the new horizons this field entailed. Perhaps the most dramatic manifestation of nanotechnology has been the proliferation of "nanoparticles" – atomic aggregates displaying a variety of compositions and shapes. This "new frontier" of scientific imagination, creativity, and synthetic acumen has yielded a remarkable variety of nanostructures (spheres, cubes, rods, stars, and many others), and physical properties and phenomena associated with the structural features of these nanometer-scale particles. Admittedly, an incentive to write this book has been my desire to acquire a comprehensive understanding of this multidisciplinary field – the types of nanoparticles, their compositions and how the relationship between the atomic constituents affects their properties, as well as potential practical applications of nanoparticles. Indeed, the enormous scope of nanoparticle science and technology became even more apparent to me on researching material for this book and during the writing process. Moreover, the explosive growth of this field, which continues unabated to this day, has meant that I came across many new discoveries and interesting reports on nanoparticles as the book progressed; deciding which systems to include and which to leave out has been especially difficult. My hope is that the final product will endow the reader with a sound knowledge base and new perspectives on nanoparticles, and encourage further exploration of this exciting field.

This book has benefited from significant help and support of several people. First and foremost I want to thank my family for their patience with me spending many hours working on the book through the nights and over weekends … I am also grateful to my graduate student Alex Trachtenberg for his help and efforts putting together many of the figures in the book. Special thanks to the Y Café in the Nachlaot neighborhood, Jerusalem, and the Austrian Hospice Café in the Old City of Jerusalem for the hospitality, many hours of inspiration, great coffee and beer that contributed to the realization of this book.

Contents

1 Introduction

For a young scientific discipline which burst on to the stage less than 30 years ago, *nanotechnology* has had a tremendous impact on both fundamental research and development of technology. *Nanoparticles* (NPs) have been among the most visible facets of nanotechnology research. The essence of this field is the realization that the properties of matter are often significantly altered as one ventures into the nanoscale. Indeed, the unique properties of NPs are due in large part to their *nanometer* (10^{-9} meter) dimensions. The interest and activity in this field have led to dramatic contributions in diverse fields of science and technology – chemistry, physics, biology, electronics, and others. In fact, NPs can be considered both products and promoters of the "nanotechnology revolution", as attested by the huge body of work on the subject.

Although the precise definition of NPs may be somewhat fluid, this book focuses on atomic and molecular aggregates which are generally smaller than *tens of nanometers*. While the term "nanoparticle" often evokes an image of a small *spherical* particle, this book is not limited to spherical NP configurations. The discussion rather spans the diverse structural universe of nanoparticles, including (*nano)wires, rods, stars, cubes* and various other morphologies enabled by nature, our imagination, and synthetic acumen. This book is designed to be an introductory textbook to the rapidly evolving field of nanoparticle science and technology. As such, the book aims to present different facets of nanoparticle research to readers who are not necessarily active or experts in this discipline. A scientific knowledge base, however, is quite essential for grasping many of the subjects discussed. Overall, this book aims to endow the reader with a methodical summary of the field – how concepts, synthesis schemes, and applications of NPs have been developed and implemented.

Discussion of the broad and diverse array of systems and experimental strategies is carried out primarily through presentation and analysis of studies published in the scientific literature. Starting from a historical perspective, the book has several underlying themes, including *synthetic routes* for preparing NPs; different NP *structures* and the way the structural and morphological features of the particles affect *functionalities;* novel *constructs* and *devices* utilizing NPs; and the use of NPs *beyond the nanoscale – as* building blocks in higher-order materials. Specific emphasis is placed upon the interface and relationships between NPs and *biological systems*, as important developments of biomedical applications underscore both the potential and risks associated with increased applications of NPs as therapeutic and diagnostic tools. While unique physical phenomena are intrinsic to the properties and applications of NPs, detailed analyses of the *physics* aspects of NPs are beyond the scope of this book.

Naturally it is difficult to cover all pertinent topics and aspects in a single textbook. Accordingly, this textbook will hopefully serve as a "starting point" for nanoparticle science and technologies; the reader is accordingly referred to many excellent comprehensive reviews and scientific publications, outlined in the "Further reading" section at the end of the book. Importantly, the focus here is on nanoparticles and not "nanoscale materials" as a whole. Accordingly, discussion in the text is focused mostly on "stand-alone" *synthetic* NPs self-assembled in *solutions*, rather than nano-structures produced via techniques such as lithography which can technically be perceived as parts of larger entities (e.g. surface). This book also excludes the huge field of "carbon nanomaterials"; carbon nanoparticle allotropes, such as fullerenes and carbon nanotubes, exhibit distinct properties related to the organization and binding of carbon atoms and deserve an independent textbook.

The chapters in the book are devoted to different nanoparticle *compositions* and *types*: *semiconductor NPs* (Chapter 2), of which "quantum dots" occupy a prominent position; *metal NPs* (Chapter 3), including the highly diverse applications of *gold, silver*, and *transition metal NPs; metal-oxide NPs* (Chapter 4) employed in varied technologies such as solar energy harvesting and biomedical imaging; *biological and polymer NPs* (Chapter 5), in which organic building blocks have been used to construct nanoparticles; *hybrid NPs* (Chapter 6), comprising more than one component and displaying intriguing configurations – from *core-shell NPs*, all the way to more exotic species, such as "nanostars", "nanodumbbells", nanocages, and others. A specific chapter is devoted to the effects of nanoparticles on biological entities – cells, proteins, and DNA (Chapter 7); and the last chapter focuses on the use of NPs as building blocks for larger and more complex materials (Chapter 8). A certain overlap naturally exists between topics. Thus, for example, *NP assemblies* are discussed both in a dedicated chapter (Chapter 8), as well as in individual chapters (such as solar cells comprising of *titanium oxide NPs*). Similarly, the interface between NPs and the biological world is a vast and recurring theme in several chapters; the significance of this topic is also reflected in a thorough discussion in a specific chapter (Chapter 7).

Nanoparticles have inspired the scientific and technological communities for several decades now, and the sheer activity in this field promises to continue generating new discoveries, revolutionary products, and novel physical phenomena. The remarkable progress in our understanding of NPs and the ability to control and modulate their properties will undoubtedly further expand the frontiers of chemistry, physics, material sciences, and biomedicine.

1.1 Historical context and early work

"Pornography is a matter of geography" as the saying goes; this aphorism might seem relevant to many scientific disciplines in which long-known phenomena are explained using new physical and chemical tools and new terminology. This has also been partly the case with nanoparticles. Indeed, NPs have been produced since mankind learned to manipulate materials, although the actual term (and hype ...) of "nanoparticles" was coined much more recently. One of the earliest and most famous examples of the use of NPs for everyday objects was the "Lycurgus Cup" (Fig. 1.1). Manufactured by a

Fig. 1.1: The Lycurgus Cup. Image provided by the British Museum.

Roman craftsman almost 2000 years ago from special glass speckled with "gold and silver dust", this extraordinary object changes its color depending on the position of the incident light. When illuminated from the outside, the cup appears green, however when the light source is placed inside the cup it shines red. This rather unusual property is directly related to the interplay between reflection and scattering of the light beam from metal nanoparticles embedded within the glass. The Romans did not of course know they were working with NPs, and in fact the unique mechanism responsible for the optical properties of the Lycurgus Cup was deciphered not that long ago. However, the Lycurgus Cup illustrates a notable facet of NP science and technology – that varied chemical and physical phenomena associated with NPs have, in fact, been known for quite a long time. Indeed, part of early NP research was aimed at providing a solid physical/chemical understanding of known processes and materials.

In a historical context, NP research emanated in large part from a convergence of two distinct scientific disciplines – the study of *atomic clusters*, and *colloids* research (Fig. 1.2). Clusters are loosely defined as aggregates of relatively small numbers of atoms, held together by both noncovalent and covalent bonds (Fig. 1.3). Importantly, it has been determined that clusters possess different physical properties, both compared to individual molecules, as well as in relation to the bulk material. In particular, scientists concluded that the unique characteristics of atomic clusters can be largely traced to the significantly high ratio between atoms at the surface of a cluster and its inner core. Indeed, this (high) ratio is a major determinant distinguishing clusters (and nanoparticles) from their bulk counterparts.

Colloid research is the other major preceding field which led to the emergence of nanoparticle science. Colloidal systems are defined as molecular aggregates which are usually dispersed within a more abundant substance (such as a solvent. Milk is a prime example of an aqueous colloidal suspension). Indeed, colloid dispersions are

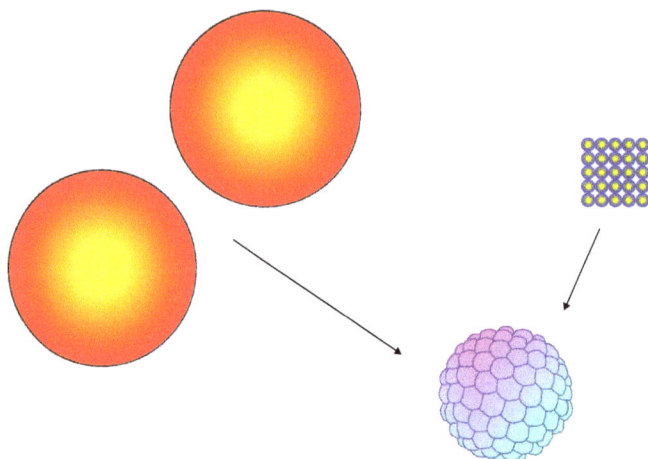

Fig. 1.2: Nanoparticle research emerged from the convergence of colloids research (*top left*) and the study of atomic clusters (*top right*).

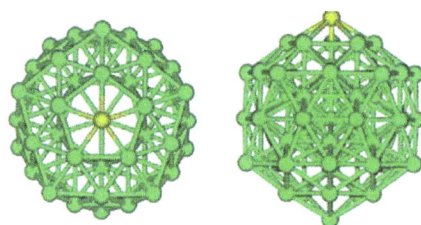

Fig. 1.3: Scheme of a metal cluster. Top view (left) and side view (right) of a Cu_{55}^- cluster. Reprinted with permission from Rapps T. et al., *Angew. Chem.* **52** (2013), 6102–6105, © John Wiley and Sons.

among the bedrocks of metallurgy and materials science in general. While colloids have been prepared routinely for millennia, the advent of science and technology has brought about the realization that the properties of colloids, particularly particle size, have intimate relationships with the overall functions and macroscopic characteristics of colloid assemblies. This link between the size of colloids and their overall material properties is one of the most important aspects of NP research, and is a fundamental phenomenon manifested in different NP systems presented throughout this book. Milk, in fact, is a case in point highlighting the significance of nanoparticles in determining material properties. The white, opaque appearance of milk is due to its composition as an *emulsion* of small colloids comprising of fats, proteins, and calcium. Moreover, it has been found in recent years that some of these colloidal species are tiny protein nanoparticles which are easily digested and contribute to the important nutritional properties of milk (Fig. 1.4).

The microscopy image in Figure 1.4 highlights another important aspect of the explosive growth of NP research – visualization. Indeed – "seeing is believing", and the advent of microscopy techniques, particularly variations of electron microscopy

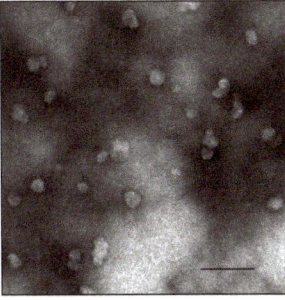

Fig. 1.4: Protein nanoparticles in milk. Transmission electron microscopy image of nanoparticles comprising casein, a major protein in milk. Scale bar corresponds to 100 nm.

and scanning probe microscopy have been pivotal to the expansion of NP science and technology. Indeed, as discussed below in detail, obtaining microscopic insights into the fine structural features of NPs, their crystallinity and atomic organization, have been among the main aspects shaping the field to this day. The development and refinement of NP synthesis schemes have been another powerful driving force. While progress in metallurgy and gold chemistry has occurred over hundreds (or thousands) of years providing tools for manipulating metallic materials, the much more recent and dramatic proliferation of NP studies and technological applications is linked to the rapidly evolving synthetic universe involving inorganic materials, semiconducting assemblies, rare earth metals, biological molecules, and others.

While it is hard to "pinpoint" the exact birth of nanoparticles as a distinct scientific discipline, the onset of research on *nanocrystals (NCs)*, particularly *semiconducting NCs*, in the 1980s, is considered (partly in retrospect ...) a prominent marker. The great interest in NCs arose from the observation that they exhibited unique physical properties remarkably different from macroscopic crystalline aggregates (e.g. the "bulk" state of the material). Indeed, NC *size* has been shown to be a fundamental parameter affecting a variety of physical features. *Quantum confinement* (Fig. 1.5), in particular, has been one of the most central experimental observations contributing to burgeoning NC research.

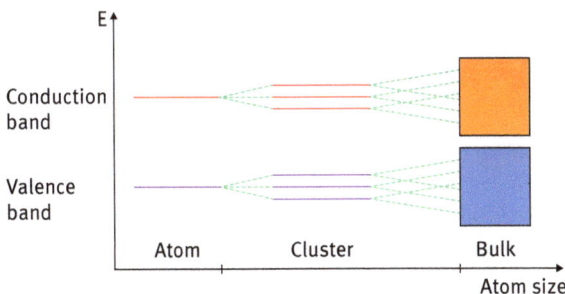

Fig. 1.5: Quantum confinement in semiconductor nanoclusters. The energy level diagram depicts the transformation of energy band-gap between atom-cluster-bulk configurations. Atomic clusters exhibit discreet energy levels.

Quantum confinement encompasses spectroscopic phenomena – essentially light absorption and emission – which directly emanate from the size regime of an NC. Quantum theory heralded the "golden age" of modern physics by predicting the existence of distinct energy levels for atoms and molecules, accurately explaining phenomena such as the discreet light absorption/emission spectral lines observed for gases. It was soon realized that while individual atoms or molecules exhibited energy level separation in larger aggregates (and in bulk materials), the levels essentially "coalesced", forming continuous energy bands. The remarkable phenomenon which came to light in the 1980s was the "intermediate" spectral properties of NCs – between individual atoms on the one hand and bulk materials on the other. Specifically, while NCs are comprised of many atoms, the physical dimensions of the NC still give rise to "quantum effects" – the discreet energy levels predicted by quantum theory. Apart from the fact that NCs provide a fine example of "quantum mechanics in action", quantum confinement has spawned major research efforts towards controlling optical properties of materials through careful tuning of NC sizes, utilization of the various atomic compositions and material classes, and exploration of practical applications and commercial targets. More detailed discussion of the quantum confinement phenomenon and its scientific and technological implications for different NPs is provided in the individual chapters.

2 Semiconductor nanoparticles

Nanoparticles and nanocrystals made of semiconducting materials were among the first to burst onto the world nanotechnology stage in the 1980s. The impact of this family of NPs has been profound, both in terms of the new scientific phenomena as well as their applications in diverse fields including solar energy, biological imaging, photonics and electronics. By common definition, semiconducting materials exhibit electrical conductivity between *metals* and *insulators*. In physics parlance, the *bandgaps* of semiconductors, e.g. energy difference between the *valence energy band*, which in semiconductors is usually fully occupied by electrons, and the empty *conduction band* at a higher energy, generally fall between insulators and conductors (Fig. 2.1).

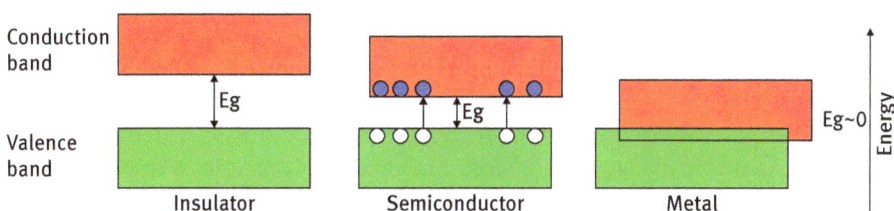

Fig. 2.1: Relative bandgaps (Eg) in insulators, semiconductors, and metals.

Significantly, the magnitude of the bandgap in a semiconductor makes excitation of electrons from the valence band to the conduction band through thermal energy, light irradiation, and other means possible, thereby creating an "electron-hole" pair (e.g. "exciton"). Exciton mobility and electron-hole recombination kinetics constitute the basis for the fundamental optical and electronic phenomena encountered in semiconductors. Another parameter pertaining to the performance and practical applications of semiconducting materials is whether the energy bandgap is *direct* or *indirect* (Fig. 2.2). Generally, indirect bandgap semiconductors exhibit slower and less efficient exciton formation and consequent light emission through electron-hole recombination, thus they exhibit more limited applicability in optoelectronic devices.

Semiconductors by themselves are often poor conductors; their conductivity can generally be enhanced through physical or chemical modifications, such as addition of foreign substances as "impurities" (e.g doping). The most common semiconductor is *silicon*. In the nanoparticle universe, the impact of silicon was smaller than that of *binary semiconductor* materials, such as "II-VI" compounds (i.e. an element from group II in the periodic table, such as cadmium, bonded to an element from group VI, such as selenium) or III-V materials, such as gallium arsenide.

Crystalline semiconductor nanoparticles (NPs) took center stage in the 1980s as striking examples of *quantum effects* – physical phenomena recorded when the di-

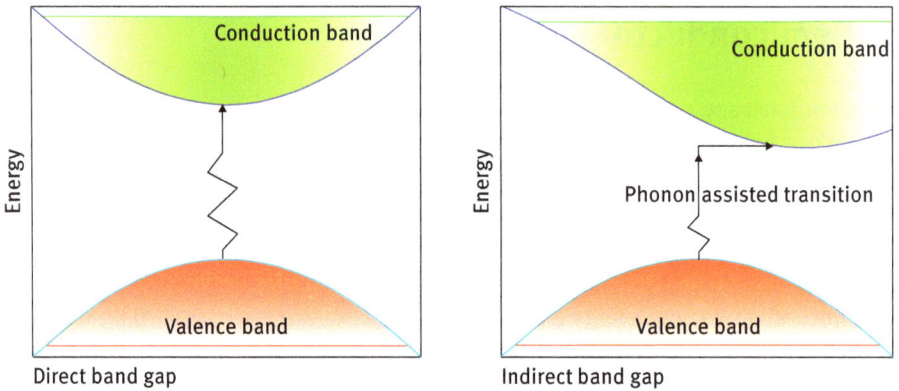

Fig. 2.2: Energy structure of (left) direct and (right) indirect band gap semiconductors.

mensions of the materials are significantly reduced, spanning a few nanometers. One of the most dramatic quantum effects in semiconducting NPs is *quantum confinement*, distinguishing the NP aggregates both from bulk materials as well as from individual molecules. As shown in Figure 2.3, quantum confinement generates discreet bandgap energy levels (and corresponding transitions) which constitute the basis for the remarkable optical and electronic properties associated with semiconducting NPs. Spherical semiconductor NPs, which were among the pioneering nanoparticle assem-

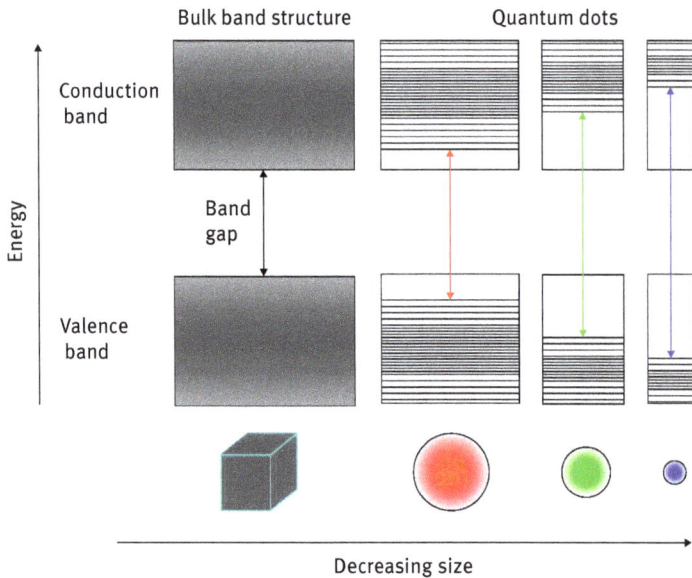

Fig. 2.3: Size-dependence of energy bandgaps in semiconductor nanoparticles (quantum dots).

blies displaying quantum confinement effects, were termed "quantum dots" (QDs). Since their inception, QDs (and their morphological relatives – *quantum wires, quantum rods*, and others, see discussion below) have experienced dramatic proliferation in varied scientific and commercial applications.

As outlined in Figure 2.3, in a bulk semiconductor material, both the valence band and the conduction band are broad, as they are comprised of a multitude of levels from all individual atoms in the material. However, when the diameter of the particle which confines the electron is at the order of (or lower than) the average distance between an electron in the valence band and conduction band (defined as the "exciton diameter" or "exciton Bohr radius"), a quantum confinement effect becomes predominant, leading to occurrence of discreet (quantized) energy levels in both the valence and conduction bands. As a consequence, the difference between the highest valence energy level and lowest conduction energy level increases compared to the bulk material, and a corresponding blue shift in the excitation energy of the electron occurs (and similarly a blue shift of the *emission energy* when an electron falls from the conduction band to the valence band).

A remarkable consequence of quantum confinement is the ability to tune the color (i.e. *luminescence wavelength*) of the quantum dot, simply by modifying the particle size (Fig. 2.4). Size-dependent QD color tuning has been manifested in numerous cases and has led to several commercial applications, mainly focused on biological imaging; see below for a more detailed discussion. Indeed, the explosive growth in QD technologies and research has been aided by the feasibility of tuning the energy levels of the particles making diverse optical and electronic applications possible.

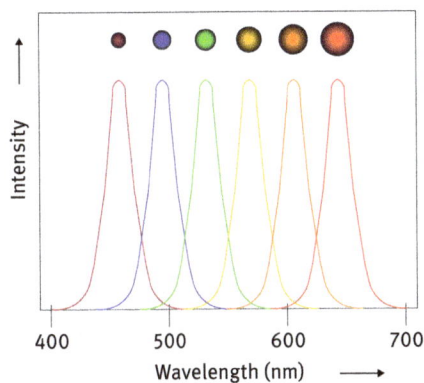

Fig. 2.4: Shifts of QD color (luminescence wavelength) are size dependent.

Diverse synthetic approaches have been developed for construction of semiconductor NPs. Synthetic techniques include "top down" schemes for QD fabrication for example "sculpting" a semiconductor film to generate NPs and surface manipulation techniques such as lithography. Many QD fabrication processes rely on *chemical vapor deposition (CVD)*. In this technique, NPs are formed on solid substrates through addition of reagents in the gas phase; chemical reactions are consequently induced on the surface, usually through high temperature conditions, leading to growth of crystalline NPs. In addition to top-down and CVD techniques, many "bottom up" methods have been introduced in which NPs spontaneously assemble in solution. Generic solution-based synthesis schemes of binary semiconductor NPs rely on mixing the NP precursors, cationic metal salts or organo-metallic compounds and their anionic counterparts (Fig. 2.5). Such reactions are usually carried out in *nonpolar solvents* (i.e. non-aqueous solutions); accordingly, stabilization of solution-synthesized QDs is usually accomplished via addition of hydrophobic agents which coat the resultant particles.

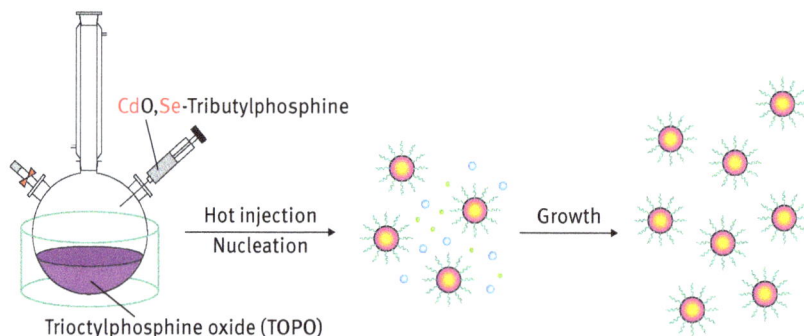

CdO,Se-Tributylphosphine

Hot injection
Nucleation

Growth

Trioctylphosphine oxide (TOPO)

Fig. 2.5: Synthesis of quantum dots in organic solution.

Many QD applications, particularly biologically-oriented, require aqueous solutions. QDs can be transferred from nonpolar solvents into aqueous solutions through substituting the coating layer of the QD with *hydrophilic* ligands. This strategy has spun numerous synthetic variations and examples are discussed below for specific QD families. An alternative route, however, is to perform the entire QD synthesis in water. This approach necessitates dissolution of the ions forming the QD in water, usually through the use of the appropriate salts, or by embedding the ions within shells of hydrophilic ligands. In a subsequent step the ions are reacted with each other, fusing into binary NPs.

Semiconductor nanowires (NWs) have attracted interest due to their significant technological potential. An important facet of semiconductor NWs is the feasibility of electron transport; furthermore, the semiconducting properties of NW building blocks open the way to diverse applications and optoelectronic/photonic devices, including transistors, light emitting diodes (LEDs), "nanolasers", and photonic wave-guides. Representative examples for some of these applications are presented below.

2.1 Metal chalcogenide nanoparticles

NPs comprising of transition metals and *sulfides* (S^{2-} ions), *selenides* (Se^{2-}), and *tellurides* (Te^{2-}) have been among the most widely studied semiconductor nanostructures due to their remarkable size-dependent optical and electronic properties. Transition metal chalcogenides such as CdS and CdSe, particularly in quantum dot configurations, have played central roles in pioneering studies in this field, displaying unique physical properties and applications that were later encountered in other types of semiconductor NPs. Development of facile synthetic methods for preparing QDs in controlled dimensions and compositions, and in high yield have greatly contributed to their broad applicability. Most synthetic schemes utilize high-temperature reactions of organometallic precursors (for example *dimethylcadmium*, $(CH_3)_2Cd$, providing the cationic Cd^{2+} ion) with their anionic counterparts (such as *alkylphosphine-selenide*, yielding the *selenide* ion) in nonpolar (organic) solvents. Numerous variations of this basic approach have been introduced, with an emphasis on developing pathways in milder reaction conditions, particularly in low temperatures.

QDs comprising two or more metal chalcogenide components have been introduced as well, often exhibiting superior optical and chemical properties as compared to the single component species. *"Core-shell" structures*, in particular, are prominent members of a sub-class of metal chalcogenide NPs containing different semiconductor compounds at the core and shell, respectively (Fig. 2.6). For example, core-shell NPs with shells comprised of wide bandgap group II-VI semiconductors such as CdS or ZnS were found to exhibit stronger light emission compared to single-component

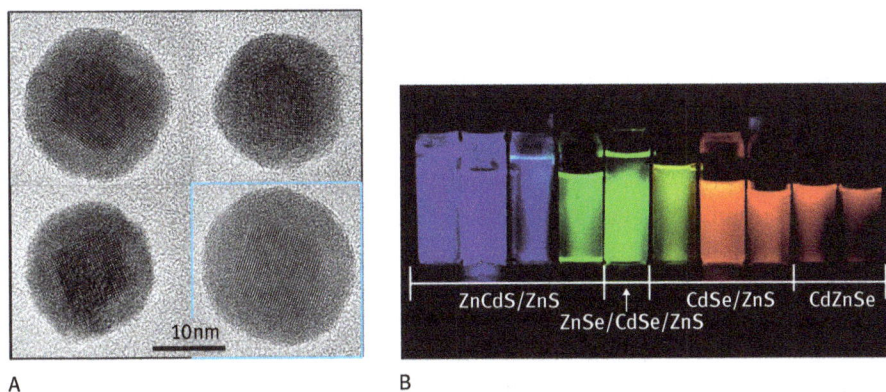

Fig. 2.6: Core-shell semiconductor nanoparticles. **A:** high-resolution transmission electron microscopy images of PbTe@PbS core-shell NPs. The darker crystalline core is clearly distinguished. Reprinted with permission from Ibanez et al., *ACS Nano* **2013** *7*, 2473–2586, ©2013 American Chemical Society. **B:** different colors (fluorescence emission upon excitation at 365 nm) observed for different core-shell NPs; core and shell compositions as indicated. Reprinted with permission from Anikeeva et al., *Nano Lett.* **2009** *9*, 2532–2536, © 2009 American Chemical Society.

QDs. This phenomenon has been ascribed to longer exciton lifetimes in core-shell NPs, presumably through "confinement" of the excitons at the shell and/or reduction of the concentration of crystal defects within the QD. Longer exciton lifetimes reflect lower energy losses occurring through electron-hole recombination processes (i.e. "exciton dissipation"), and therefore result in greater light emission/excitation efficiencies. Also noteworthy is the fact that mixing different semiconducting species in core-shell NPs provides another vehicle for color tuning (Fig. 2.6B).

2.1.1 Quantum dots in biology

QDs have enjoyed broad usage in biology and biomedicine, with commercial applications being introduced in these fields early on. The appeal of QDs largely stems from the tunable luminescence properties of the particles and their use as imaging agents. QDs offer distinct advantages for bio-imaging as compared to conventional fluorescent dyes (historically the main workhorses for microscopic analysis of cellular environments). Specifically, synthetic protocols for production of QDs in narrow, pre-selected size distributions have been developed and refined in the past couple of decades. Controlling the particle size and composition provides powerful tools for tuning the fluorescence emission, e.g. color, of the QDs – from the ultraviolet to infrared regions in the electromagnetic spectrum. Moreover, the intrinsic optical properties of QDs are attractive for microscopic imaging applications – brightness, narrow emission spectra, and low photo-bleaching (e.g. photochemical degradation of the fluorescent marker, widely encountered in many conventional fluorescence dyes).

Another notable QD feature which contributes to their biological applicability is the diverse synthetic routes which have been developed to couple QDs with (bio-) molecular recognition elements. Such QD functionalization capabilities are crucial since they permit covalent display of molecular units which could latch onto various targets: proteins, DNA, and larger biological entities such as cells, viruses, and bacteria. Figure 2.7 depicts a representative experimental scheme for bioconjugation of QDs. Specifically, the binding of hydrophilic molecules to the QD surface (necessary for the solubilization of the QDs in aqueous solutions) is carried out with *thiol* (e.g. sulfhydril, S-H) moieties. A molecular "linker" such as mercaptoacetic acid (MAA; Fig. 2.7) can be further coupled with the thiol residues and employed for covalent binding and the display of additional biological units such as proteins and other recognition elements.

While thiol-derivatization schemes have been among the more popular methods of rendering QDs water soluble (and thus amenable to biological applications), other reaction pathways utilizing amines, phosphonic residues and other reactive groups have been introduced for anchoring hydrophilic residues onto QDs. Single-step water-based synthetic routes have attracted interest as "greener" alternatives to the conventional synthesis methods carried out in organic solvents. A method for aqueous synthesis of mixed ZnCdTe QDs displaying bio-recognition units on their surface is de-

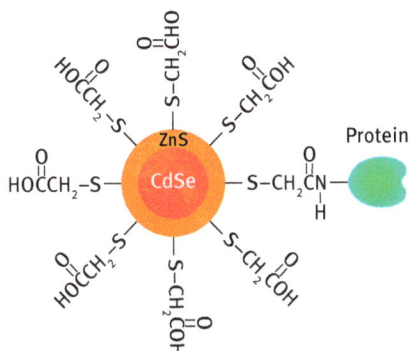

Fig. 2.7: Bioconjugation of a core-shell quantum dot with a protein recognition element.

picted in Figure 2.8. In this approach, developed by H. Xu and colleagues at the China Petroleum University, the short *argnine-glycine-aspartate* peptide (RGD in the single letter amino acid code), has been interspersed with the ionic precursors (Cd^{2+} and Zn^{2+}) prior to mixing with the tellurium counter-ion. The sequence of the co-added peptide further included a *cysteine* residue which supported chemical linkage to the QD surface. The reaction yielded CdTe QDs which were coated with the RGD ligands. Modulating the reaction parameters such as the acidity of the aqueous solution and the incubation time provided effective means for varying the QD sizes and thereby tuning the colors of the QDs as highlighted in Figure 2.8. The RGD caps were particularly useful for cellular imaging applications, as this short peptide is known to recognize and bind cell-surface proteins.

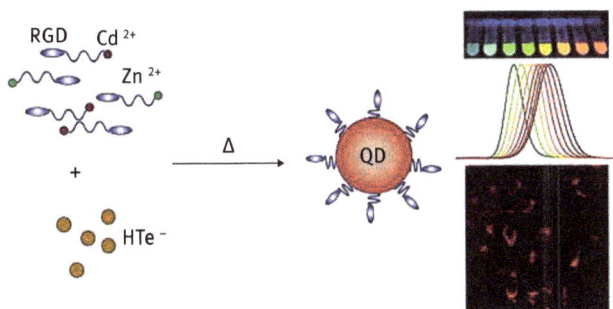

Fig. 2.8: Single-step synthesis scheme for peptide-displaying quantum dots. ZnCdTE QDs coated with the peptide recognition element arginine-glycine-aspartate (RGD) prepared by mixing the ionic precursors and a peptide sequence containing the RGD motif. The colors of the bio-conjugated RGD-QDs could be tuned through modifying the experimental conditions (affecting the QD sizes). The RGD-QDs could be used for cell imaging (fluorescence microscopy image on the bottom right) as the RGD motif binds cell surface proteins. Reprinted with permission from He et al., *ACS Appl. Mater. Interfaces* **4** (2012), 6362–6370, © 2012 American Chemical Society.

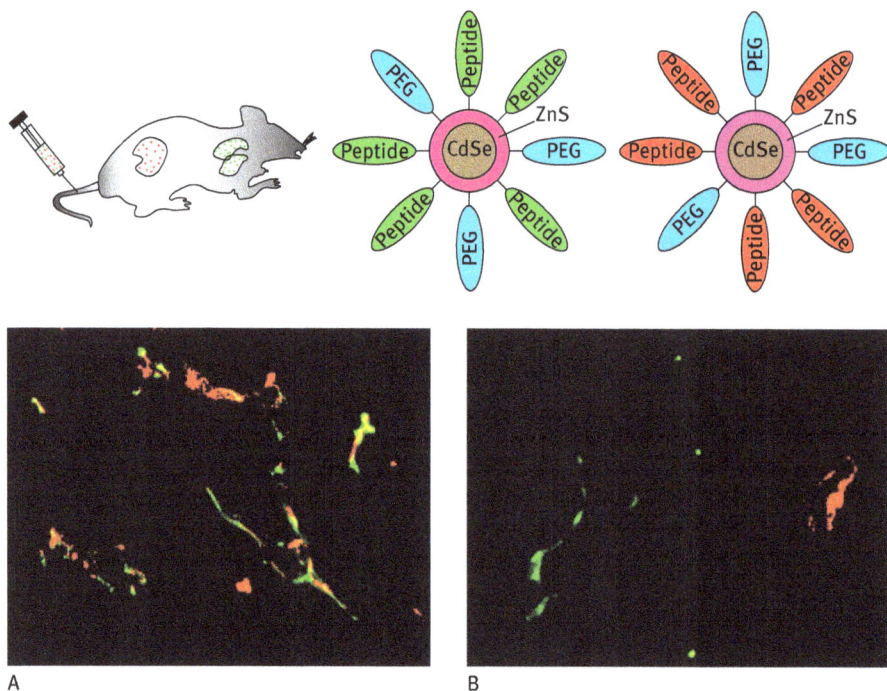

Fig. 2.9: In vivo imaging using peptide-derivatized quantum dots. **Top:** Schematic depiction of the experiment. Core-shell quantum dots displaying peptides targeted to different tumor regions were prepared and injected into the mice. **Bottom:** Fluorescence microscopy images demonstrate selectivity: **A:** QDs functionalized with a peptide recognizing blood vessel tumors (red) co-localized with a blood-vessel specific fluorescent dye (green); **B:** QDs derivatized with a peptide recognizing lymphatic vessel tumors (red) are localized differently compared to a blood vessel-specific fluorescent dye (green). Magnification is X400 (**A**) and X600 (**B**). Reprinted with permission from Akerman et al., *PNAS* **99** (2002), 12617–12621, © 2002 National Academy of Sciences, U.S.A.

Numerous experiments have demonstrated the use of QDs for cellular imaging, both *in vitro* (cells outside their physiological environment) and *in vivo* (within the live organism). Figure 2.9 presents an example of an in vivo imaging experiment using peptide-derivatized QDs, carried out by E. Ruoslahti and colleagues at the Burnham Institute, California. The researchers prepared two types of QDs which were coated, respectively, with peptides targeting different tumor cells, one peptide recognizing blood vessel tumors, the other honing onto lymphatic vessel tumors (the coating layer also comprised of polyethylene-glycol (PEG) moieties to improve water solubility and minimize nonspecific binding). Following injection of the bio-conjugated QDs into mice implanted with tumors, their vascular systems were visualized with fluorescence microscopy. To identify the precise localization of the bio-functionalized QDs, the mice were also injected with a fluorescent dye known to attach to blood vessels. The fluorescence images in Figure 2.9 provide a convincing visual depiction of tumor targeting by the QDs.

Specifically, while the QDs derivatized with the *blood vessel* tumor-recognizing pep-tide (appearing in red in Fig. 2.9A) were co-localized with the blood vessel-specific dye (shown in green in Fig. 2.9), the QDs displaying the peptide recognizing the *lymphatic vessel* tumor (red area in Fig. 2.9B) appeared in a different location to the blood vessel-specific dye (green domains in Fig. 2.9B).

QD-based cell imaging can be further employed for applications such as drug screening. Figure 2.10 depicts a cell imaging experiment utilizing CdSe@ZnS core-shell QDs functionalized with an antibody generated against hemagglutinin (HA) – a widely studied sugar-binding protein. R. Song and colleagues at the Institut Pas-teur Korea have shown that the anti-HA-QDs recognized cells programmed to express the HA protein on the surface of G-protein cell receptors (GPCRs), the prominent fam-ily of ubiquitous membrane-associated proteins. Crucially, addition of GPCR *agonists*, e.g. molecules mimicking the natural ligands which bind and activate the receptor, resulted in "opening" the membrane and subsequent internalization of the QDs into the cells. The fluorescence microscopy images in Figure 2.10 clearly show both the cell-surface binding of the QDs conjugated with the anti-HA antibody (middle fluorescence microscopy image in Fig. 2.10C), and cell insertion of the QDs following activation by an agonist (fluorescence microscopy image at the right in Fig. 2.10C). Importantly, this imaging concept can be employed for evaluation of potential drugs which act as ago-nists, and also for screening substances which block GPCR, e.g. *antagonists*, thereby preventing internalization of the QDs into the cells.

The biosynthesis of metal chalcogenide QDs (as well as other NP species) has attracted interest in recent years both as an alternative "green" route for NP produc-tion, and also as a means for exploring the interface between QDs and living systems. QDs have been synthesized within different microorganisms by manipulation of their metabolic pathways. Indeed, researchers have noted that genetically-engineered bac-teria, yeast, and other microorganisms could be harnessed to synthesize QDs simply through "feeding" them with salts of the QD building blocks. Biosynthetic routes for generation of QDs likely occur through production of various cell metabolites which favor nanoparticle formation and stabilization. Such molecular species might in-clude amphiphilic capping agents (i.e. fatty acids) and thiol-containing amino acids and peptides (cysteine or cysteine-containing peptides) which constitute stabilizing agents for both the ionic reagents as well as the final QD product.

Identifying specific cellular pathways which promote QD biosynthesis might provide a useful vehicle for tuning QD production through the powerful arsenal of genetic engineering. Figure 2.11, for example, vividly demonstrates this concept. In the experiment depicted, Z. X. Xie and colleagues at Wuhan University, China exam-ined two types of yeast cells – a wild-type species and a strain genetically engineered to over-express *glutathione* – a cysteine-containing tripeptide involved in various intracellular biochemical pathways. The yeast cells were incubated with cadmium chloride and sodium selenide, designed to produce CdSe QDs. Remarkably, following this treatment the two cell types could be clearly distinguished according to their

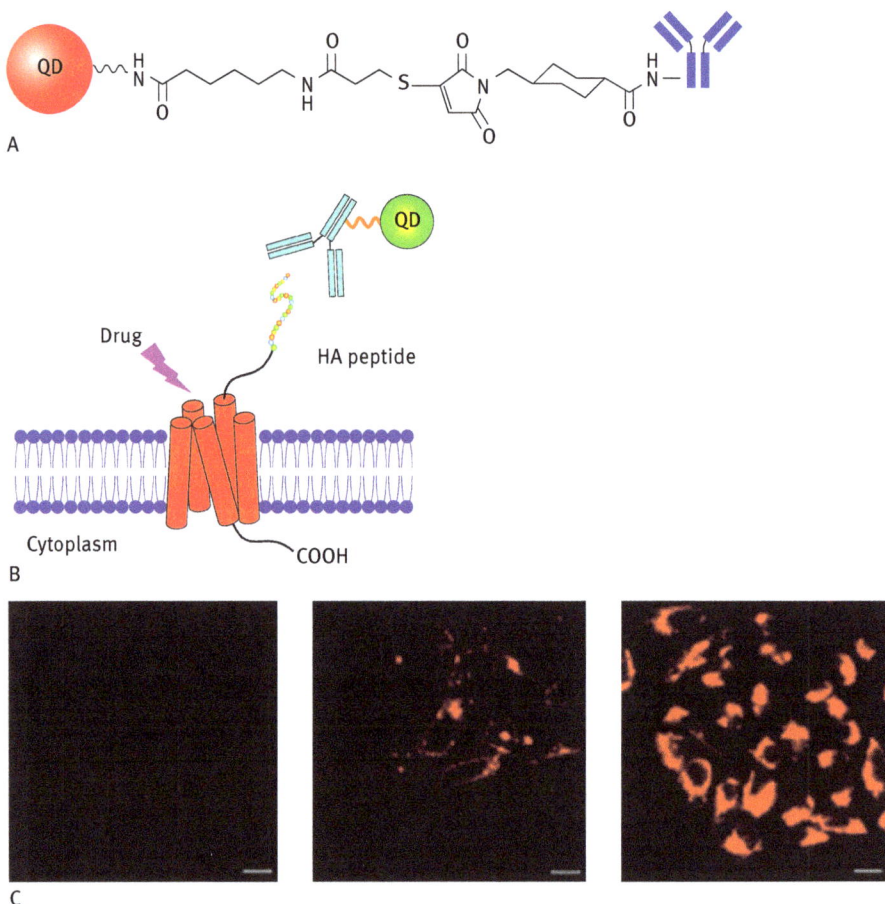

Fig. 2.10: Application of quantum dots for screening substances binding to cell-surface receptors.
A: chemical structure of the quantum dot attached to an antibody directed against hemagglutin
(anti-HA). **B:** the anti-HA-quantum dot binds to the HA peptide displayed on a membrane-associated
G-protein cell receptor (GPCR). Upon action of a drug agonist molecule, the GPCR receptor (as well
as the bound quantum dot) is internalized by the cell. **C:** fluorescence microscopy images demonstrating QD internalization into cells: cell incubated with QD that were *not* conjugated with anti-HA
(*left image*) – no staining observed; cells incubated with anti-HA-QDs (*middle image*) – the QDs
were attached to the cell surface; image recorded after addition of GPCR agonist to cells incubated
with anti-HA-QDs (*right image*) – the QDs were internalized within the cells, rendering them fluorescent. Scale bars correspond to 20 μm. Reprinted with permission from Lee et al., *ChemBioChem* **13**
(2012), 1503–1508, © 2012 John Wiley and Sons.

fluorescence emission. Notably, Figure 2.11 reveals significantly enhanced intracellular yellow-green fluorescence, ascribed to biosynthesized CdSe QDs, in case of the
yeast cells over-expressing glutathione. To explain this observation, the researchers

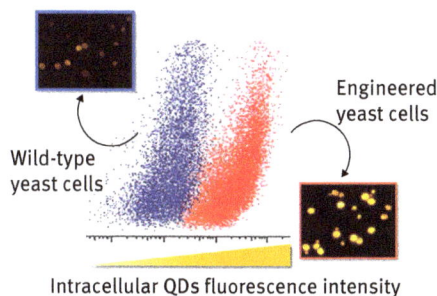

Fig. 2.11: Biosynthesis of CdSe quantum dots in yeast cells over-expressing glutathione. The genetically engineered cells producing glutathione exhibit more intense and differently-colored fluorescence (right), due to accumulation of the fluorescent CdSe quantum dots inside the cells. Reprinted with permission from Li et al., *ACS Nano* **7** (2013), 2240–2248, © 2013 American Chemical Society.

hypothesized that glutathione promoted CdSe QD formation through stabilization of both the telluride precursors and the budding CdSe QDs.

While biosynthetic approaches for QD-labelling of cells are intriguing, such techniques are limited to certain microorganisms and QD compositions. Besides biolabeling through latching QDs onto the external surface of cells (as in Figures 2.8, 2.10), insertion of QDs into the cell interior has been mostly carried out via conjugation of QDs with delivery vehicles such as cell penetrating peptides, DNA transfection reagents, and others. Progress in this field, however, has been slow; besides issues of toxicity and interference in cell processes (see Chapter 7 for a more detailed discussion of this topic), transporting QDs across the cell membrane has been found to be difficult, partly because QDs are recognized as foreign objects. Consequently, QDs are often blocked at the cell membrane or rapidly secreted from the cells after internalization. Furthermore, it was found that QDs often degraded or underwent aggregation when exposed to physiological environments in cellular uptake processes (for example acidic solutions encountered in *endocytosis*, or cell internalization processes).

Viruses have attracted interest as a potential vehicle for QD cell entry, since they possess intrinsic stealth properties and have already been used for the transport of genetic material into cell nuclei. Figure 2.12 illustrates an interesting strategy, developed by H. Wang and colleagues at the Chinese Academy of Sciences, in which QDs were incorporated within viral particles and subsequently delivered by the viruses into cells. Specifically, the QDs were conjugated to a DNA fragment corresponding to part of the viral genome. Following packaging of the genetic material, the QDs were encapsulated within viable viruses. Subsequent infection of mammalian cells by the QD-containing viruses resulted in cell uptake of the QDs (Fig. 2.12B). This approach is noteworthy because QD attachment to viral *surfaces*, which might be conceptually and technically easier, would likely have a negative impact upon viral infectivity as it might interfere with virus-cell interactions.

Light emission by chalcogenide QDs is not restricted to the ultraviolet or visible regions. In fact, *tissue imaging* applications benefit from fluorescence emission in the second *near-infrared (NIR)* spectral window (wavelengths of 1000–1700 nm), in which there is minimal adsorption/scattering by tissues, and thus deeper light penetration is expected. In addition, less *auto-fluorescence* from tissues occurs in this spectral re-

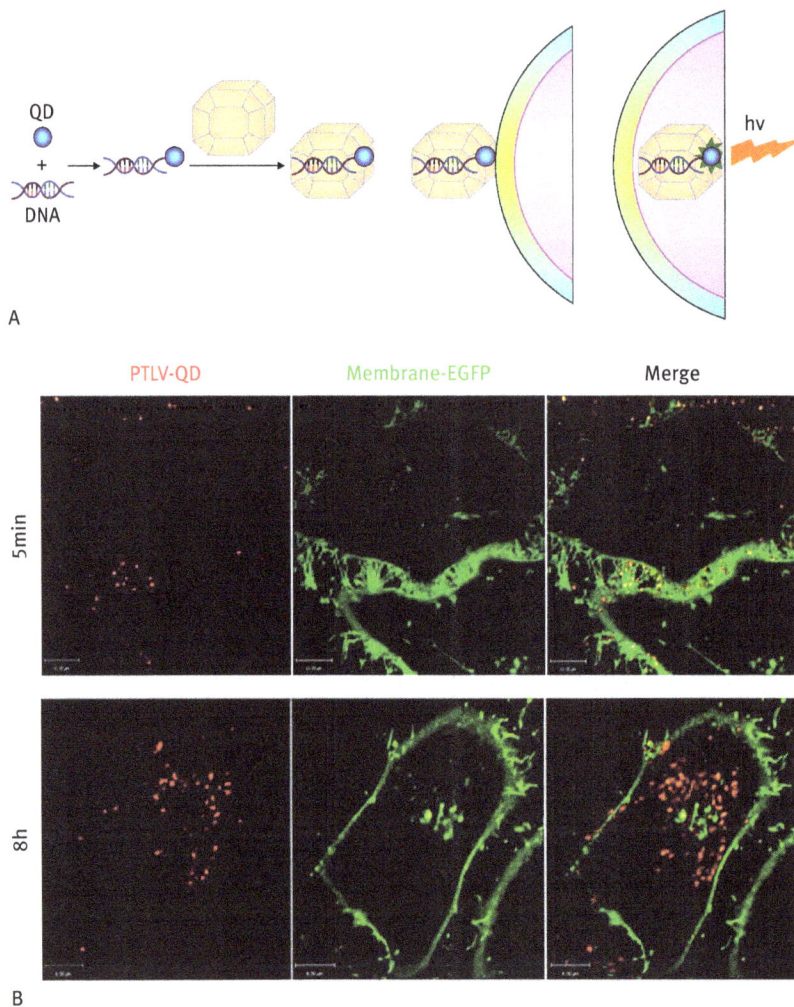

Fig. 2.12: Transport of quantum dots into cells via viruses. **A:** schematic description showing coupling of the quantum dot to viral DNA. The quantum dot is embedded within the packaged virus and is delivered into the cell following infection with the viral particle. **B:** two-color fluorescence microscopy images demonstrating quantum dot uptake. The quantum dots coupled to the viral particles are red, while a membrane-associated fluorescent dye is green. Five minutes after infection of the cells with the virus the quantum dots were still localized in the cell membrane (top row), however 8 hours after infection the quantum dots were localized inside the cells through viral insertion (bottom row). Scale bars correspond to 6 µm. Reprinted with permission from Zhang et al., *ACS Nano* **7** (2013), 3896–3904, © 2013 American Chemical Society.

gion, reducing the background emission. Ag_2S and Ag_2Se QDs have been useful in such applications since they exhibit strong emission in the NIR spectral region. Particularly important, as opposed to commonly used chalcogenide QDs comprising toxic

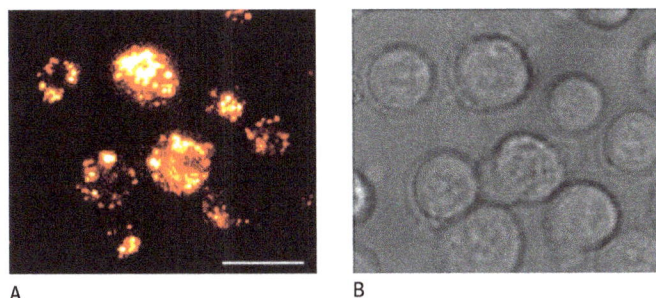

Fig. 2.13: Cell imaging in the near infrared (NIR) spectral region through attachment of Ag$_2$S quantum dots. NIR (1100–1700nm) fluorescence microscopy image (**A**) and corresponding optical microscopy image (**B**) of cells treated with Ag$_2$S quantum dots coupled to cell recognition elements. Scale bar corresponds to 25 μm. Reprinted with permission from Zhang et al., *ACS Nano* **6** (2012), 3695–3702, © 2012 American Chemical Society.

metals such as cadmium, lead, and mercury, *silver* ions are generally more biocompatible and exert minimal adverse effects upon cells. Figure 2.13 depicts a cell imaging experiment carried out by Q. Wang and colleagues at the Chinese Academy of Sciences, in which Ag$_2$S NPs were adsorbed by human cells allowing visualization in the NIR region.

The attractive photophysical properties of QDs have promoted research efforts aimed at clinical applications of QDs. The risk of cellular toxicity of QDs is, however, a major concern in this field, mostly due to the known and perceived dangers inherent in atomic constituents such as cadmium. Indeed, while some animal studies suggest that QDs do not exhibit long-term health risks, other observations, such as accumulation of QDs in the lymph nodes and the liver do raise health safety issues. In light of these observations, investigating the biological effects of QDs has become an important field of study. Parameters explored include QD composition (the semiconductor material/s comprising the QD core); size; surface charge; and type of ligands employed for solubilizing and/or directing the QDs onto their biological targets. Extensive studies in recent years have not yet provided a definitive answer as to which QD property is the most pertinent to cytotoxicity, although some studies have identified links between cell damage and the extent and density of *positive charges* upon the QD surface. Likewise, molecular units displayed on the QD surface (either as part of a molecular recognition system or other functional species) have been found to exert significant biological impact (more detailed discussion is provided in Chapter 7). Perhaps surprisingly, specific relationships between the compositions of QD cores (e.g. type of metals or chalcogenide ions) and cell toxicity have not yet been firmly established.

Various sensing and imaging applications utilizing metal chalcogenide NPs have been reported. QDs constitute an excellent molecular sensing platform for two main reasons. First, the intrinsic intense luminescence of QDs provides a sensitive signaling mechanism. Second, as indicated above, QD surfaces can be readily functionalized, making covalent attachment of recognition elements possible; binding of the respective molecular targets can, in principle, result in modulation of the QD luminescence, indicating the presence of the analytes. Many variations of this basic concept have been demonstrated.

Unlike QDs conjugated with molecular recognition units, *unmodified* QDs have also been employed in innovative sensing schemes. Figure 2.14 outlines an interesting sensing concept based upon CdTe QDs for reporting on the activity of *protein kinases*, a ubiquitous class of enzymes. The universal catalytic feature of protein kinases is the *phosphorylation* process – the addition of phosphate units onto protein targets constituting the substrates for the kinase enzyme. Z. Nie and colleagues at Hunan University, China, mixed QDs with a short peptide whose sequence contained *serine* (the phosphorylation site), *cysteine* residue for QD surface binding, and two positively-charged *arginines*. Without the presence of protein kinases in the QD solution, the positive peptides were bound to the QDs and reduced the negative electrostatic charge upon the QD surface – thereby promoting aggregation of the QDs and consequent shift of the

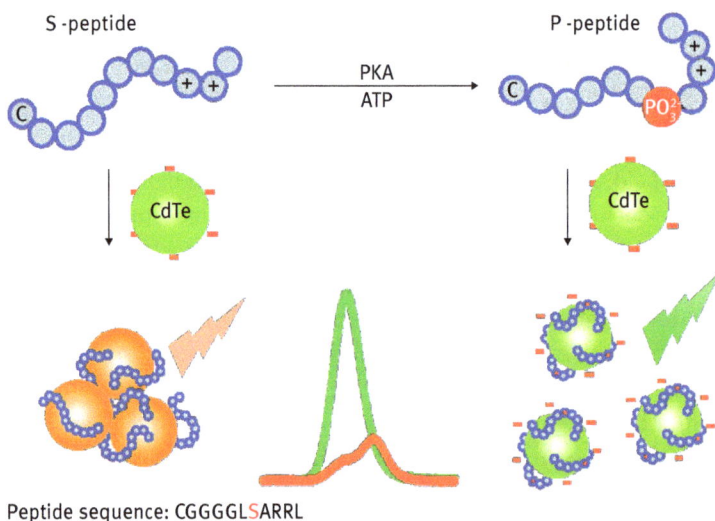

Peptide sequence: CGGGGGLSARRL

Fig. 2.14: Sensing protein kinase enzymatic activity with quantum dots. **Left:** without addition of protein kinase (PKA) the positively-charged peptide (sequence shown at the bottom) binds and neutralize the negatively-charged quantum dots, inducing aggregation of the dots and consequent red shift of the fluorescence peak (orange). **Right:** addition of PKA and adenosine triphosphate (ATP, serving as an energy source) gives rise to incorporation of negatively-charged phosphate residue onto the peptide (e.g. phosphorylation). The peptide now does not neutralize the quantum dots, which consequently repulse each other and do not aggregate, producing a fluorescence peak that is blue-shifted compared to the aggregated dots (green spectrum). Reprinted with permission from Xu et al., *Anal Chem.* **83** (2011), 52–59, © 2011 American Chemical Society.

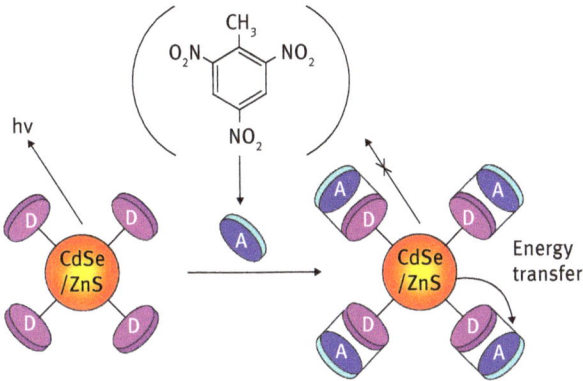

Fig. 2.15: Sensing explosives with modified quantum dots. **Left:** Fluorescence donors attached to the quantum dot's surface – fluorescence emission from the dots is recorded. **Right:** quenching of fluorescence emission following binding of fluorescence energy acceptors such as trinitrotoluene (TNT).

fluorescence emission peak to a higher wavelength (due to the larger size of the aggregates). However, upon addition of a kinase to the solution and consequent phosphorylation of the QD-anchored peptide, the negative phosphate moiety effectively countered the two positive arginine residues within the peptide, leading to *de-aggregation* of the QDs and re-appearance of their intrinsic blue-shifted luminescence. This simple approach could allow, in principle, visual detection of kinase activity and screening for kinase inhibitors.

QD-based sensing is not limited to the biological realm. A system exploiting QDs for the detection of explosives has recently been reported (Fig. 2.15). The sensor, developed by I. Wilner and colleagues at the Hebrew University, Israel, was based on attaching aromatic electron donors onto the QD surface. Binding of electron acceptors, such as the explosive trinitrotoluene (TNT), was found to quench the QD photoluminescence, presumably by capturing electrons from the QD surface. This luminescence quenching mechanism provides a generic sensing pathway for explosive analytes such as TNT exhibiting electron drawing properties.

2.1.2 Quantum dots in solar cells

Solar cell design is a major field in which QDs could be utilized. Solar cells (SCs) employ semiconducting materials as vehicles for generating electricity from light (i.e. "photocurrent"). Figure 2.16 depicts the operation principle of a semiconductor solar cell. Incident light generates excess electrons and holes in the n-type and p-type semiconductor layers, respectively; energy is subsequently harvested upon the transport of electrons from the front to the back electrode. The critical parameters affecting the per-

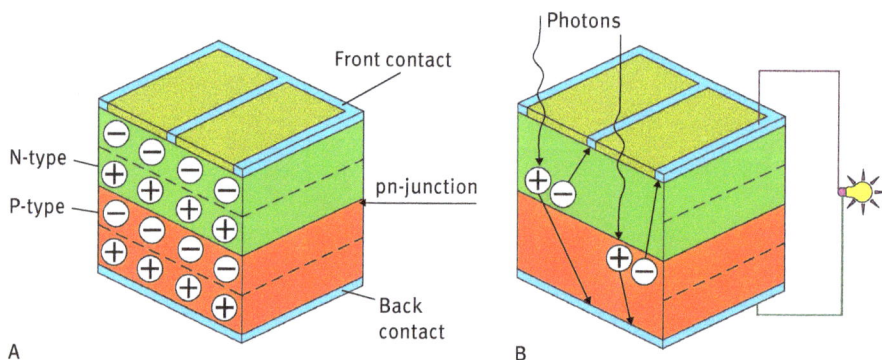

Fig. 2.16: Solar cell structure. **A:** No light irradiation; the electrons and holes equilibrate at the p-n junction. **B:** Upon illumination, light disrupts the electron-hole pair, resulting in excess electrons harvested in the front contact (electrode) and holes captured by the back contact, generating charge carrier mobility and electrical energy.

formance of an SC include the efficiency of light absorbance, transport of charge carriers induced by light irradiation, and long-term stability of the cell. QDs are generally incorporated into the "active layer" of an SC functioning as "photon harvesters" – absorbing light energy and injecting electrons (and concurrent holes) into the electrical circuit. The thrust of integrating QDs in solar cells is their tunable light absorbance (by modulating QD size), enabling uptake of light energy at different regions of the electromagnetic spectrum. In particular, QDs have been promoted as attractive candidates for solar cell applications due to their relatively simple synthesis schemes (which do not require specialized heavy-duty machines), and their relative stability in solar cell conditions. Moreover, metal chalcogenide QDs can be tuned to absorb *infrared (IR)* light – in which solar cells exhibit the most optimal efficiency for converting light to electricity.

The basic quantum-dot solar cell configuration is depicted in Figure 2.17. The QDs are usually deposited upon or within an *electron collecting layer*. In most current SC constructs, this layer is comprised of transparent nanocrystalline or microcrystalline ZnO or TiO_2 films which exhibit high surface area and efficient electron transport properties. As electrons are injected from the QDs into the collecting layer and subsequently to the electrode, the positive "holes" are transported to a hole-acceptor such as a p-type semiconducting layer through an interface referred to as the "heterojunction".

Several practical factors are usually considered in QD SC designs. Likely the most important aspects affecting cell performance are the separation and lifetime of the photo-induced electron-hole pairs generated in the QDs and their efficient transport to the respective electrodes. Accordingly, among the key design parameters in QD SCs is the interface between the QDs and the electron/hole acceptors. The optimal interface needs to facilitate rapid transfer of the photo-generated charge carriers, extending

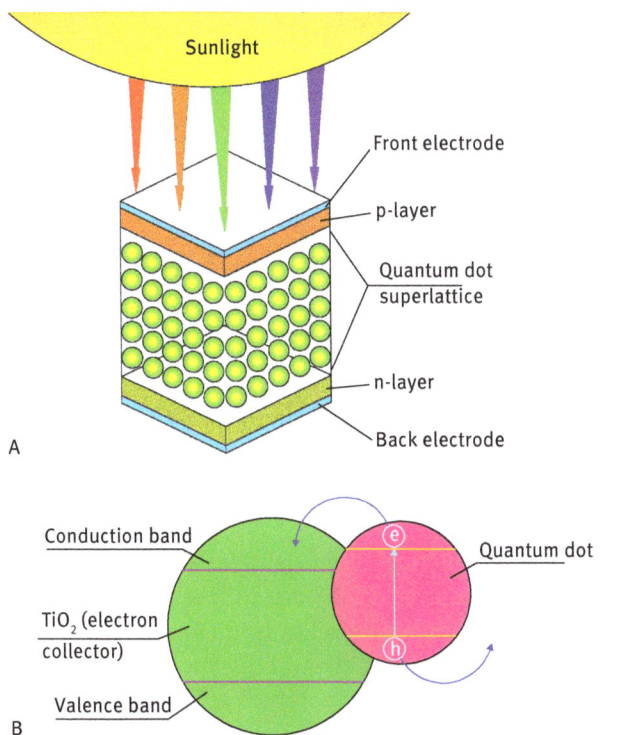

Fig. 2.17: Quantum dot solar cell. **A:** schematic drawing showing the quantum dot "superlattice" within the electron collecting layer. **B:** energy level diagram illustrating transfer of the excited electron from the quantum dot to the close energy level of the electron collector (TiO_2). At the same time the hole is transported to the hole-acceptor layer.

their lifetimes and reducing energy losses occurring through electron-hole recombination. Numerous variations of QD SCs have addressed these goals via chemical means and physical organization of the SC components.

One of the useful features of QD SCs is the possibility of using *solution-phase chemistry* for the fabrication of cell components (particularly the QD layer). Solution chemistry does not require expensive equipment and infrastructure such as clean rooms or vacuum chambers which are predominant in silicon-based fabrication techniques, thus significantly reducing SC production costs. Figure 2.18 depicts a solution-based "layer-by-layer" scheme for SC construction, based on integration of two QD species: CdSe and CdTe. This multilayer system, demonstrated by P. Mulvaney and colleagues at the University of Melbourne in Australia, was designed to exhibit several functions. First, the multilayers comprising different QD compositions were *sintered*, e.g. fused together to create a semiconductor *alloy*. Alloy formation through sintering of QDs has been shown in some instances to yield superior solar cell performance, likely due to better charge transfer throughout the semiconducting layer and enhanced electronic

Colloidal nanocrystal
synthesis

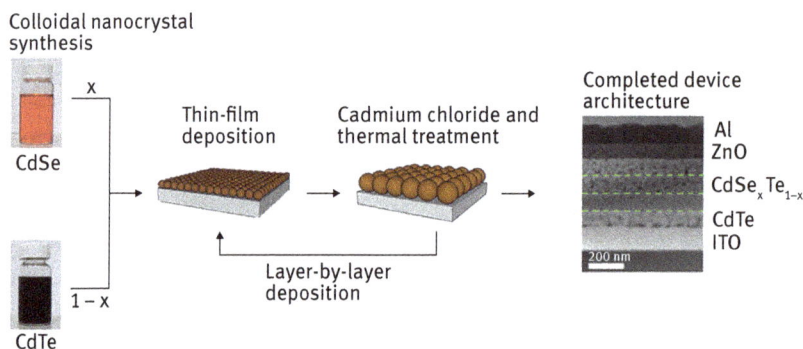

Fig. 2.18: Layer-by-layer deposition of a quantum dot solar cell. Mixed layers of CdSe and CdTe quantum dots are deposited consequentially and high temperature sintering generates a photoactive semiconducting alloy layer. Reprinted with permission from MacDonald et al., *ACS Nano* **6** (2012), 5995–6004, © 2012 American Chemical Society.

properties of the material (which, after sintering, resembles a bulk semiconducting layer). Another important property of multilayered QD SCs is the feasibility of tuning the overall energy bandgap of the SC (thus optimizing light adsorption) by modulating the compositions of the layers, i.e. the ratio between the two QD species.

An interesting SC construct in which the QD layer acts both as an electron generator and a hole transporter is depicted in Figure 2.19. This uncommon SC design, developed by L. Etgar and colleagues at the EPFL, Switzerland, using lead-sulfide QDs, obviates the inclusion of a p-type semiconductor material – required in most SC assemblies to facilitate hole mobility towards the front electrode (e.g. Fig. 2.17). The dual role for QDs is made possible because the energy levels of the 3.2 nm PbS nanoparticles allow both efficient electron injection into the TiO_2 layer as well as hole transport directly into the gold electrode (Fig. 2.19B).

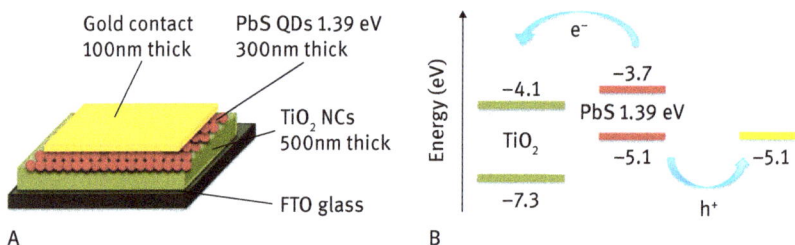

Fig. 2.19: Solar cell construct in which PbS quantum dots constitute both an electron donor and a hole transporter. **A:** solar cell scheme. **B:** energy level diagram showing the close energy values of the PbS quantum dots' and TiO_2 conduction bands (enabling electron transport), and the PbS valence band and gold electrode (enabling hole transport). Reprinted with permission from Etgar et al., *ACS Nano* **6** (2012), 3092–3099, © 2012 American Chemical Society.

2.2 Semiconductor nanowires

Semiconductor nanowires (NWs), also referred to as "quantum wires" (QWs), constitute a technologically-promising class of nanomaterials. In particular, the *one-dimensional* quantum confinement (in comparison to the *zero dimension* of QDs), endows semiconductor NWs with distinct electro-optical properties – primarily the possibility of transporting excitons over large distances along the longitudinal axis of the wire. Figure 2.20 depicts photoluminescence images of single NWs comprising a CdS core coated by a CdTe shell. The quantum wires, synthesized by W. E. Buhro and colleagues at Washington University, exhibited high luminescence and relatively uniform core formation, although some defects are apparent, for example the discontinuity in the NW shown in Figure 2.20C.

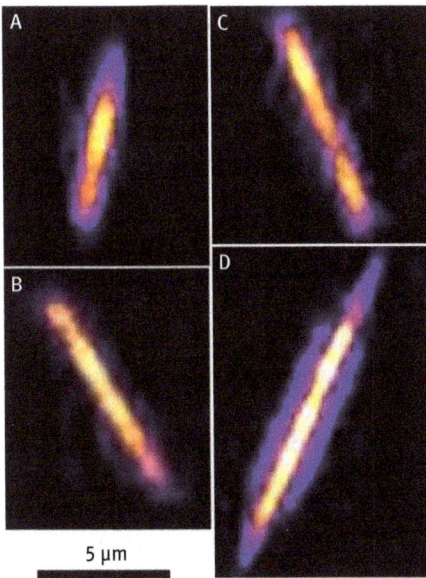

Fig. 2.20: Core-shell semiconductor quantum wires. The bright CdTe core is apparent in the CdTe@CdS nanowires. Reprinted with permission from Liu et al., *JACS* **134** (2012), 18797–18803, © 2012 American Chemical Society.

Quantum wires constitute core components in many devices. For example, field effect transistors (FETs) based on semiconductor NWs have been produced (Fig. 2.21). The current passing through the NW (placed between the "source" and "drain") is controlled by application of an electric field at the "gate". Semiconductor NWs comprising of binary compounds have been shown to exhibit good electron transport properties. NW FET designs have been applied for biological and chemical sensing, usually through coupling molecular recognition elements onto the NW surface; binding of the target analytes modulates the energy levels of the NW and consequently changes the conductivity of the wire.

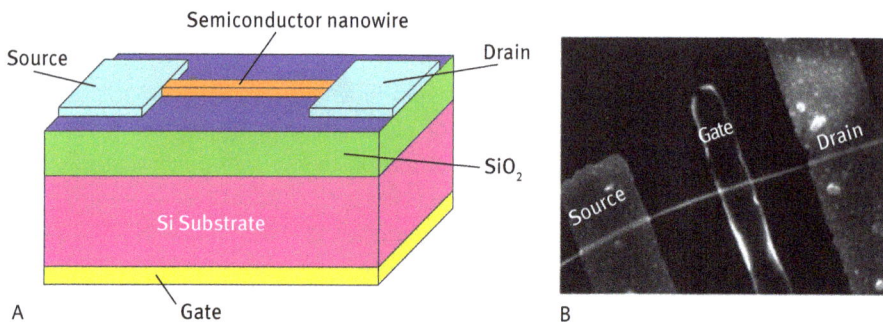

Fig. 2.21: Semiconductor nanowire field effect transistor. **A:** schematic structure of the transistor, showing the nanowire connecting the source and drain. **B:** electron microscopy image of an InAs nanowire in a field effect transistor. Reprinted with permission from Dayeh et al., *Small* **3** (2007), 326–332, © 2007 John Wiley and Sons.

Two primary synthetic routes have been implemented to produce semiconductor NWs. An early and powerful methodology, the *vapor-liquid-solid (VLS)* approach (Fig. 2.22), is based on the growth of NWs on solid surfaces through catalyzing condensation and crystallization of reagents from the vapor phase. The critical component in VLS reaction schemes is the catalyst (usually spherical nanoparticles) placed on the solid substrate. The formation of liquid droplets on the NPs lowers the activation energy for further adsorption of the NW constituents from the vapor phase; following adsorption, the solution becomes *supersaturated*, resulting in anisotropic, unidirectional crystallization of the NW. The choice of the catalyst NPs is important since their role of maintaining equilibrium with the NW building blocks in a liquid phase is dependent upon composition, shape, and size.

The scheme depicting the catalyst-induced VLS approach further shows that the diameter of the NW is essentially determined by the diameter of the spherical NP. In addition, the NW length can be controlled by the duration of reagent injection into the reaction chamber. The basic NW synthetic route can be further expanded to construct *hetero-structured* NWs (through switching between different reagents during the growth process) and core-shell *radial* NW organizations (Fig. 2.22); each of these nanostructures exhibits distinct photophysical properties. NWs synthesized via VLS technology can be utilized while still attached to the surface but can be also detached and harvested for applications in other settings.

Solution-based synthesis procedures have also been developed for the fabrication of semiconductor NWs. Figure 2.23 highlights a ZnSe NW synthesis scheme demonstrated by S. Zhou and colleagues at Sichuan University, China, based on laser-induced *ablation* (i.e. "chipping" small particulates from a larger substance) of micrometer-sized substrate material dispersed in a liquid medium. Specifically, the researchers demonstrated that irradiation of the soluble ZnSe particles with high-frequency pulsed laser resulted in formation of relatively uniform semiconducting

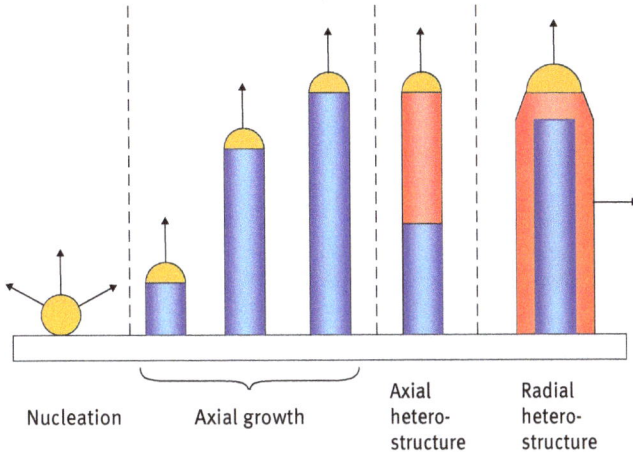

Fig. 2.22: Nanowire synthesis by the vapor-liquid-solid (VLS) method. Anisotropic growth of the nanowire is facilitated by the nanoparticle at the tip, which acts as a catalyst. The technique enables fabrication of different nanowire configurations, such as axial-, and radial-heterostructures.

NWs. According to the mechanism proposed by the researchers, energy bursts induced by the laser pulses cleaved the microparticles, generating small ZnSe nucleating seeds which subsequently fused into nanowires (Fig. 2.23).

This solution-phase procedure further enabled doping of the NWs with additional constituents, such as metal ions, co-dissolved in the solution. Doping is particularly advantageous for endowing additional properties to NWs, such as *lasing* (see discussion on nanowire lasers below). Solution-based NW synthesis routes, such as that outlined in Figure 2.23, are attractive for practical applications since they do not require expensive, heavy-duty equipment and can be carried out in nonextreme conditions of temperature and pressure. It should be noted, however, that it is often difficult to achieve sufficient product uniformity and overall reproducibility using solution-based NW synthesis techniques, as NW structures, homogeneity, and physical properties are usually affected by the presence of even minute reagent impurities.

2.3 Photonic applications of quantum dots and quantum wires

The unique electro-optical properties of QDs and QWs make these nanoparticles prime candidates for integration in photonic devices. *Light emitting diodes (LEDs;* see Fig. 2.24) are among the most commercially advanced and promising applications of QDs. An LED generally comprises of a *p-n junction* – the physical interface between an electron donor (*n-type semiconductor*) and hole generator (*p-type semiconductor*). When voltage is applied and an electron recombines with a hole at the p-n junction, light is released, the wavelength of which is determined by the bandgap

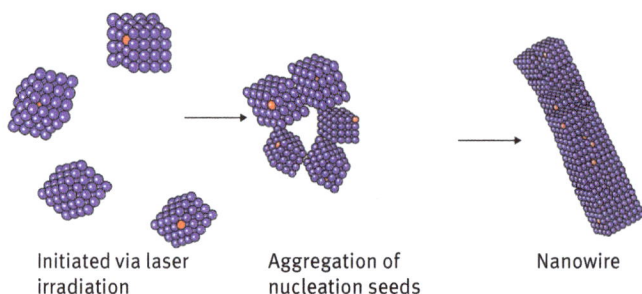

Initiated via laser irradiation   Aggregation of nucleation seeds   Nanowire

A

S4800 5.0kV 2.7mm x 100k SE(U,LA0)   500nm

B

Fig. 2.23: Semiconductor nanowire fabrication through laser ablation in solution. **A:** Small nucleation seeds aggregate and fuse to produce the nanowires; **B:** an electron microscopy image of ZnSe nanowires produced with the laser ablation technique. Reprinted with permission from Feng et al., *Nano Lett.* **13** (2013), 272–275, © 2013 American Chemical Society.

of the semiconducting substance selected. QDs constitute a powerful platform for LED fabrication, as the colors they emit (i.e. their energy bandgaps) can be tuned by controlling the particle size and composition. Furthermore, QDs exhibit narrow spectral emission ranges, facilitating the generation of pure, well-defined LED colors. Another advantage of QDs is their synthesis in mild conditions which are generally inexpensive and scalable for mass production.

Different designs have been introduced for QD-LED devices. In the most basic QD-LEDs, QDs are deposited between two electrodes; through application of an electric field, excitons (electron-hole pairs) are generated in the QD layer and photons are subsequently emitted following recombination between the electrons and the holes, a process defined as "electroluminescence". An example of a QD-LED device is shown in Figure 2.25. In this configuration, based upon a "charge injection" mechanism, both the electrons and holes are "injected" into the QD layer through an electron transport layer and a hole transport layer, respectively. Subsequent electron/hole recombination within the QD layer produces the emitted light.

Metal chalcogenides, particularly CdSe, have been the most widely investigated QDs in visible LED applications. CdSe NPs provide bright light emission in the range of 480–650 nm, and various synthetic procedures provide the means for high yield, inex-

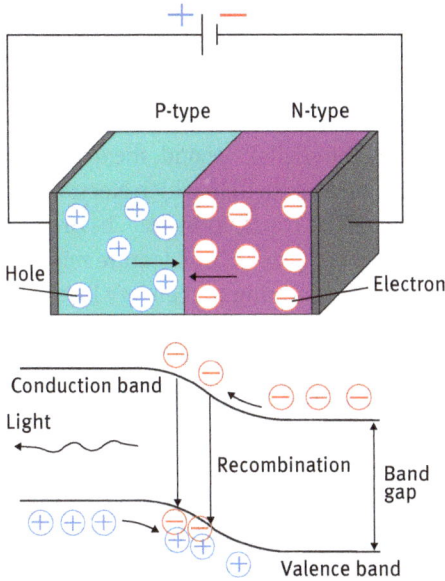

Fig. 2.24: Schematic depiction of a light emitting diode (LED). Application of an electric voltage at the p-n junction induces electron-hole recombination, producing light the energy of which corresponds to the semiconductor bandgap.

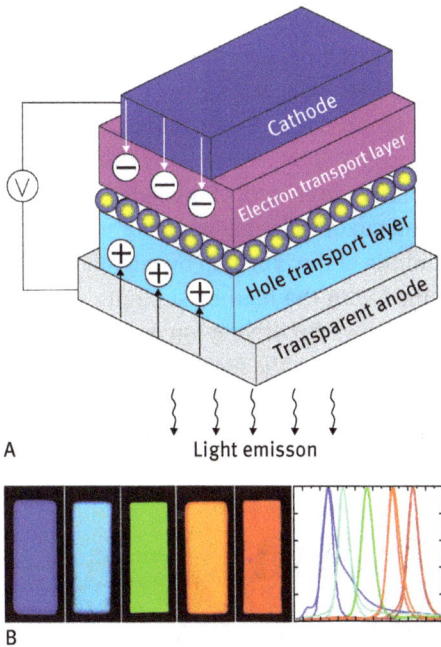

Fig. 2.25: Quantum dot light emitting diodes (LEDs). **A:** scheme of an LED operating through charge injection into the quantum dot layer. **B:** colors and corresponding electroluminescence spectra of LEDs comprising of quantum dots of different compositions. Reprinted with permission from Anikeeva et al., *Nano Lett.* **9** (2009), 2532–2536, © 2009 American Chemical Society.

pensive, and uniform CdSe NPs. Lead chalcogenide QDs (specifically PbS and PbSe), on the other hand, have been the material of choice in LEDs operating in the infrared spectral region. It should also be emphasized that core-shell nanostructures, rather than single core species, are the predominant building blocks in most current LED-QD devices due to their brightness, tunability, and stability.

Core-shell nanowires, in particular, offer distinct advantages as LEDs. Firstly, similar to QDs, light can be generated in a broad wavelength range by varying the composition of the semiconducting materials (primarily the shells). Second, the *axial geometry* provides a well-defined interface (between core and shell) in which electron/hole injection and recombination can occur. Moreover, NWs can, in principle, be prepared as "free-standing" constructs (i.e. not attached to a surface). This feature might account for higher efficiency of light emission, since scattering and energy loss to the LED environment are minimized (such scattering effects are usually encountered in the high refractive-index materials employed to embed LEDs). In addition to the above aspects, from a practical/synthetic point of view the production of core-shell NWs is relatively straightforward, accomplished via sequential deposition of the core and shell materials, for example using VLS methods, as described above.

Semiconductor NWs could also function as nanoscale *lasers*. The core phenomenon responsible for lasing is *stimulated emission*, in which high intensity, coherent light is generated by inverting an electron population from the ground energy state to an excited state through continuous pumping with incident light photons. In conventional lasers this is achieved by creating a "resonance cavity" (or "optical cavity") in which photons, initially injected from an external source or emitted through electrical stimulation of the medium electrons, are reflected between two parallel mirrors thereby producing the stimulated emission effect. Semiconductor NWs can be made into lasers through *cleavage* of the two NW ends – essentially transforming the two ends into reflective mirrors (Fig. 2.26); the *wavelength* of the light emitted by the NW laser is determined by the energy bandgap of the semiconductor material of the nanowire. Indeed, control over the magnitude of the bandgap by modulating the NW composition and/or dimensions, combined with the high crystallinity attained through existing synthetic schemes, have been powerful driving forces for the introduction of NW lasers. Since their first demonstration in the early 2000s, numerous variations of semiconductor NW lasers have been reported, both as single NW configurations as well as nanowire *ensembles*. Figure 2.26, for example, demonstrates lasing

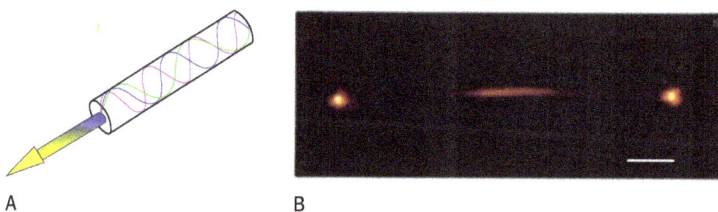

A   B

Fig. 2.26: Semiconductor nanowire laser. **A:** Scheme of the resonance cavity inside the nanowire producing the laser beam. **B:** Microscopy image of a single lasing CdSe nanowire. Reprinted with permission from Li et al., *Adv. Mater.* **25** (2013), 833–837, © 2013 John Wiley and Sons.

from a single CdSe nanowire, accomplished by Q. Yang and colleagues at Zhejiang University, China. The researchers have further shown that tuning the laser wavelength (e.g. color) can be achieved in quite a broad spectral range by changing the NW length.

2.4 III-V semiconductor nanoparticles

Binary semiconductor materials comprising of metals (or semi-metals) from group III (such as gallium and indium) and elements from group V (e.g. phosphorous, arsenic) exhibit useful physical properties, particularly high charge mobility and mostly direct bandgap configurations; as outlined in the beginning of this chapter, a direct bandgap leads to efficient electron-hole recombination and corresponding light emission exploited in light-emitting diodes and other opto-electronic applications and devices. III-V NPs have been fabricated in varied configurations including nanowires, spherical NPs, nanoplatelets, and others. Particularly noteworthy, studies have shown that tuning structural properties of III-V semiconductor nanoparticles is possible by changing the *ratio* between the atomic components.

Gallium arsenide (GaAs) is a well-known semiconductor used in photovoltaics, light emitting diodes, and other devices. GaAs NWs have been mostly synthesized via surface-induced VLS schemes; modulating growth parameters such as the ratios between the ionic precursors has been found to intimately affect the structural and photophysical properties of the NWs. A report on GaAs NWs grown via the VLS mechanism, conducted by J. Zou and colleagues at the University of Queensland, Australia, demonstrated intriguing structural features including *tapering* (e.g. bases of greater diameters than the tips), and the formation of intersecting triangular-shaped nanoplatelets at the NW base (Fig. 2.27).

Gallium phosphide (GaP) is an indirect bandgap semiconductor (in its abundant *cubic* crystalline phase), limiting its applicability in light emitting diodes because of inefficient emission following excitation of the valence electrons. In some cases, however, GaP NPs have been constructed in crystalline organizations which allow a direct bandgap configuration. *Indium phosphide (InP)* is an interesting semiconducting material yielding varied nanoparticle structures. InP NWs can be manufactured in relatively controlled and uniform dimensions (diameter and length), and exhibit high stability. Figure 2.28 illustrates a synthetic scheme developed by C. M. Lieber and colleagues at Harvard University for the production of "free standing" (as opposed to substrate-attached) InP nanowires of uniform diameters and lengths. The experimental technique was based upon the use of Au NPs as growth initiators/catalysts. Growth and elongation of the nanowires ewre subsequently carried out through laser ablation of InP substrate, giving rise to evaporation and re-crystallization of InP in a nanowire configuration. Note that control of the NWs' diameters was made possible by tuning the dimensionalities of the Au NP tips.

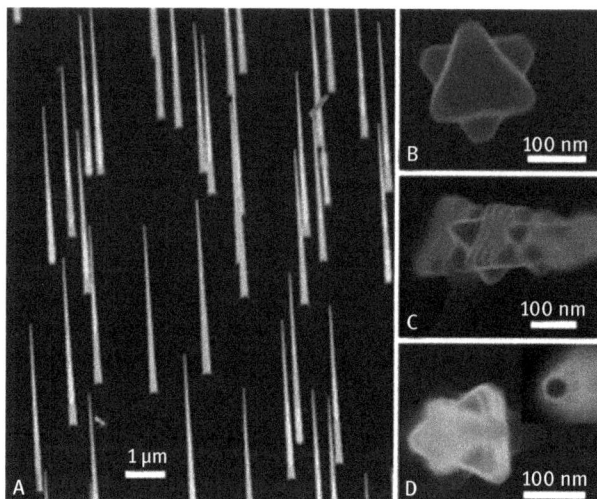

Fig. 2.27: GaAs nanowires. The nanowires were grown via the vapor-liquid-solid (VLS) method and positioned vertically at the surface. The electron microscopy images reveal nanowire tapering (*left image*). The nanowire bases are triangular (*right images*). Reprinted with permission from Zou et al., *Small* **3** (2007), 389–393, © 2007 John Wiley and Sons.

The synthetic method outlined in Figure 2.28 made isolation of single InP NWs and analysis of their optical properties possible. Indeed, individual InP NWs could be incorporated between electrodes in simple devices enabling measurement of the light-induced current (e.g. "photocurrent") from single nanowires. Remarkably, C. M. Lieber and his colleagues have shown that the InP NWs exhibit high photocurrent sensitivity which is furthermore dependent on the polarization of incident light (i.e. the ratio between the parallel and perpendicular components of the excitation light source). The high sensitivity of the InP NWs to irradiation and light polarization and their excellent charge-carrier mobility might make their use as photodetectors, optical switches and in other photonic applications possible.

Hetero-structured NWs comprising of InP doped with additional atomic and molecular components have been used as building blocks for light emitting diodes (LEDs). Similar to the metal chalcogenide NW LEDs described above, the radial configuration of NWs produces an extended interface serving as an effective p-n junction. InP-based NW LEDs have been shown to be particularly promising as components in electro-optic devices, as they can be tuned to emit light in the infrared (IR) spectral region – a highly sought-after property in photonic devices for telecommunication applications.

Figure 2.29 depicts an LED constructed by V. Zwiller and colleagues at the Kavli Institute of Nanoscience, Holland, from a hetero-structured InP/InAsP NW. As shown in the atomic force microscopy (AFM) image in Figure 2.29A, the NW was placed between the two electrodes and the "step" in the NW approximately one third of the way between the two electrodes corresponded to the location of the p-n junction. Indeed, the optical microscopy images in Figure 2.29B demonstrate that electroluminescence was generated specifically in the nanoscale "spot" within the NW – confirming the

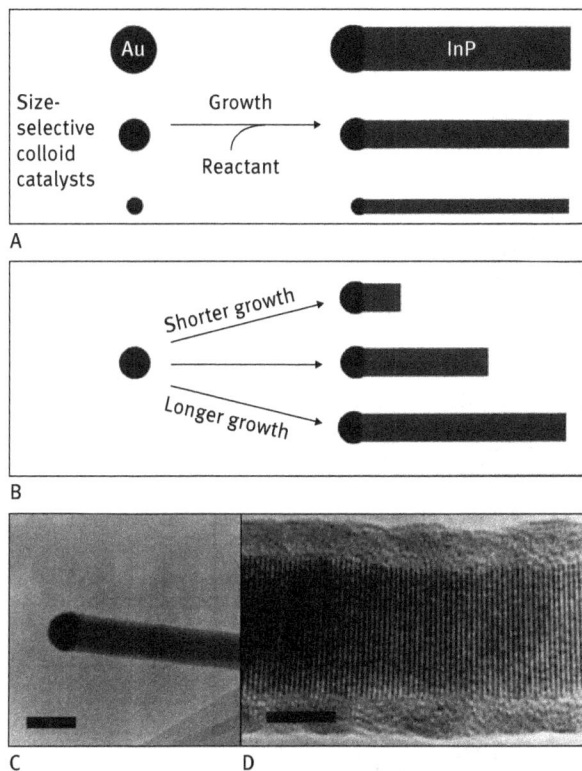

Fig. 2.28: InP nanowires. **A, B:** Methodology for growth of InP nanowires using Au nanoparticles as nucleating agents and catalysts. **C, D:** Electron microscopy images showing the nanoparticle tip (**C**) and nanowire crystallinity (**D**). Reprinted with permission from Gudiksen et al., *J. Phys. Chem. B* **105** (2001), 4062–4064, © 2001 American Chemical Society.

occurrence of electron-hole recombination and consequent light emission in the p-n junction.

III-V semiconductors were shown to adopt interesting *multishell* NWs (Fig. 2.30). These NWs are an intriguing variation of the core-shell model, providing a useful vehicle for tuning the wavelength (e.g. color) of the emitted light. Figure 2.30 depicts LED constructs comprising of a gallium nitride core (the n-type component) coated with several layers containing mixtures of gallium, indium, and aluminum nitrides serving as the p-type part of the diode. Indeed, Figure 2.30 reveals that the LEDs, synthesized by C. M. Lieber and colleagues at Harvard University, produced bright light in distinct colors (e.g. different electroluminescence wavelengths), depending upon the composition, thickness, and number of shells.

The photoluminescence of III-V semiconductor NWs could be tuned not only by modulating the NW composition or dimensions, but also through application of post-synthesis chemical modifications (Fig. 2.31). The composite nanoparticle system, de-

A

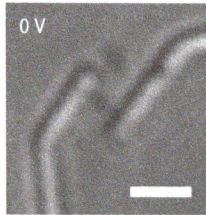

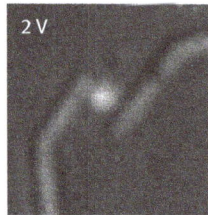

B

Fig. 2.29: Light-emitting diode (LED) based upon a single hetero-structured InP-based nanowire. **A:** Atomic force microscopy (AFM) image showing the InP/InAsP nanowire between two electrodes. The "step" corresponds to a p-n junction formed within the nanowire. **B:** Optical microscopy images showing electroluminescence generated at the p-n junction. Reprinted with permission from Minot et al., *Nano Lett.* **7** (2007), 367–371, © 2007 American Chemical Society.

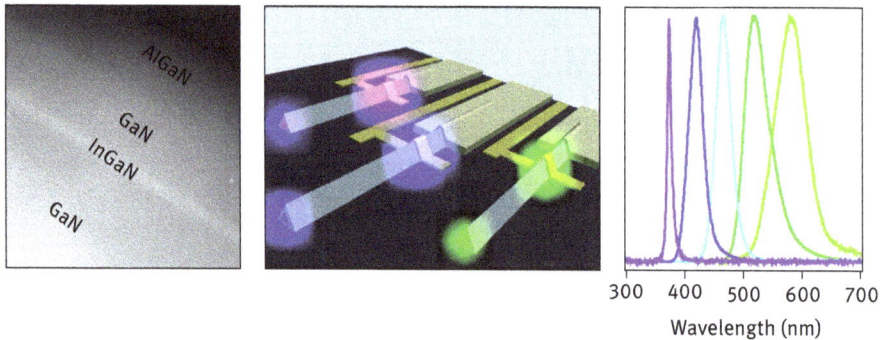

Fig. 2.30: Core multishell nanowire light emitting diodes (LEDs). **Left:** Electron microscopy image of the nanowire cross section showing that the GaN core and shells have different compositions. **Middle:** depiction of the core-shell nanowires. Note the triangular cross section of the wires. **Right:** electroluminescence spectra corresponding to the different LED colors generated. Reprinted with permission from Qian et al., *Nano Lett.* **5** (2005), 2287–2291, © 2005 American Chemical Society.

signed by M. B. Bavinck and colleagues at Delft University, Holland, comprised of an InAsP QD embedded at the tip of an InP NW, which was itself further coated with a silica (SiO_2) layer (produced through vapor-deposition). The experiments revealed that the luminescence of the silica-coated NWs could be tuned over a broad spectral range by changing the deposition conditions of the silica envelope. The pronounced photoluminescence shifts have been ascribed to *physical strain* induced by the silica coating which affects the physical properties of the semiconductor NW, particularly the

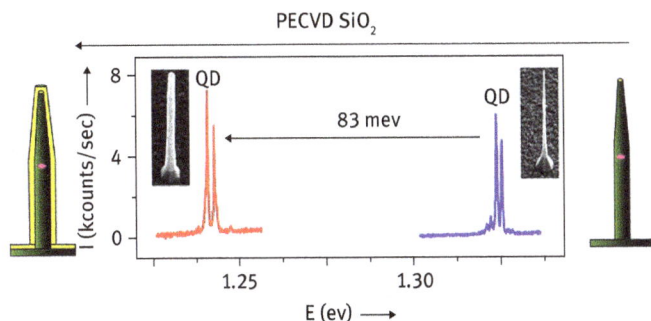

Fig. 2.31: Tuning light emission of semiconductor nanowires through silica coating. Schemes, electron microscopy images, and photoluminescence spectra of silica-coated composite nanowires. **Right:** as-synthesized InP nanowire embedding InAsP quantum dot at the tip. **Left:** significant red shift observed following coating the nanowire with a silica layer (*yellow*). Tuning of the emission color is possible by modifying the silica deposition parameters. Reprinted with permission from Qian et al., *Nano Lett.* **12** (2012), 6206–6211, © 2012 American Chemical Society.

energy bandgap. Modulating light emission properties of the NWs through silica coating points to the possibility of using externally-applied strain as a vehicle for bandgap engineering in semiconductor NPs.

2.5 Other binary semiconductor nanoparticle morphologies

Semiconductor *nanorods* (NRs) generally have aspect ratios (i.e. the ratio between length and cross-section diameter) smaller than nanowires, although the distinction between these two nanoparticle morphologies is not clear-cut. In some instances, however, semiconductor NRs exhibit distinct photophysical properties not encountered in NW systems. Particularly interesting properties have been reported in systems of *heterostructured* (or *hybrid*) NRs, i.e. NRs comprising of different semiconductor materials, such as semiconductor/metal systems. Heterostructured NRs have been mainly prepared by *seeded growth* techniques which are conceptually similar to core-shell synthesis schemes for spherical QDs. In seeded growth, preformed crystalline seeds (usually spherical) are injected into the growth solution containing the precursors for NR formation (which are distinct from the seed material). The NRs which subsequently grow embed the seeds, resulting in core-shell configurations.

Figure 2.32 presents an example of this strategy for the synthesis of CdSe@CdS NRs (i.e. CdS NRs containing embedded CdSe nanocrystals) resulting in highly uniform NR populations. In this study, conducted by L. Manna and colleagues at CNR-INFM, Italy, CdSe nanocrystal seeds were injected into a solution containing Cd^{2+} and ligand-coated S^{2-}. The CdS NRs displayed a preference for assembling on the CdSe seeds rather than forming their own CdS nuclei, since the activation energy for the

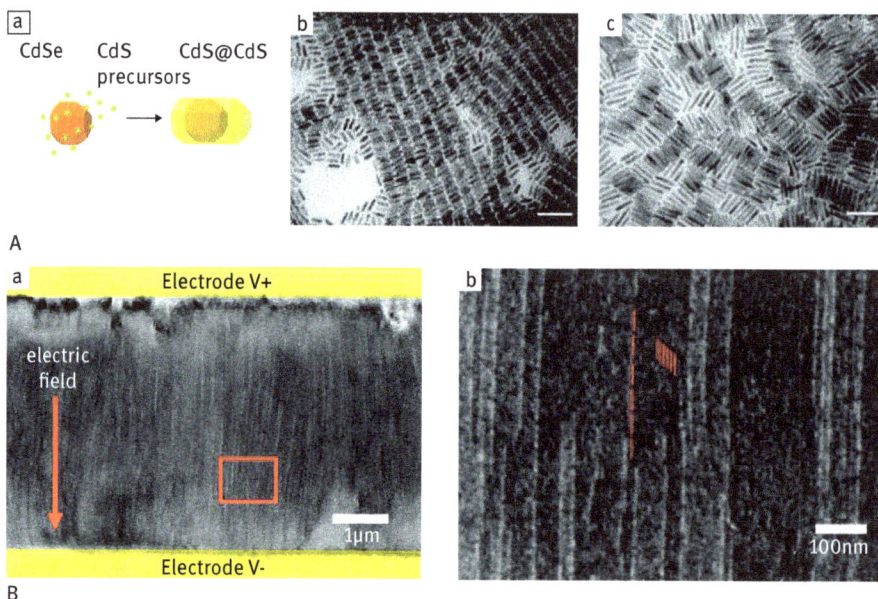

Fig. 2.32: Core-shell CdSe@CdS nanorods. **A:** schematic description of nanorod growth through coating CdSe nanoparticle seed with CdS. The electron microscopy images on the right reveal spontaneous organization of the nanorods upon drying of the solutions. Scale bars correspond to 50 nm. **B:** electron microscopy images demonstrating alignment of the nanorods through application of an electric field between two electrodes. The red arrow in the left image indicates the direction of the electric field. The magnified area in the right image highlights alignment of individual nanorods. Reprinted with permission from Carbone et al., *Nano Lett.* **7** (2007), 2942–2950, © 2007 American Chemical Society.

heterogeneous growth process is lower. Several experimental parameters were found to intimately affect the NR structures and dimensionalities, including the diameters of the CdSe seeds and solution temperature, which dictated the mean diameter of the rods and their aspect ratios, while precursor concentration and reaction times were found to determine the rod lengths.

The anisotropic structure of semiconductor NRs gives rise to interesting phenomena and practical applications. For example, the CdSe@CdS nanorods spontaneously adopted oriented macro-scale organization upon drying of the solvent, apparent in the electron microscopy images in Figure 2.32A. Externally-applied electric fields could similarly induce NR alignment over large surface areas (Fig. 2.32B). NR alignment was further exploited for the generation of *polarized light* through preferred transmission of light waves oscillating parallel to the NR alignment axis; this effect is similar to the well-known phenomenon in which polarized light is induced after passing through liquid crystals.

Other morphologies of semiconductor nanoparticles have been reported besides QDs, NWs, and NRs. Metal chalcogenide *nanoplatelets*, in particular, have attracted interest since they constitute *two-dimensional* NP systems (also referred to as "quantum wells") – exhibiting different photophysical properties than either zero-dimensional NPs (QDs) or one-dimensional species (NRs and NWs). Figure 2.33A illustrates a synthetic scheme for the production of CdSe nanoplatelets developed by B. Dubertret and colleagues at UPR5 du CNRS, France. The researchers demonstrated that nanoplatelet organizations could be achieved through dissolution of the ionic precursors (Cd^{2+} and Se^{2-}) in an organic solvent containing a weak basic salt (acetate). Importantly, modulating the nanoplatelet thickness (by modifying temperature and reagent concentrations) yielded dramatic shifts of the optical properties of the particles (e.g. changing the bandgap of the semiconductor quantum well; see Fig. 2.33B), pointing to possibilities for the use of nanoplatelets in electro-optical devices.

Nanoparticles comprising *copper-based chalcogenides* exhibit interesting structural and physical properties. The presence of copper ions, in particular, endows the nanoparticles with *localized surface plasmon resonance (LSPR)* properties. LSPR is a fairly common phenomenon encountered in metal NPs, particularly Au and Ag, due to oscillation of partly-bound electrons at the metal surface (see Chapter 3 for a more detailed discussion). LSPR is rare, however, in crystalline semiconducting NPs, as the concentrations of surface electrons are generally lower than metals. LSPR likely plays a role in the biosensing functionalities of the *copper-telluride* NPs produced by A. Cabot at the University of Barcelona through solution-based reaction of a copper salt with hydrophobically-coated telluride ions (Fig. 2.34). Interestingly, the CuTe NPs exhibited remarkable nonspherical morphologies which were dependent upon co-addition of an organic lithium-silicon reagent (which did not react with either copper or tellurium precursors). This reagent was hypothesized to activate reaction intermediate species which were subsequently transformed into NPs with distinct shapes – nanocubes, nanoplates, and nanorods.

The proliferation of research into semiconductor NPs has led to the synthesis of other anisotropic semiconductor NP morphologies exhibiting interesting structural and physical properties. *Branched nanostructures*, or *multipods*, have been reported for metal chalcogenide and other semiconductor NP systems (Fig. 2.35). In many instances, these nanostructures are comprised of different semiconducting building blocks in the core and branches (thus denoted "hetero-structured multipods"), and exhibit intriguing optical properties as their luminescence is affected by both the core material and the branches. *Dual* light emission has been observed, for example, in samples of hetero-structured multipods. The anisotropic structure of the multipods also affects the semiconductor bandgap, giving rise in some cases to broad spectral absorption windows (beneficial for applications such as solar cell design) and generation of luminescence in the infrared region.

"Seeded growth" approaches have been quite successful as means for the synthesis of branched semiconductor NPs of narrow size and shape distributions. Seeded

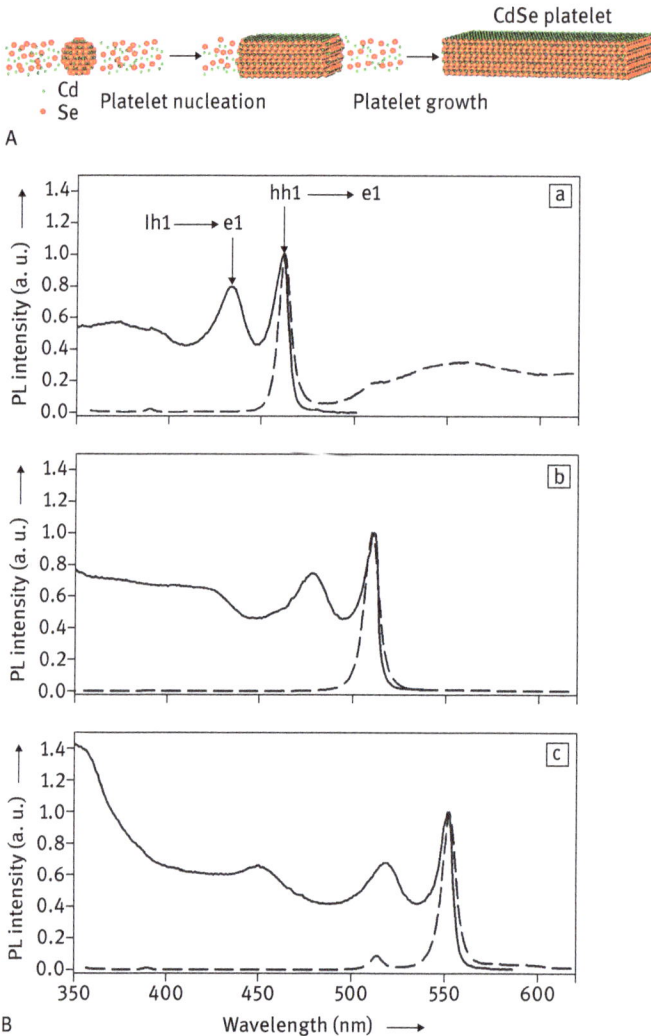

Fig. 2.33: Semiconductor nanoplatelets. **A:** schematic description of CdSe nanoplatelet synthesis. **B:** optical properties of CdSe nanoplatelets of different thicknesses (**a–c**). Significant shifts are observed in both absorbance spectrum (solid lines) and luminescence spectrum (broken lines) when thickness is modulated. The arrows indicate the two transitions expected in the visible spectrum of the crystalline CdSe nanoplatelets. Reprinted with permission from Ithurria and Duberthret, *JACS* **130** (2008), 16504–16505c © 2008 American Chemical Society.

growth techniques essentially distinguish between synthesis of the multipod nucleus (or core) and the growth of the arms. A generic seeded growth scheme, developed by L. Manna and colleagues at CNT-INFM, Italy, is presented in Figure 2.35. Initially, cubic or faceted seeds were synthesized and subsequently added to solutions containing

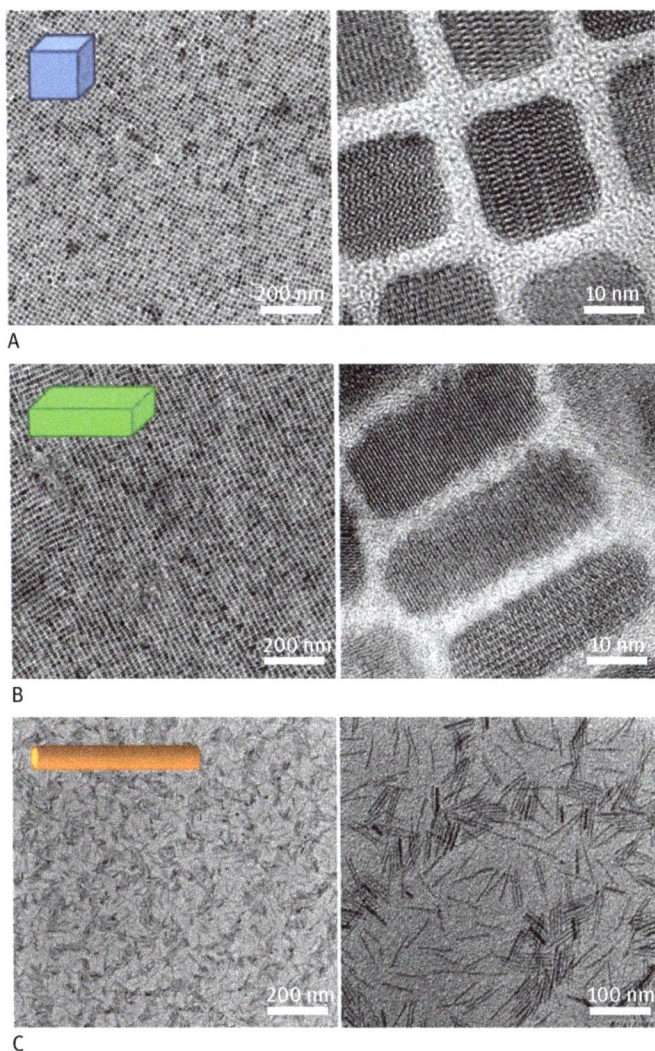

Fig. 2.34: Nonspherical CuTe nanoparticles. Electron microscopy images showing different NP morphologies. The distinct shapes were produced through modifications of the reaction conditions, including Cu:Te ratio, reaction time, and temperature. Reprinted with permission from Li et al., *JACS* **135** (2013), 7098–7101, © 2013 American Chemical Society.

the building blocks of the arms. Crystal growth of the arms was then initiated on the facets of the seeds (usually adopting different crystalline structures than the seeds), as growth of the branches is more energetically favorable compared to spontaneous nucleation in the solution. Reaction conditions such as temperature, concentration of surfactants, and concentration of the semiconducting precursors forming the crystalline arms can be altered, enabling control over structural parameters, particularly

S or Te precursors, co-injected with the seeds in TOP

Cubic sphalerite seed
(CdSe, CdTe, ZnTe)

A

Sphalerite core (CdSe, CdTe, ZnTe)/
wurtzite arms (CdS, CdTe)
Tetrapod

B

2nm

C

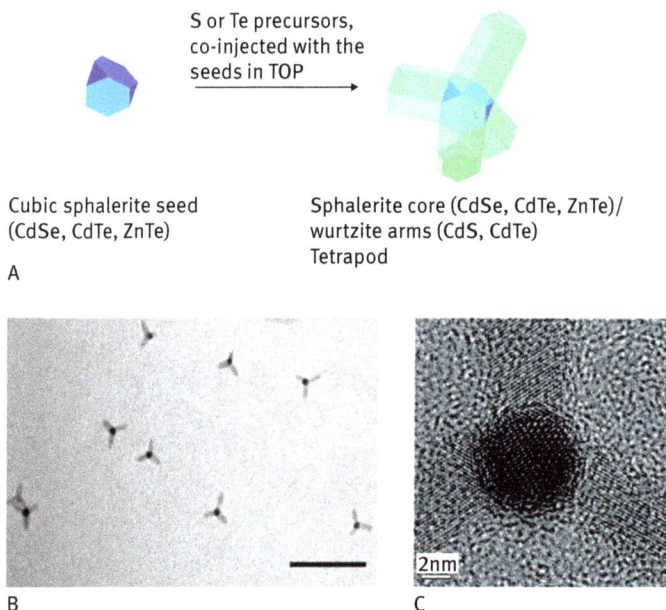

Fig. 2.35: Hetero-structured tetrapods. **A:** synthesis scheme: The core is prepared separately, and the branches are subsequently grown upon the seed's facets. **B:** electron microscopy image showing typical tetrapods; scale bar corresponds to 100 nm. **C.** A high resolution transmission electron microscopy (TEM) image demonstrating the different crystallinities of the core and branches. Reprinted with permission from Li et al., *JACS* **131** (2009), 2274–2282, © 2009 American Chemical Society.

branch length. It should also be noted that the nucleating seeds can be assembled from materials other than semiconductors; multipods grown from seeds of Au NPs and other noble metals have been reported (see Chap. 6).

Branched semiconductor NPs can also be assembled by imposing differential growth kinetics upon the crystal facets of an NP serving as the core, leading to the preferred crystal growth of "arms" protruding from a central nucleus. It has, however, been notoriously difficult to achieve uniform shapes of such nanostructures. Researchers have explored various strategies to control the morphology and structural features of branched NPs, particularly through the use of *template agents* – surfactants which promote material growth at specific facets of the core.

Other intriguing semiconductor NP morphologies have been reported, demonstrating both synthetic acumen and new functional avenues. Figure 2.36, for example, presents "split-crystal" morphologies of bismute-sulfide (Bi_2S_3) NPs. Strikingly, the sheaf-like nanostructures, synthesized in solution by A. P. Alivisatos and colleagues at the University of California, Berkeley, resemble naturally-occurring minerals. This observation suggests that crystal-splitting is probably a universal driving force for the assembly of seemingly complex NPs. Moreover, this study indicates that semiconduc-

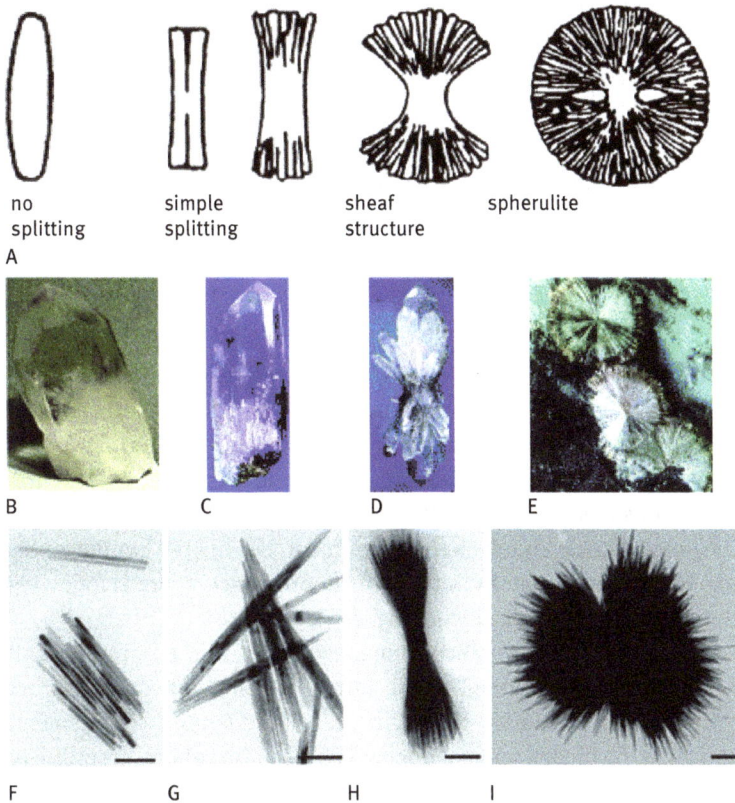

Fig. 2.36: "Split crystal" nanoparticles structurally resemble inorganic minerals. Bi_2S_3 nanoparticles of different structures (electron microscopy images in panels **F–I**; scale bars correspond to 100 nm) mimic known split-crystal organization of inorganic minerals (**B–E**). The mineral structures are schematically shown in panels **A–D**. Reprinted with permission from Tang and Alivisatos., *Nano Lett* **6** (2006), 2701–2706, © 2006 American Chemical Society.

tor NPs (and other NP compositions) with diverse and new morphologies might be inspired by nature, and synthetic schemes mimicking natural assembly processes of inorganic crystalline minerals might be harnessed to expand the NP synthesis toolbox.

2.6 Silicon and germanium nanoparticles

Silicon is the "bedrock element" of broad swaths of modern industries, including semiconductor-based microelectronics, solar energy, and others. The prominence of silicon stems from its tunable semiconducting properties, specifically following doping with small amounts of other elements, making it an n-type or p-type semiconduc-

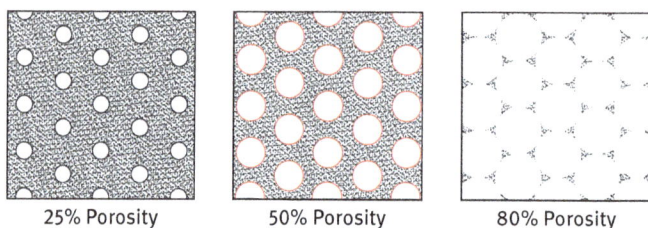

25% Porosity   50% Porosity   80% Porosity

Fig. 2.37: Silicon nanoparticles formed in porous silicon. Gradual etching of the silicon matrix leaves behind a skeleton of silicon nanowires, giving rise to photoluminescence.

tor, depending on the impurities introduced into the silicon lattice. The introduction of silicon to the "nanoparticle universe" has been slow, however; silicon is an indirect bandgap material which translates into relatively inefficient light absorbance and emission, limiting overall applications in photonics and photo-electronics. The dramatic breakthrough in this field occurred in the late 1980s with the demonstration by L. T. Canham at the Royal Signals and Radar Establishment, UK, of high luminescence from *porous silicon*. In that seminal study, intense red fluorescence emerged following gradual dissolution of crystalline bulk silicon (Fig. 2.37).

While the exact nature of porous silicon luminescence is still controversial, it is likely that the phenomenon is linked to the formation of a network of *silicon nanowires*. The "free standing" silicon NWs likely display quantum confinement, conceptually identical to the binary semiconductor NPs discussed above, resulting in a significantly greater bandgap in comparison to crystalline bulk silicon. Other Si NP morphologies have been studied, particularly colloidal spherical Si NPs. Figure 2.38, for example, demonstrates the (expected) size-dependent photoluminescence shifts of Si NPs, echoing the color modulation of binary semiconductor quantum dots accomplished by changing the particle diameters.

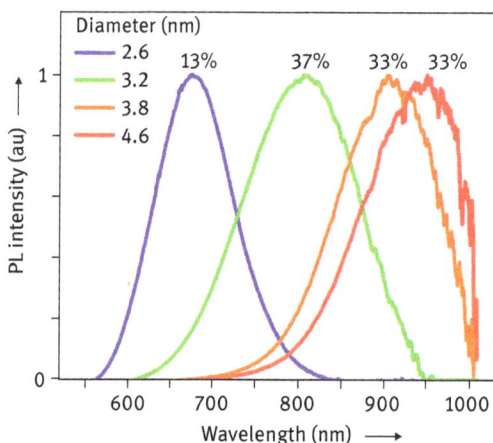

Fig. 2.38: Size-dependent photoluminescence shifts of silicon nanoparticles. The spectra were recorded for nanoparticles of the diameters indicated by the color code. The percentage values indicate the quantum yields (i.e. efficiency of luminescence emission). Reprinted with permission from Hannah et al., *Nano Lett* **12** (2012), 4200–4205, copyright (2012) American Chemical Society.

The discovery of luminescent porous silicon has precipitated an explosive growth in silicon nanoparticle research (also termed *silicon nanocrystals*). The advantages of silicon in comparison to other semiconductor materials are significant: silicon is an abundant and inexpensive material, it is not toxic (in contrast to common NP constituents such as cadmium, selenium, and lead), and importantly, silicon is compatible with current electronic and photonics technologies and their sophisticated infrastructures. As has been the case with other NP systems, this field has progressed along two somewhat parallel tracks feeding each other: the development of *synthetic techniques* for the construction of Si NPs with controlled and uniform sizes, and advances in scientific and technological applications arising from the exploitation of the photophysical properties of Si NPs.

Various methods of Si NP synthesis have been developed. Like other NP systems, the main synthetic challenges have been to develop the means of creating uniform (i.e. *monodisperse)* particles in controlled sizes and shapes, and/or to introduce effective procedures for size-separation of the NP products. These intertwined goals have led to proliferation of experimental schemes. Methodologies of Si NP synthesis can be roughly divided into three main groupings: etching of bulk silicon designed to create free-standing nanostructures (i.e. the "porous silicon" methodology); "top-down" processes employing ablation, vapor deposition, and similar techniques largely based on construction of Si NPs with the aid of bulk materials or growth-directing surfaces; and solution-based procedures aimed at utilizing spontaneous self-assembly phenomena for NP assembly.

Solution self-assembly processes rely on the reaction of various reagents to produce the pure silicon species. *Silanes* (Si_xH_y) and silane derivatives have been common precursors in many solution-based Si NP synthetic schemes. Silanes are reactive and can be derivatized with various functional units. Upon functionalization they can be dissolved in different organic solvents and can also be vaporized and used in *chemical vapor deposition* (CVD) techniques. Some experimental procedures have achieved "clustering" of silane precursors followed by hydrogen (H_2) elimination, resulting in Si NP formation.

While many Si NP synthesis techniques have been introduced, a major challenge in this field has been to achieve homogeneity of the resultant NPs (i.e. generating uniform size distributions). Figure 2.39, for example, presents electron microscopy images and optical properties of monodisperse Si NP preparations, accomplished by G. A. Ozin and colleagues at the University of Toronto, Canada, by applying advanced *centrifugation*. In this approach, uniform alkyl-capped Si NPs were obtained through a "density gradient ultracentrifugation" scheme, based on the observation that suspensions of Si NPs with different diameters possess varying macroscopic densities. The relationship between NP size and weight opens up other avenues for achieving purification of Si NPs through precipitation-based techniques. Figure 2.39B shows a photograph of Si NP suspensions, each containing different-sized monodisperse particles isolated through size-selective centrifugation.

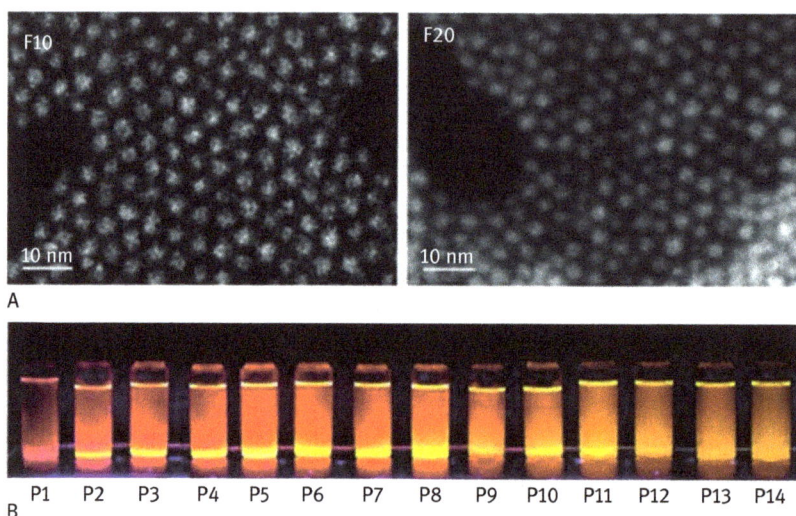

Fig. 2.39: Monodisperse silicon nanoparticles. **A:** Uniform distribution of silicon nanoparticles obtained through density-gradient ultracentrifugation; **B:** suspensions of silicon nanoparticles separated through centrifugation have distinct colors related to nanoparticle sizes. Reprinted with permission from Mastronardi et al., *Adv. Mater.* **24** (2012), 5890–5898, © 2012 John Wiley and Sons.

Framework-based (i.e. *guest-host*) synthesis has been promoted as a versatile strategy for creating uniform Si NPs. This approach aims to utilize porous materials as scaffolds for the assembly of Si NPs; accordingly, selecting host matrixes with uniform nanoscale pores might enable formation of NP guests of narrow size distributions. One of the challenges of this approach is the development of facile experimental techniques for synthesis of the guest NP species inside the host's pores and their eventual release from the host matrix. *Zeolites* have been among the framework materials examined as hosts for Si NP synthesis. Zeolites constitute a diverse family of porous aluminosilicates, exhibiting varied applications, primarily as adsorbents and catalysts. Their advantage as possible host material stems from the broad variation of pore sizes in different zeolites, and the chemical reactivity of the pore "walls".

Similar to other nanoparticle systems, stabilization of Si NPs in solution and prevention of their aggregation require capping the particles with molecular layers usually comprised of hydrophobic residues. Figure 2.40A presents a molecular model of hydrophobic-coated Si NPs comprising of a crystalline silicon core encapsulated within a layer of carbohydrate *alkyl chains*. Importantly, H. K. Datta and colleagues at Newcastle University, UK, demonstrated that such alkyl-capped Si NPs were stable in aqueous solutions and could be internalized by mammalian cells facilitating fluorescence microscopy imaging (Fig. 2.40B).

Cetyl trimethylammonium bromide (CTAB) is an amphiphilic surfactant widely used as a coating and stabilizing agent for Si NPs (and quantum dots in general). CTAB

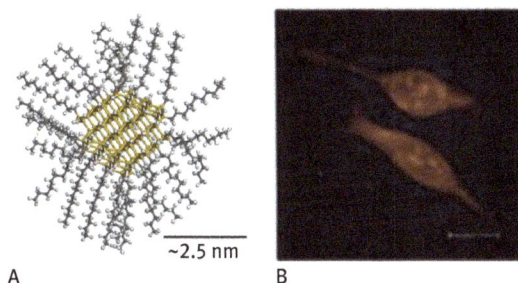

~2.5 nm

A   B

Fig. 2.40: Coated silicon nanoparticles for cellular imaging. **A:** molecular model of the nanoparticle showing the crystalline silicon core (yellow) embedded within the alkyl chain layer (grey). **B:** fluorescence confocal microscopy image of mammalian cells incubated with the hydrophobic-coated silicon nanoparticles. The nanoparticles are clearly internalized within the cells. Reprinted with permission from Alsharif et al., *Small* **5** (2009), 221–228, © 2009 John Wiley and Sons.

exhibits a long hydrophobic tail and ionic headgroup which permits water solubility. As such, CTAB both prevents aggregation and, importantly, enables transfer of coated Si NPs from organic phases (in which the NPs are usually synthesized) into the aqueous phase. Stabilization of Si NPs in water can also be accomplished through *covalent attachment* of hydrophilic residues onto the NP surface. This generic route is aided by the extensive chemical arsenal of (bulk) silicon functionalization. Moreover, residues endowing specificity and targeting capabilities have also been chemically conjugated onto Si NPs, thereby expanding their biological applicability.

Biological imaging has been one of the most dynamic fields of Si NP applications. Part of the impetus for using Si NPs for cell and tissue imaging is their perceived lack of toxicity, especially in comparison to other well-known NP markers used for bioimaging, such as quantum dots. Similar to QDs, the proliferation of Si NP imaging applications has correlated with the increased sophistication of synthetic strategies, particularly aimed at attaining nanoparticle size control required for color tuning capabilities, and surface display of biomolecular targeting units. Figure 2.41 provides an example of tissue imaging with functionalized Si NPs. The experimental strategy pursued by P. N. Prasad and colleagues at the University of Buffalo, NY, was to conjugate Si NPs with the arginine-glycine-aspartic acid (RGD, single letter code) peptide motif. This tripeptide is known to bind *integrins*, membrane-associated proteins abundant in cancerous cells. The functionalized Si NPs were injected into the bloodstream of tumor-bearing mice and, as shown in Figure 2.41, enabled tumor visualization (red luminescent regions) by targeting the integrin-displaying cells.

Silicon nanowires (Si NWs) have attracted considerable interest as conduits for electron transport and possible components in nanoelectronic and nanophotonic devices. Si NWs have mostly been produced via the vapor-liquid-solid (VLS) approach, in which slow growth of the nanowire on a solid support is facilitated through reaction with reagents injected in a vapor phase (a detailed description is provided above,

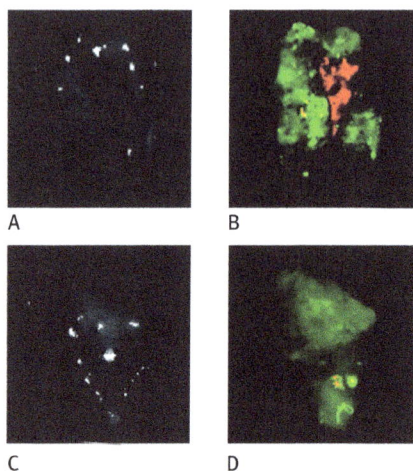

Fig. 2.41: Tumor imaging with peptide-functionalized silicon nanoparticles. Optical images (**left**) and photoluminescence images (**right**) of mice tumors labeled with silicon nanoparticles. **Top row:** addition of silicon nanoparticles conjugated with the RGD peptide motif. **Bottom row:** addition of bare silicon nanoparticles (not displaying the peptide motif). The red luminescent regions at the top image correspond to binding of the RGB-silicon nanoparticles to the tumor cells. Reprinted with permission from Erogobgo et al., *ACS Nano* **5** (2011), 413–423, © 2011 American Chemical Society.

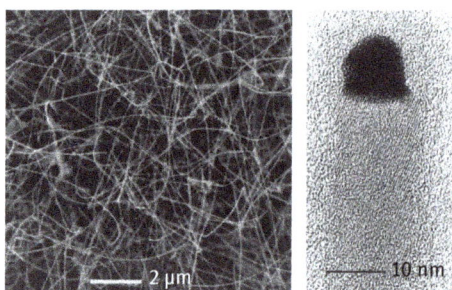

Fig. 2.42: Silicon nanowires grown via the vapor-liquid-solid (VLS) method. The electron microscopy image on the left shows a mesh of Si nanowires (scale bar corresponds to 2 μm); the high resolution transmission electron microscopy on the right shows the gold nanoparticle tip and the crystalline silicon nanowire attached to it. Reprinted with permission from Hu et al., *Acc. Chem. Res.* **32** (1999), 435–445, © 1999 American Chemical Society.

in the case of the binary semiconductor NWs). Figure 2.42 shows Si NWs produced by C. M. Lieber and colleagues at Harvard University using the VLS method. The procedure relied upon first fabricating gold nanoparticles, generated through ablation (e.g. cleavage) of a gold sheet. The Au NPs functioned as "nucleation seeds" and catalysts for the growth of crystalline Si NWs through reaction with a silicon source (silane gas).

The VLS technique also permits modulation of NW *compositions*, and thereby their physical properties through addition of *doping substances* to the vapor reaction mixture. M. A. Filler and colleagues at the Georgia Institute of Technology, for example, observed localized surface plasmon resonance (LSPR) in Si NWs doped with *phosphorous* atoms (Fig. 2.43). LSPR is usually observed in metal nanoparticles, arising from free surface electrons (see Chap. 3); surprisingly, the researchers recorded intense light absorbance in the mid-infrared spectral region in samples of the phosphorous-doped Si NWs, the peak position of which could be tuned by chang-

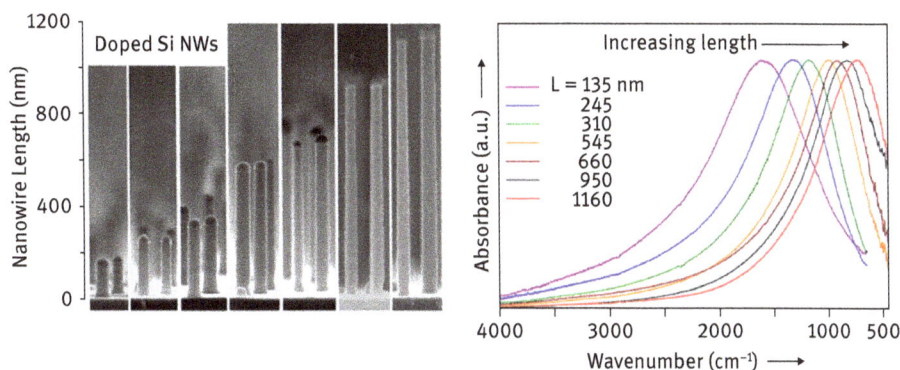

Fig. 2.43: Localized surface plasmon resonance (LSPR) in phosphorous-doped silicon nanowires. **Left:** silicon nanowires of different lengths grown via the vapor-liquid-solid (VLS) technique; phosphorous doping was carried out through addition of the reagent in the vapor phase. **Right:** optical absorbance of the nanowires recorded in the mid-infrared region, indicating LSPR effect. The shifts of the absorbance peaks are correlated to the nanowire lengths. Reprinted with permission from Chou et al., *JACS* **134** (2012), 16155–16158, © 2012 American Chemical Society.

ing the NW length, suggesting that the absorbance peak arose from a longitudinal LSPR effect (i.e. LSPR associated with the NW long axis), echoing optical effects commonly encountered in gold nanorods and nanowires (see Chap. 3). The apparent LSPR effect was ascribed by the researchers to the increased concentration of charge carriers at the Si NW surface, induced by the phosphorous dopants, which resulted in enhanced electron mobility.

An interesting and potentially broad-based application of Si NWs has been as field effect transistors (FETs). Si NW FETs exhibit an important advantage, as the use of silicon makes the conjugation of diverse molecular entities and recognition elements upon the NW surface possible. Targeting and docking of specific analytes is thereby feasible, generating a distinct electrical signal in the FET device. Figure 2.44 shows a photograph of an FET fabricated by H. Haick and colleagues at the Technion, Israel, comprising of a single Si NW. Importantly, by coating the Si NW with different

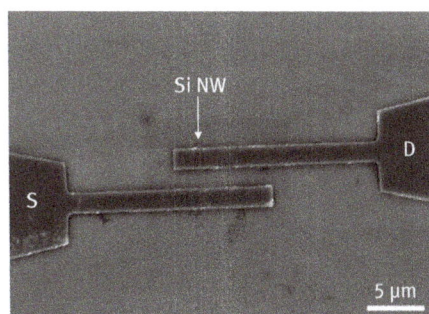

Fig. 2.44: Silicon nanowire field effect transistor (FET). A single silicon nanowire, surface functionalized to attract gas molecules, was placed between the source and drain electrodes. Adsorption of gas analytes onto the nanowire generates a distinct electrical signal in the transistor. Reprinted with permission from Wang et al., *ACS Appl. Mater. Interfaces* **4** (2012), 4251–4258, © 2012 American Chemical Society.

layers, both hydrophobic and hydrophilic, the researchers used the device for sensing a wide variety of analytes, particularly gas molecules (also referred to as *volatile organic compounds, VOCs*). Furthermore, creating *arrays* of Si NWs, in which each array element (i.e. Si NW) was differently surface-functionalized, could produce specific fingerprints of VOCs and other analytes comprising of the transistor response (i.e. conductivity) recorded upon interactions of a particular analyte with the differently-functionalized NWs.

Si nanorods (NRs) represent another silicon nanoparticle configuration, conceptually similar to NWs, which exhibit interesting photophysical properties. Figure 2.45 depicts a synthesis scheme developed by B. A. Korgel and colleagues at the University of Texas, Austin, for the production of luminescent Si NRs. The researchers employed a "solution-liquid-solid" procedure in which the silane precursor was dissolved in solution (rather than in a vapor phase, as in VLS methodologies), and the anisotropic growth was initiated on tin (Sn) clusters serving as "seeds". Intriguingly, while the as-synthesized Si NRs were *not* fluorescent (presumably due to quenching by silicon oxide impurities), treatment with a strong acid and further coating of the NRs

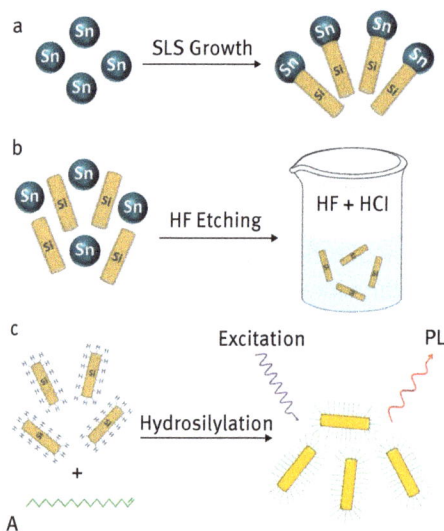

Fig. 2.45: Luminescent silicon nanorods. A: synthesis scheme: (a) "solution-liquid-solid" (SLS) growth using tin seeds as substrates; (b) acid treatment to remove the silicon-oxide layer; (c) coating with a hydrophobic monolayer to prevent re-oxidation of the nanorods.

B: Photoluminescence properties of the coated silicon nanorods: the untreated silicon nanorods are not luminescent (left bottles) due to quenching by silicon oxide, while the protected silicon nanorods are highly fluorescent (right bottle emitting red luminescence). Reprinted with permission from Lu et al., *Nano Lett.* **13** (2013), 3101–3105, © 2013 American Chemical Society.

with hydrophobic moieties resulted in high photoluminescence (Fig. 2.45B). This phenomenon has been ascribed to the removal of surface oxide species through acid etching and subsequent protection of the treated surface from further oxidation by the hydrocarbon layer.

Similar to the binary semiconductor nanoparticles discussed above, Si NPs and NWs might be used in electro-optical applications and devices. Si NP-based light emitting diodes (LEDs) have been successfully demonstrated, generating light in a broad spectral range – from the infrared to the visible spectral region. G. A. Ozin and colleagues at the University of Toronto, for example, fabricated a Si NP LED in which modification of the size of the particle made tuning the energy bandgap and thus the color emitted by the LED possible (Fig. 2.46). The data in Figure 2.46 directly demonstrates that quantum confinement effects pertinent to the Si NPs play a fundamental role in determining the optical properties of the LED, analogous to binary semiconductor quantum dots.

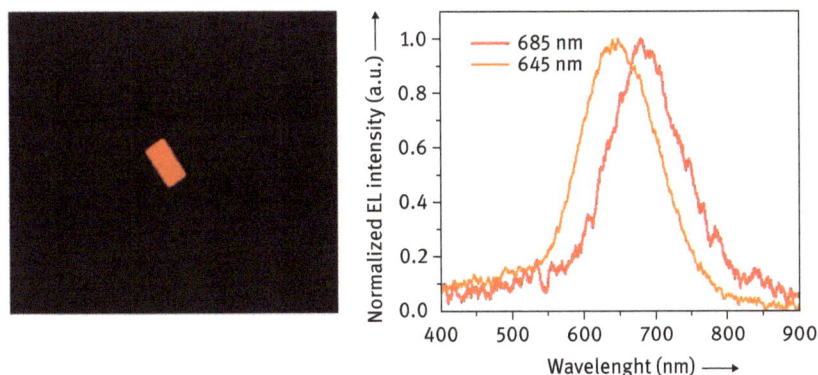

Fig. 2.46: Tunable light emitting diode (LED) fabricated from silicon nanoparticles. **Left:** photograph of the silicon nanoparticle-LED showing the electroluminescence (voltage of 15V), length of nanoparticle is approximately 3 nm. **Right:** shift of the electroluminescent signal depends upon the average size of the silicon nanoparticles. Reprinted with permission from Puzzo et al., *Nano Lett.* **11** (2011), 1585–1590, © 2011 American Chemical Society.

Silicon is the dominant element in solar cell technologies. Usage of Si NPs and NWs, in particular, might improve the performance of silicon-based solar cells through enhancing light harvesting due to the tunable energy bandgap of the NWs, as well as contributing to efficient charge transport. Indeed, some studies have demonstrated that Si NW arrays exhibit better light harvesting capabilities compared to planar silicon wafers. Such NW arrays are also easier to fabricate and require less silicon material (thus lowering the costs) compared to conventional wafers.

As is the case with other materials discussed in this book, Si nanoparticle morphologies have been fabricated which are not only spherical or cylindrical. *Silicon* nanocubes, for example, might represent useful building blocks for larger nanoparticle assemblies. As such, *nonspherical* nanoparticles offer advantages since their facets might make it possible to assemble complex "superstructures", conceptually reminiscent of "lego-block" structures. Figure 2.47 depicts Si nanocubes synthesized by J. G. C. Veinot and colleagues at the University of Alberta, Canada, via high temperature solid-state "annealing" of spherical Si NPs. The researchers discovered that prolonged annealing of spherical Si NPs induced thermodynamic rearrangements of the NP surface into planar surfaces, ultimately yielding nanocubes. The facets of the Si nanocubes could be further functionalized with hydrophobic residues designed to maintain stability and prevent aggregation in solution.

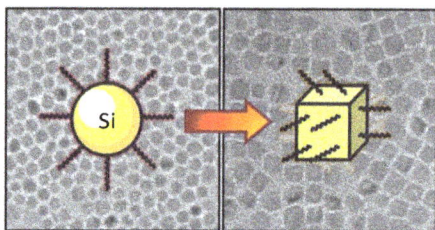

Fig. 2.47: Silicon nanocubes. The schematic figure illustrates the transformation of spherical silicon nanoparticles into nanocubes; hydrophobic moieties are attached onto the nanocube facets to prevent aggregation in solution. The background electron microscopy images depict the nanocubes formed. Size of nanocubes is approximately 10 nm. Reprinted with permission from Yang et al., *JACS* **134** (2012), 13958–13961, © 2012 American Chemical Society.

Silicon carbide (SiC) *tetrapods* (Fig. 2.48) are intriguing silicon-based NPs displaying both unusual structural features as well as notable photoluminescence properties. The NPs were produced by A. P. Magyar and colleagues at Harvard University through a chemical vapor deposition (CVD)-based scheme in which a film of polymer-embedded adamantane (a small carbonaceous molecule exhibiting a diamond-like structure) deposited upon a silicon dioxide layer was treated with a methane/hydrogen plasma vapor. The SiC tetrapods exhibit intense photoluminescence, the wavelength of which depended upon the structural properties (primarily sizes of the tetrapod arms). A rather unusual observation was the extremely narrow widths of the photoluminescence peaks (claimed to be narrower than virtually all commercially available semiconductor quantum dots), a feature which could make these NPs very useful as high-resolution biological imaging agents. Indeed, the presumed low biological toxicity of both silicon and carbon should make the SiC "nanotetrapods" particularly attractive bio-imaging agents as compared to inorganic quantum dots.

Germanium (Ge) is another potentially useful element in nanoelectronic and photonic devices. Ge NWs exhibit high charge carrier mobility and efficient light absorption due to their large surface areas. The low energy bandgap of germanium (0.67 eV,

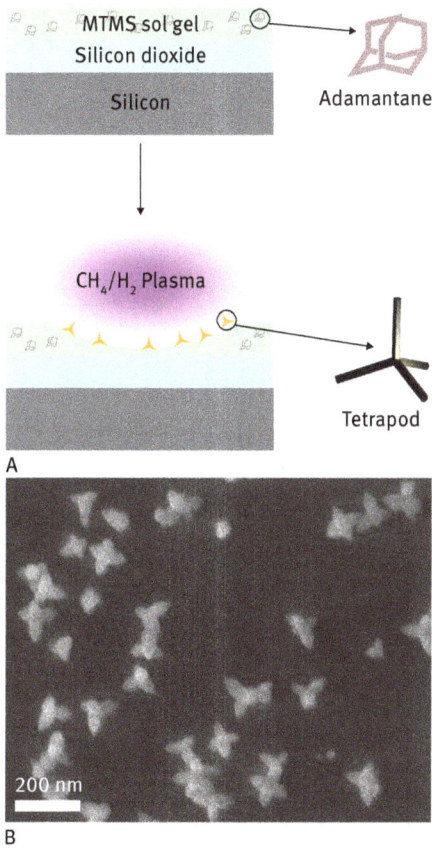

Fig. 2.48: Silicon carbide nanotetrapods. **A:** Synthesis scheme: adamantane film is plasma-treated to generate silicon-carbide tetrapods. **B:** Electron microscopy image of the nanotetrapods. Reprinted with permission from Magyar et al., *Nano Lett.* **13** (2013), 1210–1215, © 2013 American Chemical Society.

in the near infrared region of the electromagnetic spectrum) has been a primary reason for the use of Ge NWs in visible and infrared photo-detectors. Figure 2.49 depicts a photo-detector design based upon a field effect transistor (FET) utilizing a single Ge NW as the "gate" component. The detection mechanism in such FET-based devices relies upon modulation of the electrical carrier density across the gate following illumination of the semiconducting NW, enabling flow of electrons between the source and drain electrodes. In the setup shown in Figure 2.49, designed by J. Park and colleagues at the Rowland Institute, USA, a Ge NW was placed between two electrodes acting as the source (S) and drain (D); light absorption by the Ge NW generated charge carriers and "opened" the gate, enabling the passage of current between the electrodes.

Devices utilizing single NWs as active components are hard to produce in large batches, limiting their practical and commercial potential. Bundles of Ge NWs, however, have been employed in opto-electronic devices, performing similarly to devices containing individual NWs. Figure 2.50 shows a photo-detector comprising of an interconnected network of Ge NWs as the photoactive component. The system, developed by H. E. Unalan and colleagues at the Middle East Technical University, Turkey,

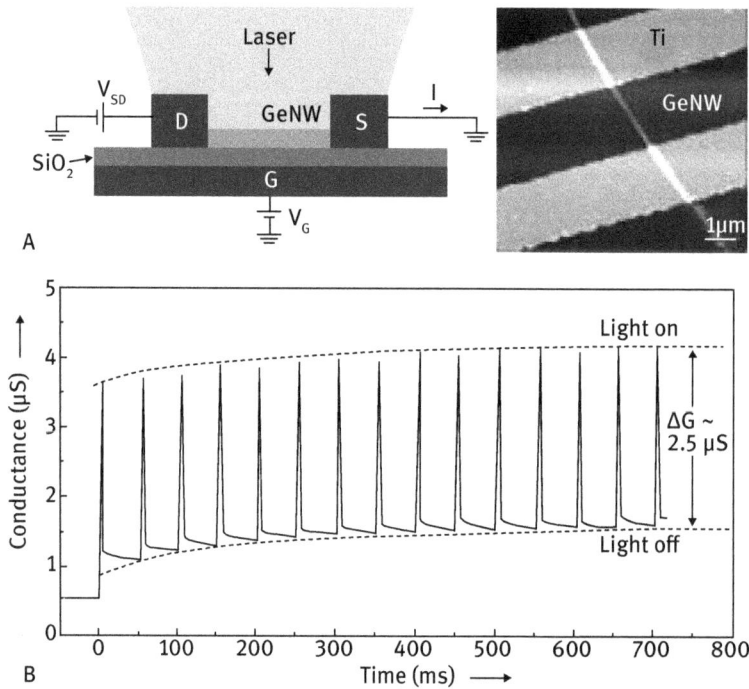

Fig. 2.49: Photo-detector based upon a germanium nanowire field effect transistor (FET). **A:** scheme of the experimental setup; changes in the electrical current passing through a germanium nanowire placed between the source and drain electrodes are induced by light irradiation. The atomic force microscopy (AFM) image on the right shows the germanium nanowire transistor. **B:** time-dependent electrical conductance response as the laser light is turned on and off. Reprinted with permission from Ahn and Park, *Appl. Phys. Lett.* **91** (2007), 162102, © 2007 American Institute of Physics.

contained two types of electrical contacts – silver nanowires and single wall carbon nanotubes (SWNT) – and exhibited good light sensitivity and photo-response parameters (low light intensity threshold and fast relaxation times). Such NW networks might be useful for potential practical applications because they can be produced via simple solution-based deposition, are optically transparent, and are stable for long time periods.

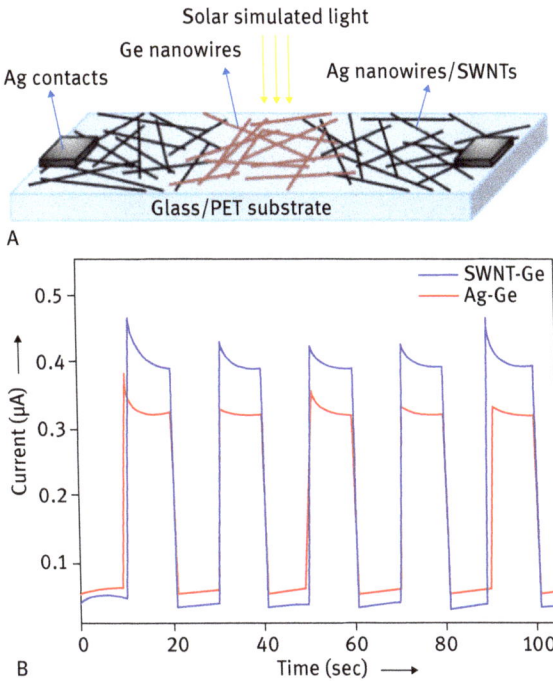

Fig. 2.50: Photo-detector comprised of multiple germanium nanowires. **A:** scheme of the device; efficient electrical contacts are maintained by placing the germanium nanowires between dispersed silver nanowires or single-wall carbon nanotubes (SWNTs). **B:** on-off response times of the photo-detector upon light irradiation. Shown are the data recorded with silver nanowires and SWNTs, respectively, as electrical contacts. Reprinted with permission from Aksoy et al., *Nanotechnology* **23** (2012), 325202, © 2012 IOP Publishing.

3 Metal nanoparticles

Metallic nanoparticles (and their compositional relatives bimetallic nanoparticles) constitute a diverse and prominent group of NPs. The physical properties and applications of metallic NPs stem largely from their (relatively) free surface electrons which endow unique electronic and optical features. This chapter discusses the main nanoparticle groups, specifically comprising of *gold*, *silver*, and *transition metals*. *Hybrid metal (bimetallic)* NPs are also discussed, reflecting their unique structural properties and functions.

3.1 Gold nanoparticles

Man's interest in manipulating gold is as old as our fascination with the metal. Indeed, the sophisticated methods of gold processing developed over millennia have provided a solid base for the huge scientific/technological field of *gold nanoparticles (Au NPs)*, which is among the most diverse in NP research. While Au NPs have become prominent members in nanotechnology research only recently, these particles have contributed to science and technology for much longer. The "Lycurgus cup" (shown in Chap. 1, Fig. 1.1) is a fitting example of the physical manifestation of Au NPs, even though Au NPs were not known to goldsmiths, artisans, or scientists of the era this ornament was produced in. However, the same properties that endeared gold to humanity – inertness, long-term stability, physical properties of color, heat and electrical transport, and its variety of synthetic pathways for size/shape manipulation – have been widely exploited in Au NP design and applications. The discussion below aims to cover the main areas of Au NP research and practical applications, but the ever-expanding scope of this field naturally means that not all facets of these fascinating nanoparticles are discussed.

As a discipline sprouting in large part from colloid science, Au NPs encompass a wide range of sizes and shapes (Fig. 3.1). Likely the most abundant and studied morphology – spherical Au NPs (often monopolizing the generic term "nanopar-

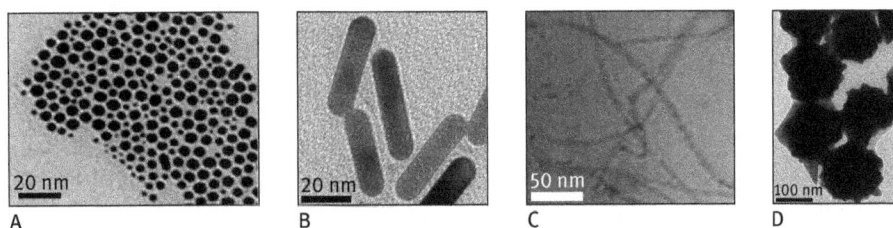

Fig. 3.1: Gold nanoparticle morphologies. Electron microscopy images showing **A:** spherical nanoparticles; **B:** nanorods; **C:** nanowires; **D:** "nanostars".

ticles" in the scientific literature) – vary in size from sub-nanometer scale all the way to hundreds of nanometers. The spherical configuration is closely related to the generic formation mechanism of Au nanostructures, discussed in more detail below, in which particles assemble through "crystallization nuclei" which subsequently induce isotropic growth of metallic gold. Au *nanorods* have also been widely studied, exhibiting unique properties arising from the *anisotropic* rod morphology. Most Au nanorod structures adopt *cylindrical* organization, and varied synthetic techniques enable tuning of the rod *aspect ratio* (i.e. ratio between nanorod length and diameter), producing Au nanorods and nanowires with diverse dimensionalities. More recent research has brought forth more "exotic" Au NP structures, including nanostars, nanoshells, nanocubes and others, some of which are discussed below.

Color has been one of the most-recognized characteristics of Au NPs, and is intimately related to particle dimensions and the aggregation state. The appearance of color is associated with one of the most fundamental properties of NPs – the *surface plasmon resonance (SPR)* phenomenon. SPR (or more fittingly *localized* SPR in the case of Au nanoparticles) occurs through interaction of incident light with electrons at the surface of the metallic (Au) NP. Figure 3.2 illustrates the SPR phenomenon in which the free electrons at a metal surface exhibit a collective oscillation – a *plasmon*. Accordingly, when the frequency of an incident light shining on a metal matches the oscillation frequency of the surface electrons, resonance occurs, manifested in a maximum (or "peak") in the absorbance (and/or emission) spectrum. For Au nanoparticles (and NPs of other noble metals), the oscillations of the electrons are constrained by the particle size – i.e. localized – giving rise to localized surface plasmon resonance (LSPR). In general, LSPR excitation (i.e. light absorbance) occurs in the visible region of the electromagnetic spectrum – thus the appearance of color.

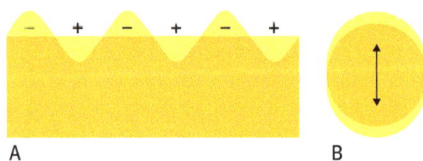

Fig. 3.2: Surface plasmons: **A:** surface plasmons occurring through oscillations of free electrons on a metal surface; **B:** localized plasmons are encountered in metal nanoparticles. Reprinted with permission from Mayer and Hafner, *Chem. Rev.* **111** (2011), 3828–3857, © 2011, American Chemical Society.

LSPR phenomena in Au NPs have been studied extensively and a detailed discussion of this important physical concept is beyond the scope of this book. Figure 3.3 highlights some of the important physico-chemical properties pertinent to nanoparticles which affect LSPR (the effects of hybrid NPs – i.e. NPs containing more than a single metal species – upon LSPR are discussed in Section 3.4). Among the most important

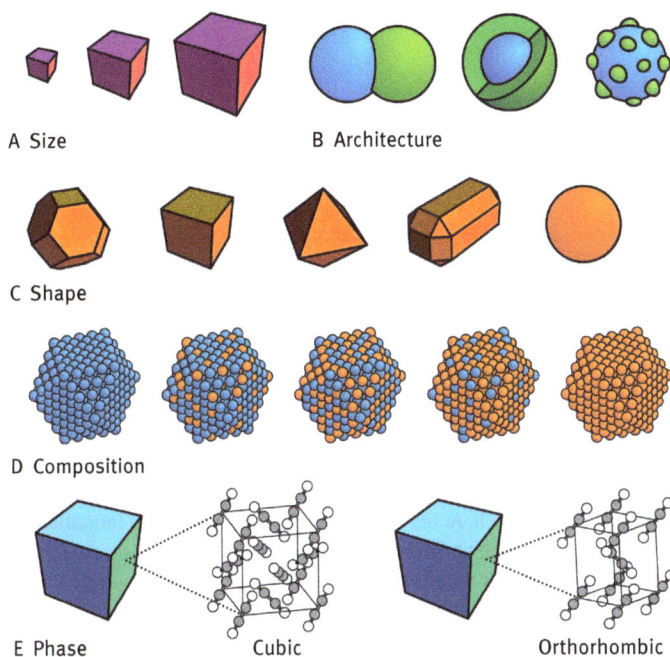

Fig. 3.3: Physico-chemical properties of nanoparticles affecting the localized surface plasmon resonance (LSPR). Reprinted with permission from Preston and Signorell, *Acc. Chem. Res.* **45** (2012) *45*, 1501–1510, © 2012, American Chemical Society.

facets of LSPR exploited in numerous Au NP applications have been the distinct relationships between the structural parameters of the individual NPs and NP assemblies and LSPR wavelengths. Examples of such relationships are discussed throughout the book.

3.1.1 Spherical gold nanoparticles

Spheres have been by far the most abundant and studied structures of Au NPs, and in most cases "Au NPs" generally refer to *spherical* Au NPs. The spherical shape is a direct outcome of the synthetic routes employed for the fabrication of Au NPs. While many synthesis procedures for Au NP production have been developed, most processes rely on solution-phase reduction of Au^{3+} ions to metallic Au^0 (Fig. 3.4A). The Au^0 "seeds" induce further reduction of gold ions, effectively giving rise to isotropic deposition of metallic gold resulting in overall spherical particles. While this basic particle growth mechanism is conceptually simple, there has been a heated debate as to the intermediate constituents in the synthetic pathways. Figure 3.4B, for example, depicts a proposed route for the formation of medium-diameter (around 7 nm) Au NPs. This

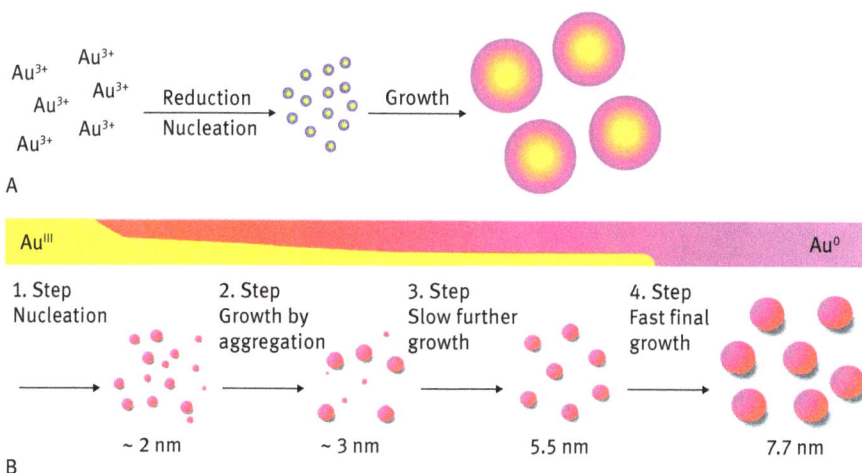

Fig. 3.4: Gold nanoparticle synthesis and growth. **A:** Generic scheme of Au nanoparticle formation, involving reduction of Au^{3+}, creating Au^0 nuclei, and nanoparticle growth. **B:** Proposed model for formation of medium-size Au nanoparticles. Reprinted with permission from Polte et al., *JACS* **132** (2010), 1296–1301, © 2010, American Chemical Society.

"four-step" model, outlined by R. Kraehnert and colleagues at the Universitaet Rostock, Germany, combines ion attachment to nucleation seeds with subsequent aggregation of individual Au NPs, ultimately producing a suspension of spherical Au NPs with a highly uniform size distribution.

The *Brust method* has been a popular synthetic approach for production of Au NPs (Fig. 3.5). The technique utilizes *chloroauric acid (HAuCl$_4$)* – a widely available reagent – as the source for the Au^{3+} ions. Chloroauric acid is dissolved in an organic solvent in the presence of two types of co-solutes: *reducing agents* (sodium borohydride, $NaBH_4$, sodium citrate, and others) responsible for transferring the electrons to the Au ions to form metallic gold (Au^0), and *stabilizing agents* (such as alkyl ammonium salts) to prevent agglomeration of the nanoparticles after reduction and growth.

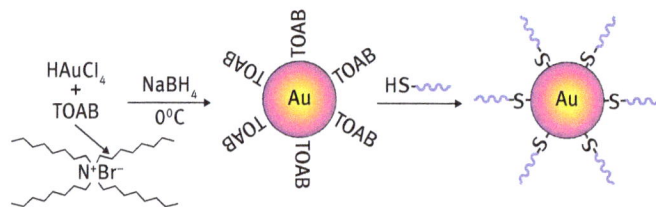

Fig. 3.5: Generic synthesis route of gold nanoparticles. Au^{3+} is reduced to Au^0 and adopts spherical morphology. The nanoparticles are stabilized in solution by surfactants such as tetraoctylammonium bromide (TOAB, *middle*), or by thiolated residues (*right*).

Many variations of this generic scheme have been introduced. A common alternative route involves substitution of the alkyl ammonium surfactants with *thiolated* residues (Fig. 3.5). In comparison with ammonium-based surfactants, which exhibit a relatively weak electrostatic attraction to the gold surface, thiol-containing molecules bind semi-covalently to the gold surface through the sulfur atom, yielding more stable and chemically-resilient nanoparticle coatings.

Thiol-based chemistry has been one of the major factors contributing to the explosive growth of Au NP applications. *Bi-functional* thiolated substances can be synthesized, displaying on one side sulfydryl moieties for binding to the Au NP surface (usually denoted as "linker" units), while diverse chemical units are attached to the other side of the molecule, endowing desired functionalities to the NPs. The NP surface-displayed moieties can provide *recognition capabilities*, or they might enable modulation of overall NP *surface properties* – for example through increasing/decreasing electrostatic charge. Numerous Au NP systems utilizing surface modification schemes have been developed based on thiol functionalization.

Thiolated residues do not only contribute to *post*-synthesis applications and properties of the Au NPs. In fact, thiolated ligands might participate in the synthesis itself. Burgeoning research has explored the effect of thiolated compounds on the structural properties of NPs, including dimensionalities, size uniformity, and growth kinetics. J. B. Tracy and colleagues at North Carolina State University, for example, have studied synthesis of Au NPs in the presence of thiolated ligands of different sizes. The experiments revealed that "bulky ligands" co-incubated with the Au seeds affected NP size in an *inverse* manner – i.e. larger ligands gave rise to smaller NPs and vice versa. As illustrated in Figure 3.6, the bulkier ligands presumably prevent gold seeds from "penetrating" to the NP core, thereby limiting the overall size of the NP ultimately formed.

3.1.2 Applications of gold nanoparticles in sensing

Au NPs constitute a powerful platform for molecular sensing, particularly *bio*sensing. First and foremost, the considerable diversity of thiol-based synthetic pathways means that varied functional units and recognition elements can be displayed on Au NP surfaces, making targeting desired analytes possible. Secondly, the nanometer *size regime* of Au NPs is comparable to many biological molecules (for example proteins), enabling efficient binding between the NPs and their (bio)molecular targets. Au NPs exhibit other practical advantages – high stability of the particles and surface-displayed ligands even in elevated temperatures, salinity, and acidity; chemical inertness (endowed through appropriate surface functionalization of the NPs); and high sensitivity, as a large number of particles can be synthesized in each experimental batch.

Color changes, associated with localized surface plasmon resonance, LSPR, have likely been the most widely-used intrinsic properties of Au NPs employed for sensing

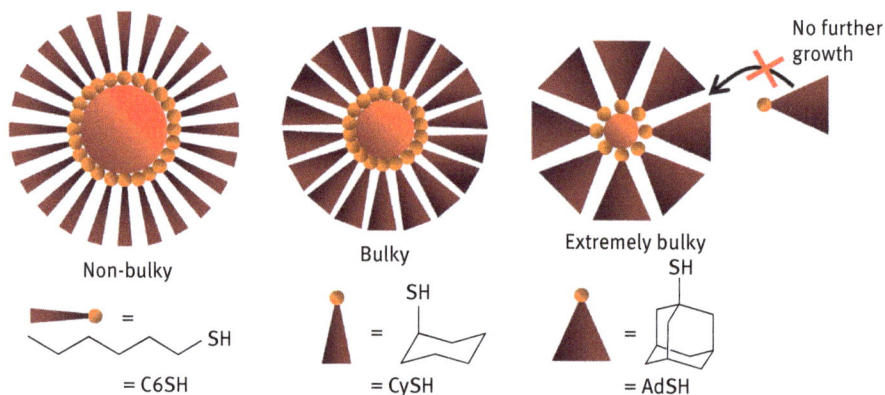

Fig. 3.6: Effect of the size of thiolated ligands upon gold nanoparticle growth. The experiment demonstrates that bulkier ligands induce formation of smaller Au nanoparticles, likely through restricting access to the growing nuclei. Reprinted with permission from Krommenhoek et al., *ACS Nano* **6** (2012), 4903–4911, © 2012, American Chemical Society.

applications. LSPR is among the "signature" physical properties of Au (and other noble metal) NPs. As indicated above, this phenomenon occurs through interaction of light with oscillating surface (conducting) electrons. For quite a wide diameter range of Au NPs, the resonating photons are confined within the particles, producing light emission at around 520 nm (appearing red-purple to the eye). LSPR changes are induced when chemical processes and/or molecular transformations occur at, or close to the NP surface. As such, LSPR shifts depend on several physical parameters, including NP size and shape, the refractive index of the medium surrounding the NP, and particularly the chemical environment in the proximity of the NP surface. The latter has been a primary factor in the broad utilization of Au NPs for biosensing, as even slight changes affecting the Au NP surface, such as type of coating (when implemented), binding of molecules through surface-displayed recognition elements, or NP aggregation, can induce detectable optical shifts.

LSPR-based sensing using Au NPs often relies on the display of molecular recognition elements on the NP surface; binding of target analytes to the functionalized NPs modulate the plasmon resonance signal, thereby providing a reporting tool on the presence of the analytes (Fig. 3.7). A broad range of binding events have been detected through Au NP-mediated LSPR, including antigen-antibody interactions, protein-protein and protein-carbohydrate recognition, DNA hybridization, and others.

Induction of NP *aggregation* was one of the early examples for the use of Au NPs for LSPR-based sensing. When NPs are located sufficiently close to each other (i.e. in aggregates or interlinked networks), electronic coupling results in extension of the oscillations over a greater distance and hence shift in the plasmon resonance frequency. The early implementation of this concept for colorimetric sensing of oligonucleotides (DNA) by C. Mirkin and colleagues at Northwestern University is depicted in Figure 3.8.

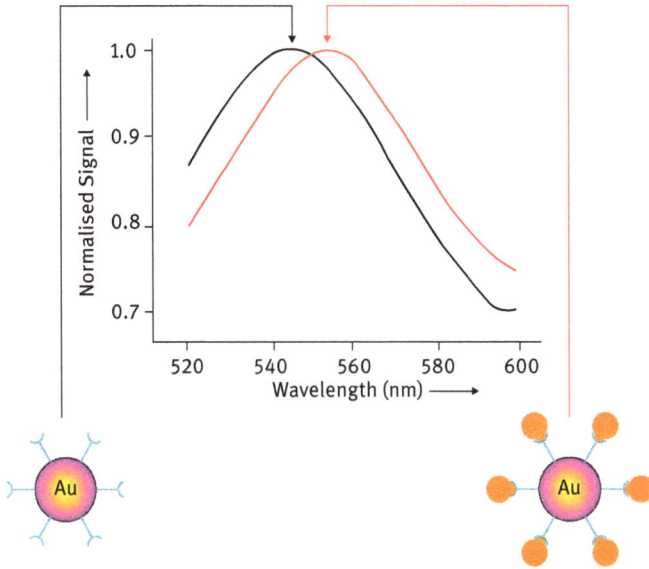

Fig. 3.7: Shift in localized surface plasmon resonance (LSPR) induced by molecular interactions occurring on the Au nanoparticle surface.

The Au NPs were chemically functionalized with two distinct thiolated oligonucleotide strands designed to recognize two domains of a target DNA strand through base-pair complementarity – the fundamental mechanism of DNA recognition. When binding did not occur between the NP-displayed oligonucleotides and its target sequence, the Au NPs appeared red as the LSPR of individual NPs was responsible for absorption of the short-wavelength component of incident light. However, when the target

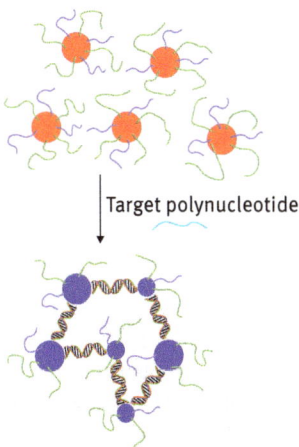

Target polynucleotide

Fig. 3.8: Color sensing of DNA through gold nanoparticle aggregation. The Au nanoparticles display two DNA strands recognizing different sequences within the target DNA. In the absence of the target strand, the Au NPs stay apart and the solution color is red. However, when the target DNA is present it binds adjacent nanoparticles through base-pair complementarity with the two distinct domains, generating an extended Au NP network and corresponding blue color.

oligonucleotide strands were present, sequence complementarity between the Au NP-displayed oligonucleotides and the two "sticky ends" of the target strand resulted in cross-linking between adjacent NPs and formation of larger nanoparticle networks. The close proximity of the NPs in this configuration led to a significant shift of the plasmon resonance to higher wavelengths – giving rise to the appearance of blue color (reflecting the complementary color observed by the eye). This simple colorimetric concept provided high sensitivity, enabling detection of minute quantities of target DNA.

While many Au NP sensing applications have exploited the complementary nature of DNA for generating signals in sensing applications, other experiments relied on *nonspecific* interactions between DNA strands and bare Au NPs. An example of such a sensing scheme is highlighted in Figure 3.9. The experiment by Z. Nie and colleagues at Hunan Univeristy, China, was designed to demonstrate the use of Au NPs for colorimetric detection of enzymatic cleavage or oxidative damage of single-stranded DNA (ssDNA – a DNA sequence not bound to its complementary strand). The sensing mechanism was based on the observation that longer ssDNA fragments bind to Au NPs at a much slower rate compared to shorter ones. Accordingly, the noncleaved ssDNA examined by the researchers did not adsorb onto Au NPs, which instead aggregated in the presence of high salt concentration making the solution blue. However, when enzymatic cleavage (or oxidative damage) occurred, the smaller DNA fragments

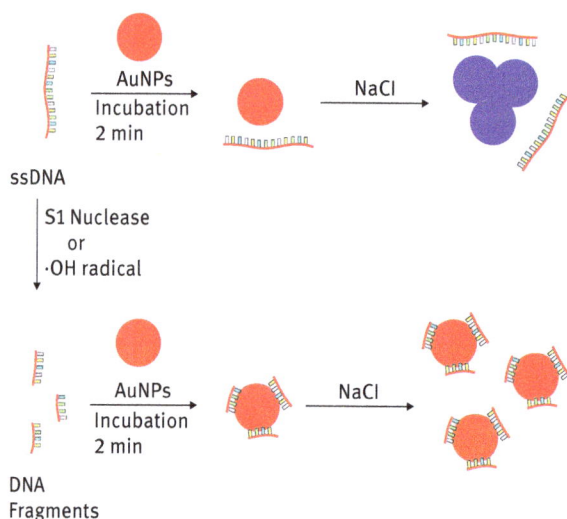

Fig. 3.9: Color detection of DNA damage through gold nanoparticle aggregation properties. Long DNA strands (*top row*) do not bind Au nanoparticles, which undergo salt-induced aggregation and generate a blue color. In contrast, shorter DNA fragments, produced through enzymatic cleavage or oxidation damage, attach to the surface of the Au NPs and prevent their aggregation. The solution color remains red.

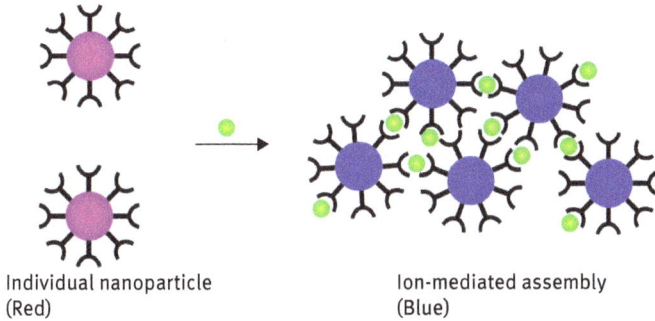

Individual nanoparticle
(Red)

Ion-mediated assembly
(Blue)

Fig. 3.10: Colorimetric ion sensing by inducing gold nanoparticle intercalated networks. The gold nanoparticles are functionalized with chelating agents which specifically recognize metal ions. The target ions are simultaneously bound to two chelators on different nanoparticles, generating a nanoparticle network and the corresponding red-blue color change. Reprinted with permission from Saha et al., *Chem. Rev.* **112** (2012), 2739–2779, © 2012, American Chemical Society.

rapidly bound onto the Au NPs surface, increasing the negative charge of the NPs – thus blocking their aggregation and leaving the solution red.

Colorimetric sensing schemes based on Au NP aggregation have been demonstrated for numerous analytes. Figure 3.10 illustrates a strategy for detecting metal ions through formation of Au NP networks. Linking Au NPs could be achieved, for example, through attachment of *chelating agents* onto the nanoparticles' surface. Metal ions are subsequently captured by the chelating residues forming an interlinked network of Au NPs, generating the red-blue color transformation. The key requirement underlining the sensing approach depicted in Figure 3.10 is the selection of recognition elements (the chelating agents) which allow *co-binding* of adjacent nanoparticles.

The transitions between Au NP *aggregation* and *disaggregation* and the corresponding colorimetric transitions have also been employed for monitoring environmental conditions such as temperature change. Figure 3.11 depicts an interesting Au NP system in which reversible *thermo-responsive* colorimetric transitions could be achieved through temperature-induced association/dissociation of Au NPs coated with a specific ligand. The essence of the experiment, devised by Y. Yin and colleagues at the University of California, Riverside, was the observation that the charge of the ligand employed [bis(p-sulfonatophenyl)-phenylphosphine (BSPP)] is temperature-dependent. Accordingly, the overall *surface charge* on the Au NPs was modulated following temperature increase/decrease, thereby affecting the degree of electrostatic attraction/steric repulsion between the NPs. This feature gave rise to reversible transformations between dis-assembled Au NPs (yielding red color), and Au NP aggregates (blue solutions).

Surface-enhanced Raman scattering (SERS) is another sensing technique demonstrated for Au NPs, which (similar to LSPR) arises from motion of free electrons on the metal surface. SERS is based on light scattered from surface-adsorbed molecules

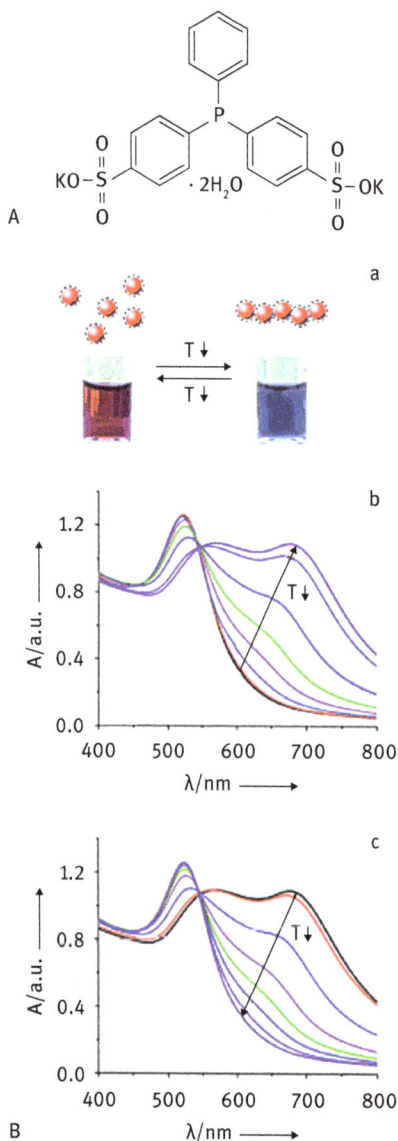

Fig. 3.11: Thermo-responsive gold nanoparticles. The nanoparticles are coated with the temperature-sensitive ligand bis(p-sulfonatophenyl)-phenylphosphine (BSPP; depicted in **A**). **B:** Details of thermo-chromic sensing. The electrostatic charge of the ligand-coated nanoparticles changes upon temperature increase or decrease, modulating the attraction/repulsion forces between the nanoparticles, which consequently result in aggregation/dis-aggregation and corresponding visible color changes (**a**). The graphs in **b–c** depict the reversible temperature-associated spectral shifts of the localized surface plasmon resonance, particularly the absorbance at around 700 nm which increases upon lowering the temperature (thus producing the blue solution color). Reprinted with permission from Liu et al., *Angew. Chem Int. Ed.* **51** (2012), 6373–6377, © 2012, John Wiley & Sons.

(i.e. Raman scattering), in which the scattering frequency is dependent on the modes of vibrations and rotations of the molecule tested. SERS provides significantly better signal intensity in comparison with conventional Raman spectroscopy through utilization of gold or silver *colloidal surfaces* upon which the examined molecules are placed. The enhanced sensitivity is likely due to localized electric fields induced by the morphology of the metal surface. The impact of nanoparticles on SERS is particularly significant, since the electric fields close to the NP surface are considerably enhanced (compared to *macroscopic* metal surfaces). Indeed, many Au NP systems have been explored as sensing platforms, their associated LSPR phenomena (and their use as signal transduction mechanisms) have contributed to SERS capabilities.

Another useful sensing avenue utilizing Au NPs exploits *fluorescence quenching* induced by the nanoparticles. This phenomenon is believed to arise from rapid energy transfer between fluorescent dyes and metal (i.e. gold) surfaces present in their vicinity – thus quenching the fluorescence emission. Indeed, such photophysical processes have made possible development of simple sensing schemes in which docking of fluorescent analytes onto Au NPs leads to quenching of the fluorescence signal, providing a generic detection mechanism. Figure 3.12 demonstrates an Au NP-induced fluorescence quenching concept implemented for DNA detection. In that study, carried out by A. J. Libchaber and colleagues at Rockefeller University, Au NPs were conjugated to a "hairpin" DNA sequence (i.e. a DNA sequence which folds onto itself through base-pair complementarity between head and tail) onto which a fluorescent dye was attached. In the initial state, no fluorescence was emitted due to hairpin folding and consequent proximity of the fluorophore and the NP, resulting in fluorescence energy transfer from the dye to the Au NP (Fig. 3.12, left). However, in situations where the target DNA analyte (i.e. the complementary sequence to the hairpin) was present in solution, hybridization (e.g. binding) between the two complementary sequences resulted in a rod-like double-helix formation with concurrent placing of the fluorescent dye away from the Au NP. No quenching occurred in this configuration and consequently high fluorescence was recorded. Remarkably, this detection scheme could even detect single base-pair *mismatches* in the target sequence, underscoring the sensitivity (and specificity) of the sensing concept.

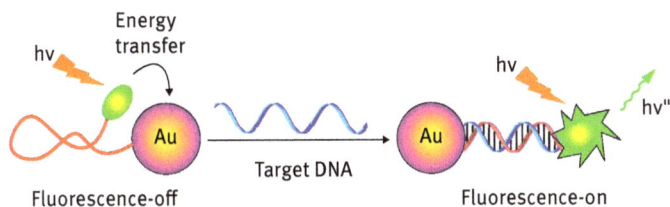

Fig. 3.12: DNA sensing through gold nanoparticle-induced fluorescence transformations. **Left:** Initial state: the fluorescent dye (green oval) is attached to a folded DNA hairpin. Fluorescence quenching occurs due to the close proximity of the dye to the Au NP surface. **Right:** Hybridization with the target DNA sequence puts distance between the dye and the nanoparticle's surface, giving rise to fluorescence emission.

Cancer detection and monitoring are among the most lucrative goals in biosensor development, and Au NPs have been extensively studied for their potential contributions to this field. An array-based cancer screening strategy utilizing Au NP-induced fluorescence quenching, introduced by V. Rotello and colleagues at the University of

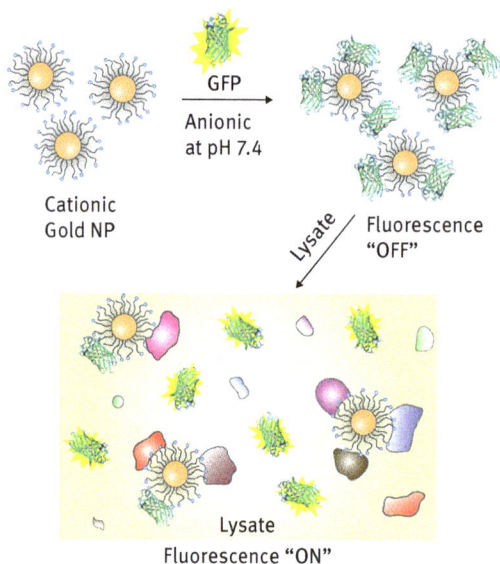

Fig. 3.13: Biological monitoring through modulation of fluorescence quenching by gold nanoparticles. Negatively-charged green fluorescent protein (GFP) is initially attached to cationic Au nanoparticles and its fluorescence is thus quenched. Following addition of cell lysate, proteins with greater affinity to the Au nanoparticles displace the GFP molecules, recovering their fluorescence emission. Creating an array using different cationic ligands displayed on the Au nanoparticles' surface produces a fluorescence fingerprint reflecting the pool of different proteins in cancer cells' lysates. Reprinted with permission from Rana et al., *ACS Nano* **6** (2012), 8233–8240, © 2012, American Chemical Society.

Massachusetts, is outlined in Figure 3.13. The signal mechanism was based on the capture/release of *green fluorescence protein (GFP)*, a widely used fluorescent marker, by the Au NPs. The NPs were chemically-functionalized with an array of ligands, each having distinctive positively-charged headgroups. Binding of the negatively-charged GFP onto the surface-functionalized Au NPs gave rise to quenching of the GFP fluorescence. However, competitive binding of proteins present in tissue lysates (solutions comprising of the molecular constituents of mechanically-disrupted tissue/cell mixture) induced release of the Au NP-bound GFP into the solution and a consequent increase in fluorescence, the intensity of which was dependent on the concentration of the displacing proteins in the lysate and their affinity to the ligands displayed on the Au NP surface. Importantly, by utilizing arrays of Au NPs functionalized with different ligands, the researchers found that proteins in lysates of cancerous tissues produced distinguishable fluorescent "fingerprints" compared to normal lysates – pointing to potential use of the technique as a diagnostic tool.

The color and fluorescence quenching properties of Au NPs can be combined in applications somewhat different to "pure" sensing. Figure 3.14 depicts an elegant

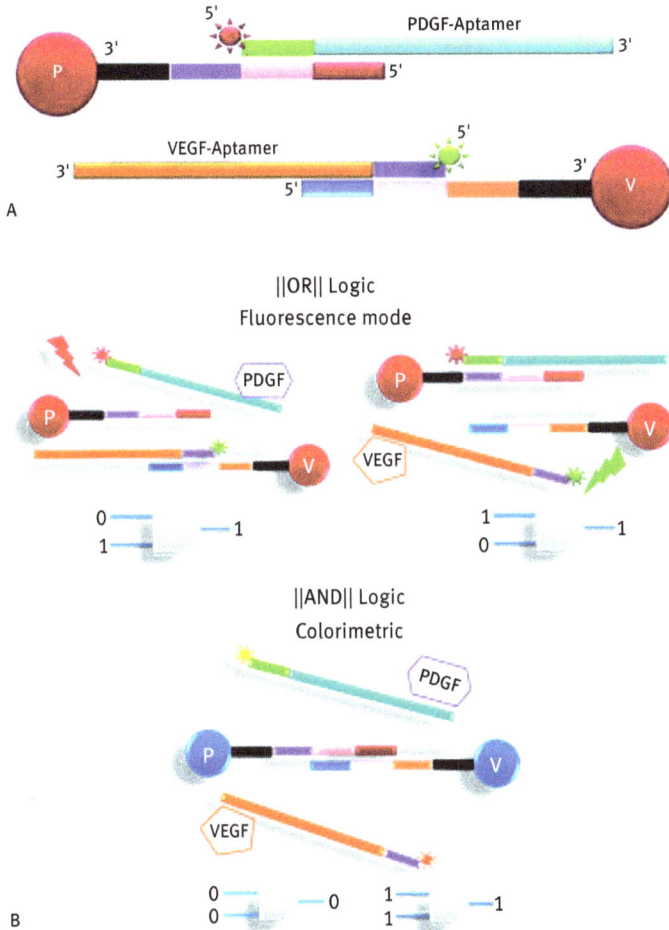

Fig. 3.14: Boolean operations using colorimetric transitions and fluorescence quenching by gold nanoparticles. **A:** The experimental system: two types of Au nanoparticles display DNA single strands which are complementary. The Au nanoparticles are initially separated, as two short complementary strands bind to the NP-displayed sequences. The short complementary strands further contain fluorescent dyes (of which the fluorescence is quenched by the Au NP), and two aptamers targeting two cancer-associated peptides VEDF and PGDF, respectively. **B:** Logic operations: OR operation is demonstrated through addition of either VEDF or PGDF; either peptide binds its respective aptamer, thereby releasing the short strand from the Au nanoparticle and inducing fluorescence of the (un-quenched) dye. The AND operation is achieved by adding both VEDF and PGDF. Release of the two short strands from the Au nanoparticles enables their linking through complementarity of the two displayed sequences - giving rise to blue color (corresponding to Au nanoparticle association). Reprinted with permission from Shukoor et al., *ACS Appl. Mater. Inter.* **4** (2012), 3007–3011, © 2012, American Chemical Society.

"Boolean logic" application developed by W. Tan and colleagues at the University of Florida, which both utilized color transformations due to Au NP aggregation/disaggregation, as well as Au NP-induced fluorescence quenching. The Au NPs were functionalized with single strand DNA sequences which were complementary to each other. The binding of two NPs through DNA complementarity was initially blocked, however, since each NP-displayed strand was separately coupled to NP-free oligonucleotides exhibiting sequence complementarity to the NP-displayed DNA (Fig. 3.14A). These complementary sequences had two important characteristics. First, fluorescent dyes were chemically attached to the DNA strands; the fluorescence was quenched upon binding to the NP-attached oligonucleotides due to the proximal NP surface. Second, the specific oligonucleotide sequences selected were *aptamers* – short DNA sequences which recognize and bind tightly onto target biomolecules, particularly proteins.

The two aptamers employed in the study were specific to well-known cancer-associated proteins – vascular endothelial growth factor (VEGF), and platelet-derived growth factor (PDGF). As shown in Figure 3.14, the color and fluorescence properties of the NP solutions were determined by the interplay between DNA complementarity and aptamer/protein binding. In the absence of the target proteins, the Au NPs move freely in solution. In this scenario the visible color of the solution is red (e.g. no aggregation), and there is no fluorescence (due to quenching of the fluorophores by the Au NPs). However, binding between the aptamers and their target proteins releases the dye-containing strands from the Au NP, resulting in observable fluorescence as the quenching effect is eliminated.

The Boolean operations based upon the Au NP/DNA system in Figure 3.14 were achieved when one or both of the target proteins were present in solution. As shown in Figure 3.14, when either VEGF or PDGF were co-dissolved, the respective aptamers were released from their complementary oligonucleotides on the Au NPs. These events resulted in *fluorescence emission*, conforming to an *OR* Boolean logic operation (fluorescence emitted when either of the target proteins was added). Moreover, in a scenario in which *both* VEGF *and* PDGF were present, the two aptamers detach from the Au NPs, resulting in binding between adjacent Au NPs through the complementary DNA strands, and the consequent blue color corresponding to Au NPs in close proximity. Accordingly, this *AND* Boolean logic operation can be distinguished by the colorimetric transformation output.

The color transitions and fluorescence quenching of Au NPs are not the only useful properties available for biosensing applications. As they comprise of metallic gold, Au NPs allow electron transport, making their use as components in electrically- and electrochemically-based sensors possible. An important class of biosensing applications in which Au NPs could play a role involves the use of *redox enzymes* as the signal transduction element. Enzymes catalyzing electrochemical reactions have been used as core components in many biosensing applications. However, a recurring practical hurdle in such systems concerns how to enable efficient electron transfer from the

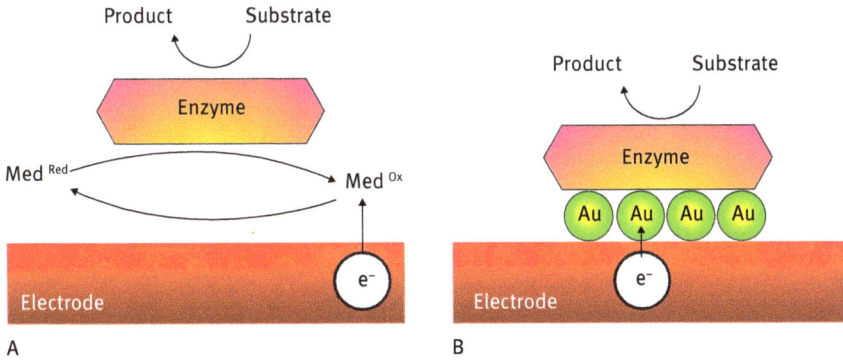

Fig. 3.15: Gold nanoparticles in electrochemical sensors based on redox enzyme activity. **A:** Conventional sensor design: enzymatic catalysis involves electron transfer by conversion between the oxidized and reduced states of a "mediator" molecule ("Med" in the figure) placed on the electrode surface, thus increasing sensor complexity. **B:** Au nanoparticle-based design in which the Au nanoparticles could substitute the mediator, facilitating efficient electron transport to the redox center of the enzyme.

redox center of the enzyme onto the *electrode*, thus generating the electrical signal recorded by the sensor.

Figure 3.15 illustrates the use of Au NPs in enzyme-based electrochemical sensors. The operation principle of such sensors is based upon recognition of a specific substrate by the enzyme, which catalyzes its conversion into a product. In redox-active enzymes, the transformation of substrate to product is accompanied by electron transfer, and the generated electrons need to be part of an electrical circuitry – transferred to/from electrodes coupled to the enzymes. In this crucial part of the biosensor (the interface between the electrode surface and the enzymes), Au NPs could play an important role, providing a "conductive medium" for electron transfer (Fig. 3.15B) which in principle is more efficient than conventional *molecular-mediated* electron transfer (Fig. 3.15A).

3.1.3 Gold nanoparticles coupled to biomolecules

While conjugating Au NPs with biological molecules acting as recognition elements has been a prominent avenue for the use of Au NPs in sensing, other interesting routes employing Au NP/biomolecule conjugates have also been followed (a discussion of *adverse effects* of the NP/biomolecule interface is provided in Chapter 7). For example, integration of Au NPs and DNA recognition has been examined as an intriguing concept to manipulate organization of *macro-scale* structures. Figure 3.16 highlights an ordered Au NP system designed by C. Mirkin and colleagues at Northwestern University. In the experiment, *noncomplementary* oligonucleotide strands were covalently

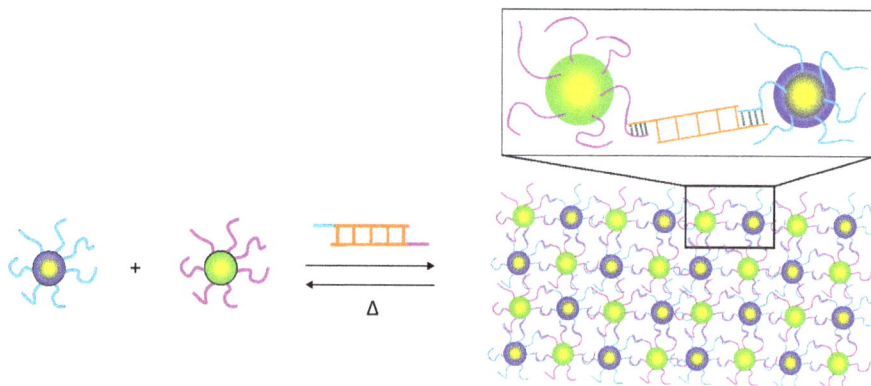

Fig. 3.16: Gold nanoparticle association mediated by DNA complementarity. Two types of Au nanoparticles are coated with different single-stranded DNA (*left*); addition of a *double-stranded DNA* with two "sticky ends" corresponding to sequences complementary to the nanoparticles-displayed strands results in binding of adjacent NPs; optimization of the double-stranded DNA concentration produces an organized three-dimensional Au nanopatricle array.

bonded to the surface of two populations of Au NPs through thiol-based chemical linkage. While the Au NPs did not interact with one another, binding of the two types of DNA-functionalized Au NPs could be achieved through co-addition of a DNA double strand exhibiting free "sticky ends" at each end, corresponding to the respective complementary sequences of the two NP-attached oligonucleotide strands. The "double hybridization" process (i.e. binding of complementary strands on each side of the DNA duplex) resulted in the formation of Au NP networks which could be optically monitored (through red-blue color transitions associated with Au NP aggregation; see Fig. 3.8). This DNA-mediated macro-scale assembly represents a general scheme for construction of hierarchical Au NP structures in a controlled manner, and can be exploited in sensing applications and material design.

Chemically linking Au NPs and oligonucleotide fragments (DNA or RNA) has led to the development of potentially powerful *gene regulation* vehicles. Identifying molecular platforms which could facilitate efficient delivery of genes and genetic regulation elements from the cell exterior onto their DNA/RNA targets inside the cell has proven to be a formidable challenge to the broad use of "gene therapy", once considered one of the "holy grails" for possible new therapies of many incurable diseases. C. Mirkin and colleagues have shown that Au NPs displaying short oligoucleotide strands could function as effective gene regulation agents (Fig. 3.17). DNA fragments tested in early experiments by researchers were designed to inactivate specific elements in the protein translation machinery by binding to the complementary sequences inside the cell (known as "antisense oligonucleotides"). Accordingly, proper functioning of the Au NP platform stipulates efficient penetration through the cell membranes, sufficient

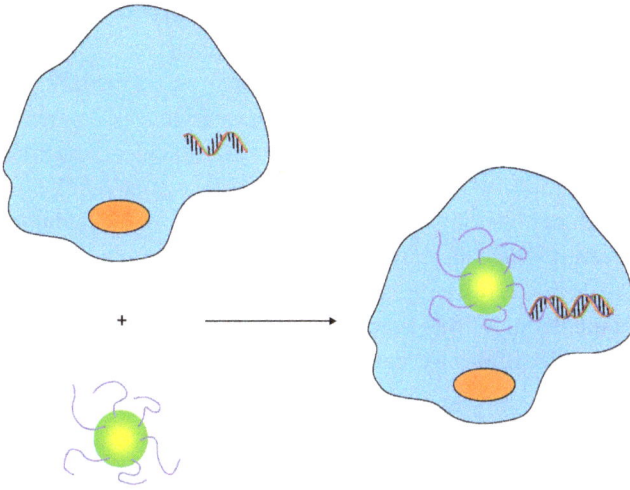

Fig. 3.17: Gold nanoparticles for gene regulation applications. Au nanoparticles displaying specific DNA fragments are taken up by cells, where they hybridize with their DNA target, thereby interfering with gene regulation pathways.

stability in the intracellular environment, and facilitating binding of the NP-displayed strands to their target sequences.

Indeed, experiments have demonstrated that oligonucleotide/Au NP hybrids easily entered cells, were not significantly degraded by intracellular nucleases (enzymes which break up oligonucleotide fragments), and appeared to exhibit very low cell toxicity. Particularly significant, gene regulation affected by the oligonucleotide/NPs was found to be more effective compared to application of free oligonucleotides. These results underscore the dual role of the NPs – on the one hand constituting an efficient delivery vehicle for displayed genetic residues and capable of overcoming formidable structural and biochemical cellular barriers. On the other hand, the chemical attachment to the NP surface apparently did not adversely affect the *functionality* of the gene regulation elements.

This duality points to other potential uses of Au NPs as biological modulators. Indeed, the use of Au NPs as inert chemical platforms, albeit still retaining the biological functionality of surface-displayed molecular elements, has opened avenues for *immunological* applications. Creating good vaccines depends on how effective the *antigen* (i.e. the foreign molecule which elicits production of antibodies) is at triggering an immune response. This issue is intimately related to effective delivery and presentation of the antigen, facilitating its recognition by the immune host cells. Antigen presentation has become central to vaccine design since it was found that *adjuvants* (substances added to the antigens in vaccines) play a crucial role in enhancing the immune response, often through enabling effective presentation of the antigens to the host cells.

Au NPs could serve as a useful platform for vaccine development for several reasons. First, versatile synthetic avenues enable the display of antigens on the NP surface, essentially providing an intrinsic presentation mechanism. In that regard, Au NPs can be co-functionalized with chemical modalities other than antigens, particularly recognition elements for targeting the NPs onto specific physiological locations and/or immune cells. Indeed, "multifunctionalization" of Au NPs can be readily exploited in vaccine design (such particles are sometimes referred to as "multimodal Au NPs"). Figure 3.18 shows a nanoparticle displaying both a peptide antigen and a "targeting peptide" designed to dock onto specific receptors on cell surfaces or antibodies. The nanometer-scale diameter of Au NPs is another critical feature favoring their potential therapeutic use, as it facilitates NP passage through lymphatic capillaries and efficient delivery of the antigen cargo to immune cells. The small size of the NPs presents an additional advantage since small particles are taken up very efficiently by *dendritic cells*, the main "work-horse" of the adaptive immune system.

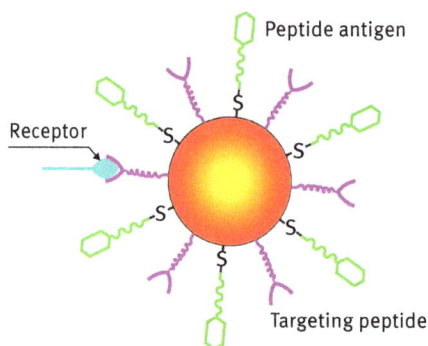

Fig. 3.18: Multifunctional gold nanoparticles for vaccine applications. Different biological residues are displayed on the surface of a single Au nanoparticle. Each entity provides a distinct functionality.

3.1.4 Gold nanoparticles in catalysis

Gold has long been considered an *inert* noble metal, and indeed this property has greatly contributed to its "preciousness". In this context, a seeming "paradox" in Au NP science and technology is the broad spectrum of *chemical reactions* involving Au NPs, particularly the varied thiol-based reaction pathways discussed above. Another facet of this reactivity has been the observation that Au NPs constitute a powerful platform for *chemical catalysis*. Research in this field benefited from the discovery in the early 1990s that Au NPs promote oxidation of carbon monoxide (CO) to carbon dioxide (CO_2) – a critical process with significant environmental and commercial implications. Numerous Au NP-induced catalytic pathways have been identified since then, including oxygenation of carbonaceous compounds, deoxygenation (oxygen removal) of organic compounds, hydrogenation (hydrogen formation), cyclization (ring formation) reactions, polymerization through carbon-carbon bond formation, and various other processes. Catalytic applications utilizing Au NPs generally require deposition of the NPs on active surfaces; metal oxide surfaces, common catalytic substrates, have often been employed in such applications.

The underlying mechanistic aspects of Au NP catalysis are not fully understood and different factors probably play roles in different reactions. It has not been fully established, for example, what the actual catalytic sites on Au NPs are. While the obvious "candidates" would be metallic Au(0) domains, some proposals point to partly-oxidized Au or ionic gold species, such as Au(I) and Au(III) as the atomic constituents activating certain reactions upon the NP surface. The latter proposal stems from the occasional use of Au ions as catalysts (known as "homogeneous catalysts") in solutions containing the reactants. It also seems likely that Au NPs play synergistic roles, together with additional active sites on the catalytic surfaces the Au NPs are deposited on.

Figure 3.19 depicts a proposed mechanism for the catalytic action of Au NPs deposited on hydrotalcite (HT, $Mg_6Al_2(OH)_{16}CO_3 \cdot nH_2O$) in selective deoxygenation of epoxides to generate alkenes, as outlined by K. Kaneda and colleagues at Osaka University, Japan. According to the proposed reaction scheme, docking of hydrogen molecule (the reducing agent) occurs between Au NPs and basic sites at the HT surface, followed by cleavage of the H–H bond. The surface-attached proton and Au-H⁻ hydride subsequently extract the oxygen from the epoxide reactant, yielding an alkene (C=C – containing carbohydrate) and water. Supporting this catalytic mechanism is the known instability of gold hydrides (Au-H⁻) compared to other metals often

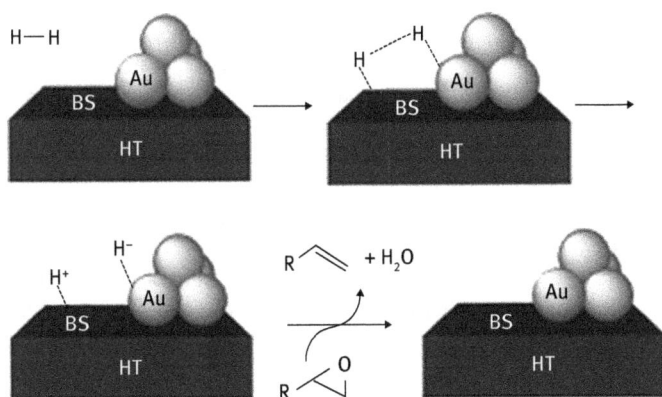

Fig. 3.19: Catalytic activity of gold nanoparticles. Proposed model for deoxygenation (oxygen removal) of epoxides in the presence of H_2 producing alkenes and water, catalyzed by Au nanoparticles deposited on hydrotalcite (HT) surface. The H_2 molecules are initially bound between the Au nanoparticle and surface basic sites (BS). Subsequent formation of highly reactive Au-H⁻ unit is the main catalytic factor for the reaction. Reprinted with permission from Noujima et al., *Angew. Chem. Int. Ed.* **50** (2011), 2986–2989, © 2011, John Wiley & Sons.

employed in catalytic applications, such as palladium and platinum. This instability should lead to high reactivity of the putative hydride, resulting in efficient oxygen removal from the epoxide reagent and water production.

While most catalytic applications of Au NPs to date have involved NPs deposited on surfaces, other configurations have been reported. E. Doris and colleagues at CEA, France, recently reported that a "corn-like" construct comprising carbon nanotubes (CNTs) coated with multiple layers of Au NPs (Fig. 3.20) exhibited notable catalytic enhancement of *silane oxidation* reactions (e.g. transformation of compounds containing Si-H residues to Si-OH). The Au NP layer was maintained by electrostatic attraction to ammonium moieties coating the CNTs, and the researchers observed that this organization yielded dramatic acceleration and yield enhancement of the oxidation of several silane-containing substrates. The catalytic activity was more pronounced in comparison to individual Au NPs, or CNTs not coated with Au NPs, attesting to the combined activity of the conjugated Au NP/CNT system. Similar to many other Au NP-based catalysts, however, the precise mechanism of the catalytic action is not entirely clear; presumably, abundance of binding sites within the CNT-stabilized Au NP layer contributes to silane adsorption and reactivity of the material.

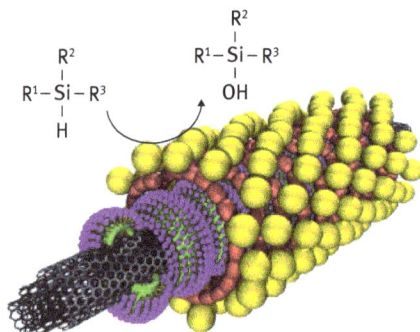

Fig. 3.20: Gold nanoparticle / carbon nanotube construct as a catalyst for silane oxidation. The Au nanoparticles are deposited on the carbon nanotube surface catalyzing the Si-H→Si-OH transformation. Reprinted with permission from John et al., *Angew. Chem. Int. Ed.* **50** (2011), 7533–7536, © 2011, John Wiley & Sons.

Au NPs have not just played roles in organic catalysis; several studies have demonstrated that Au NPs also catalyze biological reactions. Specifically, experiments have demonstrated intriguing catalytic effects on reactions involving biological molecules induced by Au NPs. An example of Au NP-catalyzed protein assembly is provided in Figure 3.21. In this work, S. Tomita and colleagues at NAIST, Japan, explored the structural consequences of adding Au NPs to a protein (trp RNA-binding attenuation protein, TRAP) in which a single *lysine* residue was substituted with *cysteine*. In normal conditions, 11 TRAP monomers form a circular, ring-like structure. However, addition of Au NPs to the protein solution resulted in rapid formation (in less than a minute) of spherical giant "capsids" (e.g. shell-like structures). While the exact mechanistic aspects of this process have not been deciphered, it is likely related to binding of the NPs to a few protein monomers by the thiol moieties of the cysteine residues. As a con-

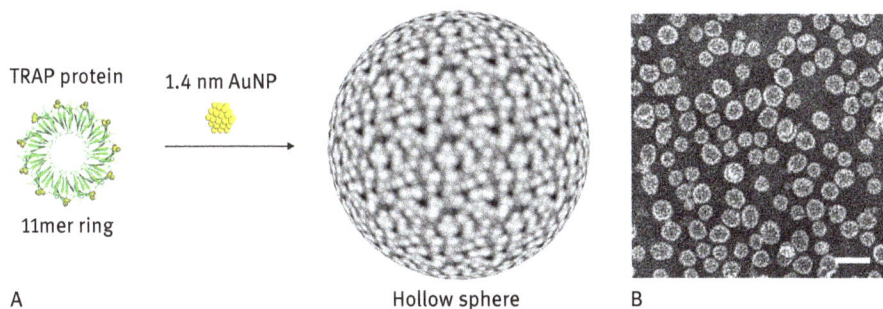

Fig. 3.21: Gold nanoparticles catalyze protein assembly. **A:** Scheme of the assembly process: Au nanoparticles catalyze formation of large hollow spheres from protein building blocks. **B:** Electron microscopy image showing the spheres. Scale bar corresponds to 40 nm. Reprinted with permission from Malay et al., *Nano Lett.* **12** (2012), 2056–2059, © 2012, American Chemical Society.

sequence, a spherical rearrangement of the monomer-bound Au NPs is energetically-favored.

While the field of Au NP catalysis is still in relative infancy, its scope and potential are significant. Research is currently being carried out with the aim of establishing better understanding of catalysis mechanisms of Au NPs, and also improving performance in industrial settings. One of the primary goals is the control of NP sizes used in catalysis, known to affect reaction yields. Also important is to characterize the exact composition of Au NPs in order to determine the nature of the catalytic sites – metallic gold, Au(0), partially ionic and completely ionic Au species, or perhaps metallic impurities embedded within the NPs throughout the synthetic processes.

3.1.5 Other applications of gold nanoparticles

The unique properties of Au NPs have been employed in applications other than sensing and catalysis. Figure 3.22 portrays a solar cell technology relying on the surface plasmon properties of Au NPs to generate photocurrent. As described in more detail in Chapter 2, a core physical event determining the performance and efficiency of solar cells is the effectiveness of generating electron-hole separation (upon which the electrons and holes are subsequently transported through the two opposing electrodes, producing energy before recombining). In the system illustrated in Figure 3.22, U. Bach and colleagues at Monash University, Australia, fabricated Au NPs directly on a commonly-used electron-transport TiO_2 layer. Following light irradiation, significant electrical current was recorded. The mechanism proposed by the researchers for this phenomenon links the photocurrent to oscillations of the free surface electrons upon the Au NPs (i.e. surface plasmon resonance). Such oscillations result in electron transfer to the TiO_2 and the base electrode, with simultaneous transfer of the holes to

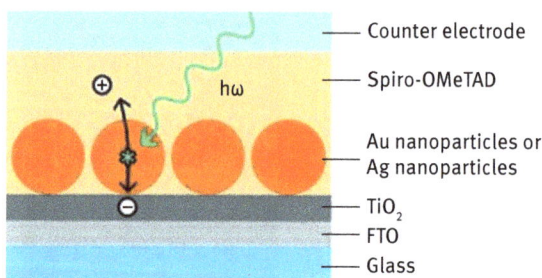

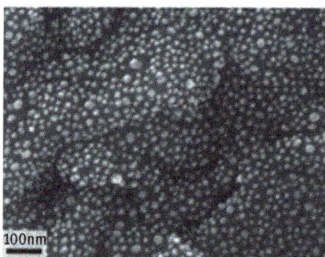

Fig. 3.22: Solar cell design based on surface plasmon resonance of gold nanoparticles.
A: Schematic drawing of the cell. Light irradiation is adsorbed by the Au nanoparticles through their surface plasmon resonance, generating an electron (transferred to the TiO_2 layer) and a corresponding hole transported through the hole-transfer layer (Spiro-OMeTAD) to the electrodes. Photocurrent is generated as a result. **B:** Electron microscopy image showing the Au nanoparticles deposited on the TiO_2-covered base electrode. Reprinted with permission from Reineck et al., *Adv. Mater.* **24** (2012), 4750–4755, © 2012, John Wiley & Sons.

the counter-electrode through the hole transport layer deposited on the Au NPs. The simple solar-cell design depicted displayed a reasonable energy performance.

Functionalized Au NPs have been employed in a recent innovative *fingerprinting* technique (Fig. 3.23). The technology is based on the well-known feature of fingerprints on paper, which largely comprise of hydrophobic (lipid) residues adsorbed onto the paper surface. Indeed, this fact is one of the reasons fingerprints are often not efficiently detected by conventional dyes, since the chemicals employed as fingerprint markers do not bind well to hydrophobic residues. Y. Almog and colleagues at the Hebrew University in Israel overcame this limitation with a technique which does not target the lipid traces themselves, but the paper background instead. Specifically, the researchers synthesized Au NPs exhibiting high affinity to the surface of plain paper (e.g. paper free of the lipid traces originating from fingerprints) by coating the NPs with thiol-based ligands whose heads were designed to tightly bind to the paper. After spreading the functionalized Au NPs on paper, a processing step akin to the old-fashioned silver-base black-and-white film development was carried out, resulting in deposition of silver on the Au NP areas and dramatic enhancement of the fingerprint

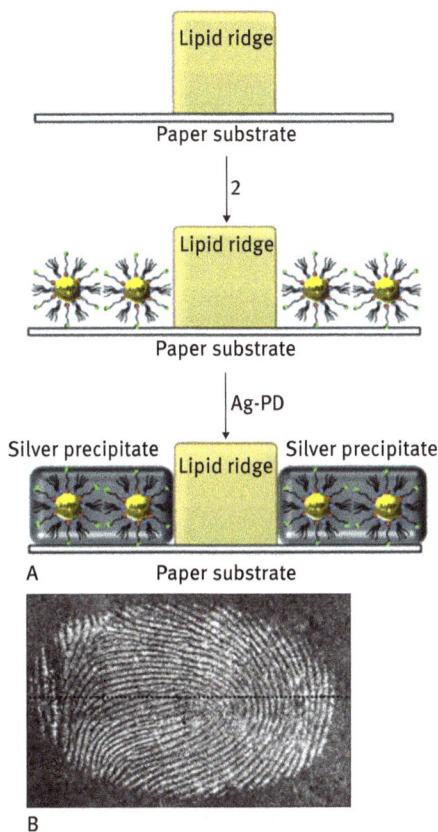

Fig. 3.23: Fingerprinting using gold nanoparticles. **A:** Scheme showing application of the technology: The Au nanoparticles are coated with hydrophilic residues and adsorbed onto the paper areas not containing lipid traces left by contact with a finger. Subsequent preferred deposition of silver upon the Au NP-containing areas reveals the fingerprint traces. **B:** Fingerprint processed using the technique. Note that the fingerprint traces are white. Reprinted with permission from Jaber et al., *Angew. Chem. Int. Ed.* **51** (2012), 12224–12227, © 2012, John Wiley & Sons.

features. Importantly, unlike conventional technologies, the actual fingerprint traces (in which lipid molecules were present) remained white, providing an intrinsic contrast mechanism (Fig. 3.23).

3.1.6 Gold nanorods

Au nanorods (Au NRs) form an interesting and technologically-promising branch of Au NP research. In particular, the *rod* configuration opens up various applications due to the effects of the anisotropic shape on physical properties. Early studies of Au NRs were constrained by the synthetic avenues available; NRs were initially produced inside porous templates which limited the range of NR dimensionalities. Only on introduction of solution-based "seeded growth" methodologies in the early 2000s did the field truly "take off". In NR solution synthesis procedures, pre-formed seeds (comprising of silver or gold) are synthesized separately and added to growth solutions containing reducing agents and stabilizers which maintain preferential crystal growth

A

B

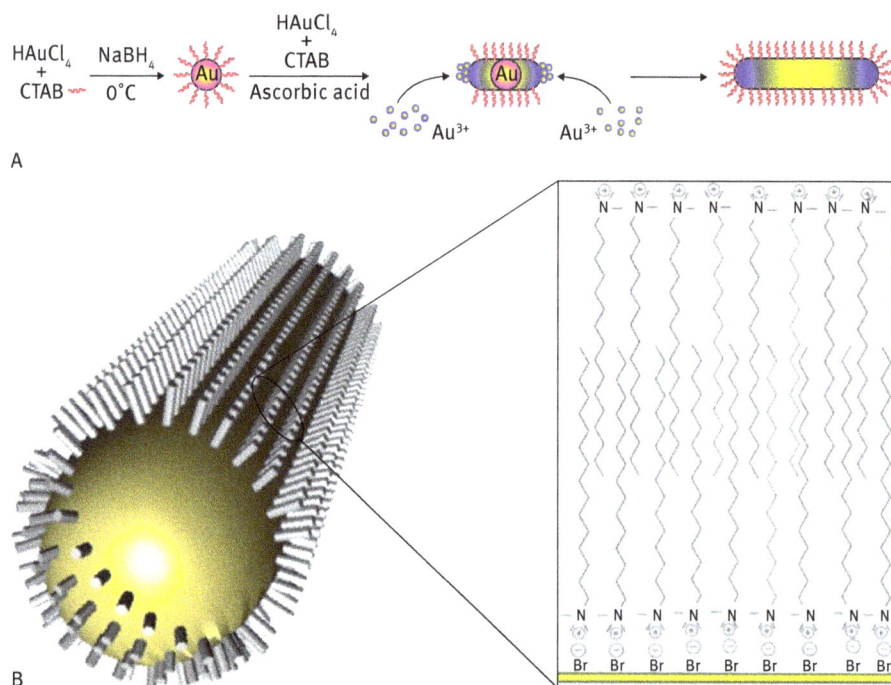

Fig. 3.24: Gold nanorod synthesis. **A:** "Bottom-up" assembly route employing pre-synthesized Au nanoparticle seeds dissolved in a solution containing reducing agents and surfactants which direct growth of the nanorod along the longitudinal axis. The Au^{3+} ions are preferentially reduced at the tips of the nanorod. **B:** Schematic structure of Au nanorod showing the surfactant layer stabilizing the cylindrical morphology. Reprinted from *Adv. Drug Delivery Rev*, vol. 64, L. M. Alkilany et al., *Gold nanorods: Their potential for photothermal therapeutics and drug delivery, tempered by the complexity of their biological interaction*, pp. 190–199, © 2012, with permission from Elsevier.

at the two rod tips (Fig. 3.24A). This function is usually carried out using surfactants such as cetyl-trimethylammonium bromide (CTAB), which preferably assemble on the side facets of growing seeds, dictating their elongation only at the tips (i.e. along the cylinder axis).

It should be noted, however, that on an atomic level crystalline Au NRs do not form "perfect" cylinders, but rather constitute prism-like structures within which the surfactants are attached to the planar facets. Figure 3.25 presents a high-resolution transmission electron microscopy (HR-TEM) image recorded by L. M. Liz-Marzan and colleagues at the Universidad de Vigo, Spain, of a cross-section of an Au NR, clearly showing the facets of an octagon-shaped particle. Interestingly, the octagonal structure has been observed in Au NR systems grown via the seeded-growth approach using diverse reaction conditions and surfactant compositions, suggesting that the eight-facet structure (although not symmetrical) is a result of a generic growth mechanism

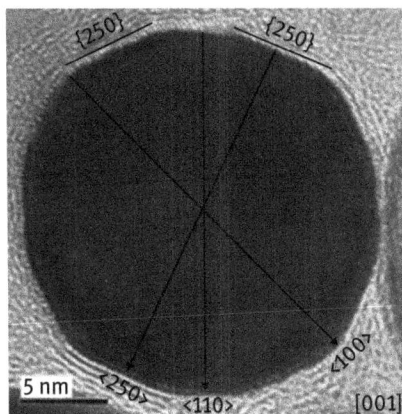

Fig. 3.25: Faceted morphology of gold nanorods. High resolution electron microscopy image of an Au nanorod cross-section showing the octahedral morphology and crystallographic directions of the facets. Reprinted with permission from Carbo-Argibay et al., *Angew. Chem. Int. Ed.* **49** (2010), 9397–9400, © 2010, John Wiley & Sons.

determined by the binding of the surfactants to molecular surfaces formed during the seeded growth process.

Many variations on the basic Au NR synthesis scheme shown in Figure 3.24 have been reported, contributing to the substantial scientific and technological activity in this field. Excellent control over NR size and aspect ratio can now be achieved, important in tuning the optical properties of the nanoparticles; see below. Figure 3.26, for example, depicts remarkably mono-disperse NR preparations, achieved through the use of a binary mixture of two surfactants (rather than only CTAB). The tunable and highly uniform NR dimensions, synthesized by C. B. Murray and colleagues at the University of Pennsylvania, were ascribed to different affinities of the two surfactants to the NR surface, resulting in distinct growth kinetics when the ratio between the two surfactants was varied. Interestingly, by careful tuning of the surfactant mixture composition the researchers constructed quite thick Au NRs which clearly showed the "facet-like" NR structure (Fig. 3.26B).

Nanorod synthesis schemes which do not rely on spontaneous particle growth in solution have been also introduced. An elegant template-directed technique enabling excellent control over NR dimensions is depicted in Figure 3.27. In this procedure, developed by Y. Yin and colleagues at the University of California, Riverside, Au (and other noble metal) NRs were grown inside a porous silica nanotube template. An initial, key synthetic step was to create tubular silica "shells" displaying amine moieties on the inner surface. Subsequently, $AuCl_4^-$ ions were attracted to the amine groups forming crystalline seeds which promoted slow deposition of Au rods within the silica nanotube framework. The final step in NR synthesis was the annealing, or "etching", of the silica shell and release of the encapsulated Au NRs. This procedure, while multistep in nature and involving careful optimization of reagent concentration, generated uniform NRs, the dimensions of which were essentially determined by the silica host matrix. It should be emphasized, however, that despite the remarkable synthetic feats highlighted in Figures 3.26 and 3.27, the exact mechanism/s of NR self-assembly in so-

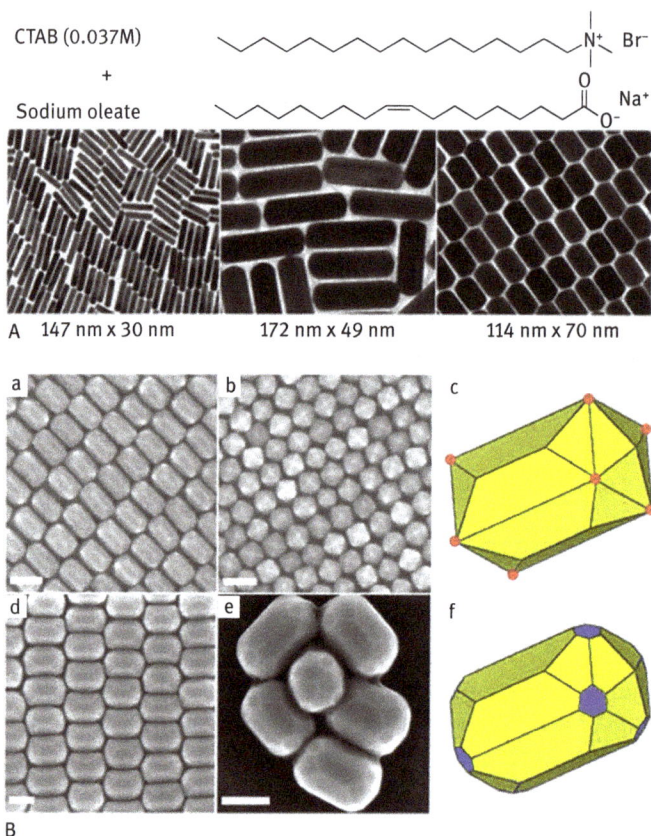

Fig. 3.26: Uniform gold nanorods synthesized using binary surfactant mixtures. **A:** Chemical structure of the two surfactants and some Au nanorod samples prepared using mixtures with different ratios between the surfactants. **B:** Electron microscopy images and models of Au nanorods showing the faceted morphology. Reprinted with permission from Ye et al., *Nano Lett.* **13** (2013), 765–771, © 2013, American Chemical Society.

lution and the factors affecting NR properties, particularly length and aspect ratio, are yet to be fully elucidated and understood.

Similar to the spherical Au NPs discussed above, *surface functionalization* of Au NRs is usually required for practical applications. Unlike spherical Au NPs, however, which exhibit bare surfaces available for chemical manipulation or surfaces coated with lightly-attached surfactants, Au NRs are generally encased within a denser coating layer comprising of tightly bound molecules. This feature presents difficulties for further surface chemical treatment and attachment of other residues. *Thiolated molecules*, for example, which have been abundantly used for functionalization of spherical Au NPs, do not readily take the place of CTAB on Au NR surfaces, thus limiting the efficiency for displaying functional units on NR surfaces.

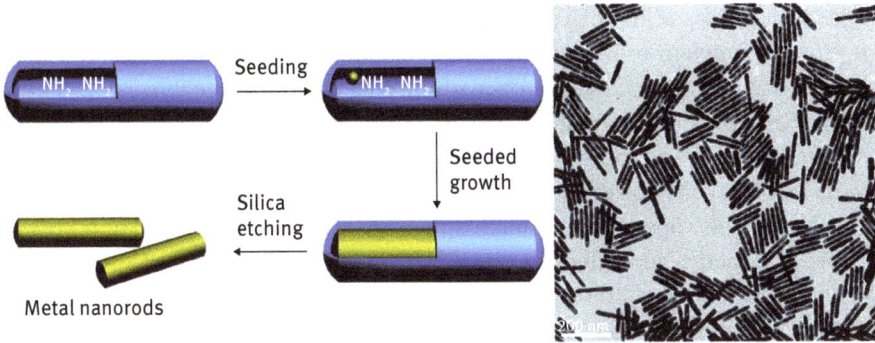

Fig. 3.27: Production of gold nanorods using a silica nanotube template. The synthesis utilizes porous silica nanotubes displaying amine residues on the internal surface. Au^0 seeds are produced by binding and reduction of Au^{3+} ions at the nanotube inner surface, followed by a "seeded growth" process. An etching process removes the silica template, releasing the Au nanorods. The electron microscopy image on the right depicts the highly uniform distribution of the nanorods. Reprinted with permission from Gao et al., *JACS* **133** (2011), 19706–19709, © 2011, American Chemical Society.

As a consequence of the challenges encountered in linking functional groups to Au NRs through thiol-gold bonds, other chemical derivation techniques have been developed. Figure 3.28 outlines a strategy in which the surfactant coating layer was not substituted, but rather employed as the substance for attachment of additional surface-displayed units. In that study, E.C. Goldsmith and colleagues at the University of South Carolina used a "layer-by-layer" approach in which polyelectrolytes were bound to CTAB-coated NRs by electrostatic attraction. The technique exploited the *positively-*

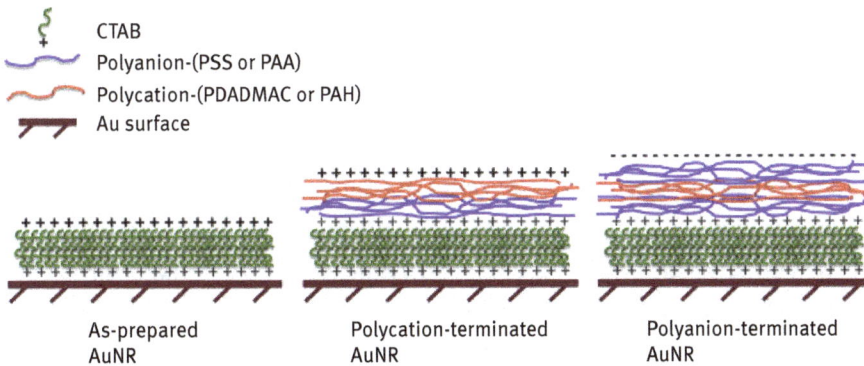

Fig. 3.28: Surface modulation of gold nanorods using a "layer-by-layer" approach. The Au nanorods are initially coated by a surfactant (CTAB). Subsequent attachment of positive and negative polyelectrolytes (polycations and polyanions) generates tunable positive or negative charge upon the nanorods' surface. Reprinted from *Biomaterials*, vol. 30, C. G. Wilson et al., *Polyelectrolyte-coated gold nanorods and their interactions with type I collagen*, pp. 5639–5648, © 2009, with permission from Elsevier.

charged headgroups of the CTAB molecules as "anchors" for polyelectrolytes which were either poly*anions* or poly*cations*, enabling "tuning" of the electrostatic charge on the NR surface. The charged surface of the NRs could then enable electrostatic-binding of additional functional elements.

A fundamental and technologically important difference between Au NRs and spherical Au NPs concerns the dependence of the surface plasmon on the anisotropic nature of the NRs. Specifically, unlike spherical Au NPs which display a single, isotropic SPR mode (dependent on the size and surface chemistry of the NP), Au NRs exhibit two (or more) resonances arising from the broken symmetry of the rod structure. As highlighted in Figure 3.29, light-illuminated Au nanorods generate both *transverse* plasmon resonance associated with electron cloud oscillations along the short axis of the nanorod (i.e. rod diameter), and *longitudinal* SPR arising from oscillations along the rod axis. The longitudinal SPR, in particular, has been shown to be highly sensitive to the aspect ratio of the nanorod, and also depends on NR length. In fact, tuning the aspect ratio makes it possible to vary the position of the SPR maximum all the way from the visible region of the electromagnetic spectrum (wavelength of 500–600 nm) to longer wavelengths in the infrared (IR) range (1500–1800 nm). The capacity of Au NRs to absorb IR light is significant since this feature facilitates *phototherapy* applications – IR-induced localized heating of NRs in specific tissue locations; see detailed discussion below.

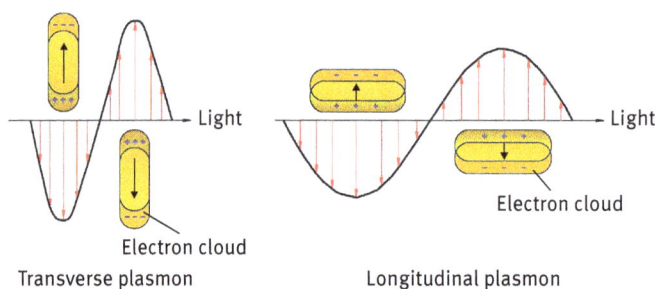

Transverse plasmon   Longitudinal plasmon

Fig. 3.29: Longitudinal and transverse surface plasmon resonance in gold nanorods.

Similar to spherical Au NPs, Au NRs have been employed as core components in optical sensors, taking advantage of the sensitivity of the SPR phenomenon to minute environmental changes occurring at the nanorod surface, particularly through adsorption of target analytes. Figure 3.30 depicts an experiment in which binding of a metal ion (the target analyte) induced a significant shift in the longitudinal LSPR signal. Specifically, M. L. Curi and colleagues at CNR-IPCF in Bari, Italy, embedded a chemical receptor for mercury ions (N-alkylpyrazole) within the CTAB layer coating the Au NRs. The mercury receptor was preferentially located at the tips of the NR; accordingly, binding of Hg^{2+} ions simultaneously to receptors in adjacent Au NRs resulted in assembly of

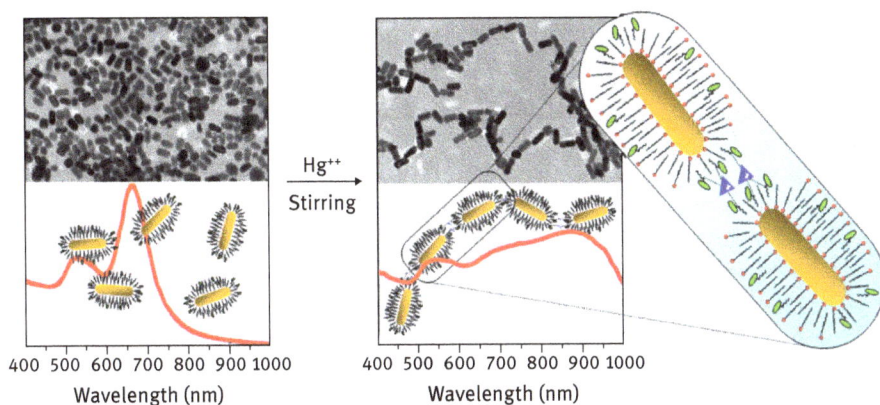

Fig. 3.30: Hg^{2+} sensing using gold nanorod chain formation. The Au nanorods are coupled to specific ion receptors which are more abundant around the nanorods' tips. In the presence of Hg^{2+} ions the nanorods are linked together through the tips, forming long chains. As a consequence the longitudinal surface plasmon resonance is shifted, providing a sensitive reporting mechanism for Hg^{2+} ions in solution. Reprinted with permission from Placido et al., *ACS Appl. Mater. Interfaces* **5** (2013), 1084–1092, copyright (2013), American Chemical Society.

elongated "chains" and consequent shift of the longitudinal SPR signal. The sensitivity observed in this experimental setup was impressive – few parts-per-billion (ppb). Indeed, in many instances Au NR-based sensors exhibit greater sensitivity than spherical Au counterparts, likely related to the more pronounced shifts of the longitudinal LSPR in response to surface modifications.

Several applications unique to Au nanorods' dimensionalities have been described, including single molecule sensing (Fig. 3.31). In such experiments, the nanoscale size offers the means to modulate the surface plasmons by single molecules attracted to the surface (compared to alteration of the SPR signal induced by population of analytes attaching to a wider surface area in a conventional sensor). In another potential sensing mode, Au NRs can be used to amplify the fluorescence signal of surface bound or adjacent single molecule fluorophores. These strategies stand in contrast to the fluorescence quenching observed in the case of spherical Au NPs, as in the nanorod the collective energy of the conductive surface electrons can be transferred to the attached analyte molecule – conceptually similar to a "lightning rod" effect. Specifically, when the excitation energy is coincident with the conductive electron's oscillations (i.e. the plasmon resonance condition), the transferred energy could lead to significant amplification of the fluorescent signal of even a single molecule.

There have been many reports of plasmon-based sensing applications using Au NRs as the transduction platform. Figure 3.32 depicts a sensing approach combining Au nanorod-based plasmonic detection with *molecular imprinting* – a potentially powerful concept for specific detection of known target analytes. In molecularly-imprinted

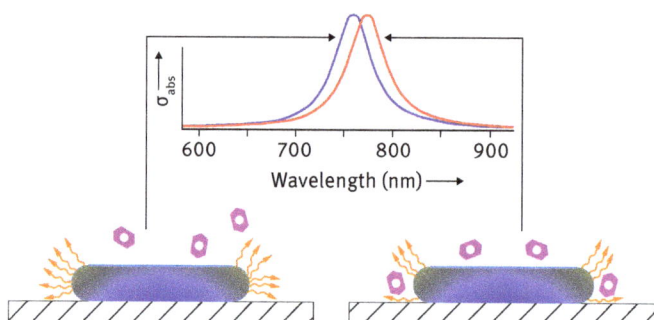

Fig. 3.31: Single-molecule sensing on gold nanorods. Shift of the longitudinal surface plasmon resonance (SPR) of the Au nanorod induced by adsorption of an analyte close to the tip. Figure inspired by Zijlstra et al., *Nature Nanotech.* **7** (2012), 379–382.

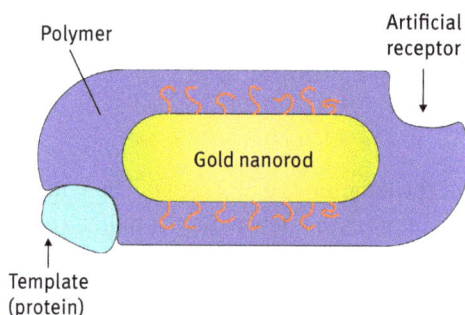

Fig. 3.32: Gold nanorod coated with an imprinted polymer. "Molecular imprinting" employed to create artificial receptor sites using protein templates. Docking of the protein target induces shifts in the nanorod's plasmon resonance. Figure inspired by Abbas et al., *Adv. Funct. Mater.* **23** (2013), 1789–1797.

systems (most such systems utilize molecularly imprinted polymers, or MIPs) the analyte molecule is used as a *template* to create binding sites (e.g. "imprinting"). The imprinting process is usually carried out by polymerization around the analyte, followed by its removal, leaving behind binding sites which specifically trace the structural contours of the analyte. Subsequent binding of the analyte is highly specific, making molecular imprinting a potentially powerful sensing concept. In the experiment depicted in Figure 3.32, S. Singamaneni and colleagues at Washington University imprinted a protein analyte within the polymer coating of Au nanorods. Docking of the protein targets onto the imprinted "receptors" on the polymer layer induced *shifts* in the LSPR signal due to changes in the refractive index of the coating layer. Moreover, the optical shifts provided the means to monitor the progression of the imprinting reaction and polymerization process *in situ* (e.g. in "real time"), providing a valuable tool for control and optimization of sensor properties.

Like their spherical counterparts, Au NRs have attracted significant interest as possible conduits for biological applications. For example, antigenic proteins have been displayed on nanorods' surfaces, eliciting a strong immune response. In the context of vaccine design, studies have specifically focused on the rod shape as a po-

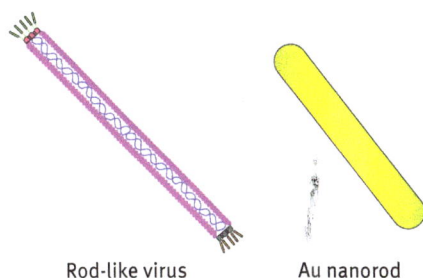

Rod-like virus   Au nanorod

Fig. 3.33: Shape and size similarities between gold nanorods and rod-shaped viruses (e.g. filamentous viruses).

tentially useful feature for therapeutic (or biomimetic) applications. To that end, Au nanorods coated with viral proteins have been examined as potential vaccines. Indeed, this strategy arose from the resemblance of Au NR shape (and occasionally also size) to viral species, specifically rod-shaped viruses (such as "filamentous viruses"; Fig. 3.33). This morphological similarity could be exploited to elicit an immune response, for example through the use of Au NRs presenting viral proteins on their surfaces as immunogens (substances generating immune response). In this sense, Au NRs are, in principle, more attractive than other adjuvants since they resemble viral particles more closely. Indeed, Au NR-based vaccines are already in the pipelines of pharmaceutical companies to compete against more established platforms.

Conjugating Au NRs with biomolecular recognition elements has opened up intriguing biomedical possibilities. R. Kopelman and colleagues at the University of Michigan have reasoned that Au NRs coupled to antibodies which recognize tumor-specific antigens could function as effective and simple "contrast agents" in x-ray computer tomography (CT). Although CT is one of the most widely-used medical imaging technologies, the technique is limited when applied to molecular- or tissue-specific imaging, since cell- or tissue-selective CT contrast agents are rare. Gold, in fact, is well-known for its capacity to attenuate x-rays (as it possesses a high atomic number), and Au NPs essentially constitute a dense volume of gold, which is expected to block incoming x-rays in a similar manner to the way bones become visible in x-ray images. In a proof-of-concept experiment for CT-oriented application, Au NRs were chemically conjugated to antibodies generated against cancer cells. Following injection into mice, the Au NPs accumulated in cancerous tissues, making those tissues appear extremely bright in a light scattering experiment mimicking the CT response (Fig. 3.34). It should be noted that Au NRs are useful for physiological imaging applications in comparison to spherical Au NPs partly because they exhibit significant light absorption in the infrared region (see below), making combined imaging and photo-thermal applications possible. Key issues need to be addressed, however, before broad adoption of Au NR-based diagnostic technologies, including assessment of the risks of Au NRs toxicity, the extent of nonspecific absorption, and technical parameters such as the available spatial resolution and image contrast.

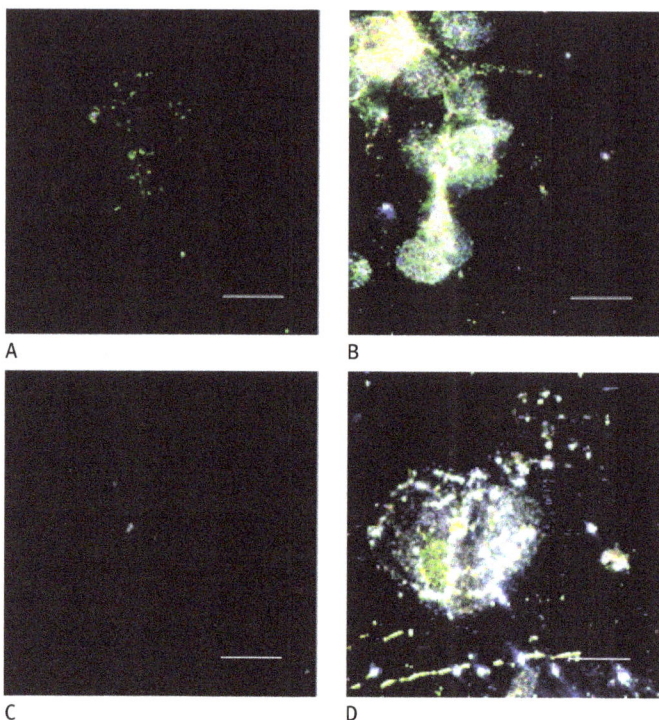

Fig. 3.34: Gold nanorods for image enhancement in x-ray computed tomography (CT). Light scattering experiments demonstrating antibody-conjugated Au nanorods accumulating in cancer cells. The light microscopy images on the right demonstrate high scattering recorded in cancer cells incubated with Au nanorods coupled to antibodies directed against specific cancer cell surface receptors. The images on the left show low scattering from cells incubated with Au nanorods coupled to nonspecific antibodies. Scale bars correspond to 10 μm Reprinted with permission from Popovtzer et al., *Nano Lett.* **8** (2008), 4593–4596, © 2008, American Chemical Society.

Au NRs have attracted significant interest as possible conduits for *photothermal therapy*. This approach utilizes absorption of light by nanoparticles embedded in target tissues or cells (usually *cancerous*), and subsequent release of the light energy as heat, thereby destroying the surrounding tissue environment. The challenges in photothermal therapy generally concern the construction of entities which will absorb light energy in spectral regions different from the absorbance of the physiological environment. In addition, the emitted heat has to dissipate only in target areas, while not damaging healthy tissues.

Au NRs could successfully address these challenges. For one thing, the diverse arsenal of gold chemistry enables functionalization of the NR surface, displaying recognition elements designed to target the NRs to specific body locations – organs, tissues, and the like. Furthermore, the spectral absorbance of nanorods (e.g. plasmon resonance) can be tuned by controlling the particle dimensions, particularly NR length and aspect ratio. Crucially, this feature makes the design of Au NRs which absorb light in the "water window" – the near infrared (NIR) spectral region (light wavelengths between 700–1200 nm) – possible in which body tissues exhibit minimal light absorption. Accordingly, Au NRs *inside* the body can, in principle, absorb energy through irradiation with NIR light; the absorbed energy then dissipates, turning the NRs into tiny "nanoheaters" which can inflict thermal damage on surrounding tissue. Remarkably, Au NR-associated photothermal effects can lead to temperatures of hundreds of degrees in the proximity of the NRs, making the particles a potentially powerful therapeutic tool.

While Au NR-based phototherapy has usually been touted as a useful platform in biomedicine, there have been other uses for the phenomenon. Figure 3.35 depicts an experiment designed by M. S. Wong and colleagues at Rice University, in which enzyme-functionalized Au NRs have been employed as a photothermal platform for induction of *biocatalysis*. Different than photothermal therapy, in which the heat generated at the NR surface is usually aimed at destroying surrounding tissue, in this case heat is used to activate a *thermophilic enzyme* covalently attached to the gold surface. Thermophilic enzymes function in elevated temperatures (sometimes up to 100°C) and thus light-induced heating through the Au NRs provides a remote-control "turn-on" mechanism for the enzyme. As shown in Figure 3.35, the Au NR/enzyme conjugates were embedded in a porous alginate matrix designed to insulate the enzyme and maintain uniform heat distribution around the enzyme molecules. Crucially, NIR irradiation generated localized heating by the Au NRs, consequently inducing enzyme action. The system depicted in Figure 3.35 points to the possible use of Au NR-induced localized heating as a noninvasive triggering mechanism for biochemical reactions in specific physiological locations.

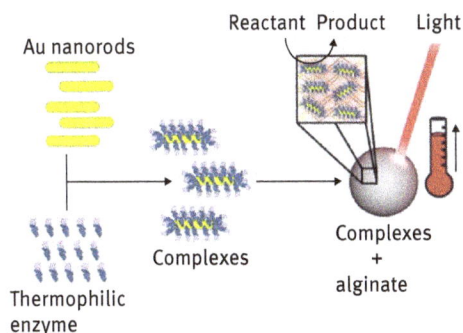

Fig. 3.35: Light-induced gold nanorod heating applied for biocatalysis. The Au nanorods are coated with a thermophilic enzyme and the enzyme-functionalized nanorods are then embedded within porous transparent alginate beads. Irradiation of the encapsulated nanorods with near infrared light generates localized heating which triggers enzyme action. Reprinted with permission from Blanckschien et al., *ACS Nano* **7** (2013), 654–663, © 2013, American Chemical Society.

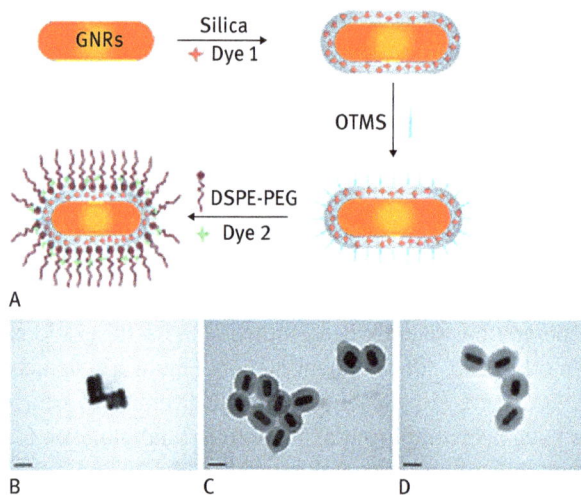

Fig. 3.36: Multifunctional gold nanorods. **A:** An experimental approach designed to construct multifunctional Au nanorods: the nanorods are first coated with a silica layer which also hosts a fluorescent dye. Subsequent coating of the nanorods with a lipid layer contributes to the nanorods' biocompatibility and allows embedding of a second fluorescent dye. **B–D:** Electron microscopy images depicting the Au nanorods encased with silica layers of different thicknesses. The scale bars correspond to 50 nm. Reprinted with permission from Zhang et al., *Angew. Chem. Int. Ed.* **52** (2013), 1148–1151, © 2013, John Wiley & Sons.

Multifunctional Au NR systems for biological applications have also been introduced (Fig. 3.36). The "multimodal" Au NRs synthesized by S. He and colleagues at Zhejiang University, China, were designed to perform several tasks in physiological environments. Specifically, the NRs were coated with silica layers (of tunable thickness) aimed at providing protection against salt species (such as NaCl), withstanding broad pH variations, and generally endowing the NRs with enhanced resilience. The silica layer also hosted a fluorescent dye functioning as a photosensitizer (i.e. light-induced trigger) to generate toxic, reactive oxygen species (ROS) in target tissues (usually cancer tissues; this technique is known as *photodynamic therapy*). The silica/Au-NRs were further coated with a *polymeric lipid* sheath hosting a second fluorescent dye employed as an imaging agent; the silica barrier maintained a distance between this dye and the NR surface, preventing fluorescence quenching. The Au NR design in Figure 3.36 underscores the range of synthetic manipulations and endowed functionalities of Au NR systems.

Similar to the spherical Au NPs discussed above, one of the most significant obstacles to expansion of biomedical and therapeutic applications of Au NRs is their toxic risk. While Au NR toxicity has been widely investigated, a consensus is yet to emerge on the nature of putative toxic agents (the Au core of the NR, the surfactant layer, specific molecules displayed upon the surface). Interestingly, some researchers have proposed that the possible toxicity profile of Au NRs might actually be utilized as a therapeutic tool, turning the Au NR itself into a "drug". A rather more conventional therapeutic avenue for Au NRs is as a vehicle for drug delivery. In this context, Au NRs offer distinct advantages. First, the relatively large surface area (or more accurately the high surface-to-volume ratio) provides an effective means for loading relatively large concentrations of drug molecules and also displaying recognition and targeting elements. Second, the combined charged gold surface and hydrophobic surfactant layer around the Au core constitute another handle for incorporating guest molecules and therapeutic cargo onto the NR surface. The hydrophobic domains of surfactant layers, for example, have been used to sequester and transport small therapeutic molecules in the bloodstream.

A related major fundamental and technical challenge encountered in biological applications of Au NRs concerns the *release* of the surface embedded cargo. In some systems, this can be accomplished by means of the photothermal effect described above. For example, drug molecules can be encapsulated within thermo-sensitive layers; heating of the Au NRs through irradiation with near IR light can then trigger release of the embedded drugs. Other release mechanisms reported relied on external stimuli such as pH changes, enzyme action, or molecular exchange with the external solution.

3.1.7 Gold nanowires

The distinction between Au nano*wires* and nano*rods* is somewhat blurred and primarily related to the higher aspect ratio of Au nanowires. In the realm of possible electronic applications, Au nanowires are naturally considered more useful compared to nanorods, for example as components in electrical/optical devices. Practical hurdles for such applications are mainly related to the somewhat limited technical routes of isolating and controlling the macroscale organization of Au nanowires prepared by "bottom-up" synthetic techniques (compared to the common lithography methods employed to produce elongated gold features and nanostructures).

Chemical synthesis of Au nanowires usually follows the same basic routes as for Au nanorods, with modifications aimed at extending the nanowire length, for example through longer incubation times in the solution containing the surfactants and reducing agents. Nanowire fabrication using lithography has also been reported due to the wide availability and sophistication of lithography technologies (specifically *nano*lithography) and patterning/etching instruments. In such schemes, generally referred to as "top-down" synthesis, surface patterning techniques such as photo-lithography or electron-beam lithography are applied to create wire patterns, subsequently acting as templates for the molding of Au nanowires. Such top-down approaches make large-scale production of uniform nanowires with pre-determined dimensions possible.

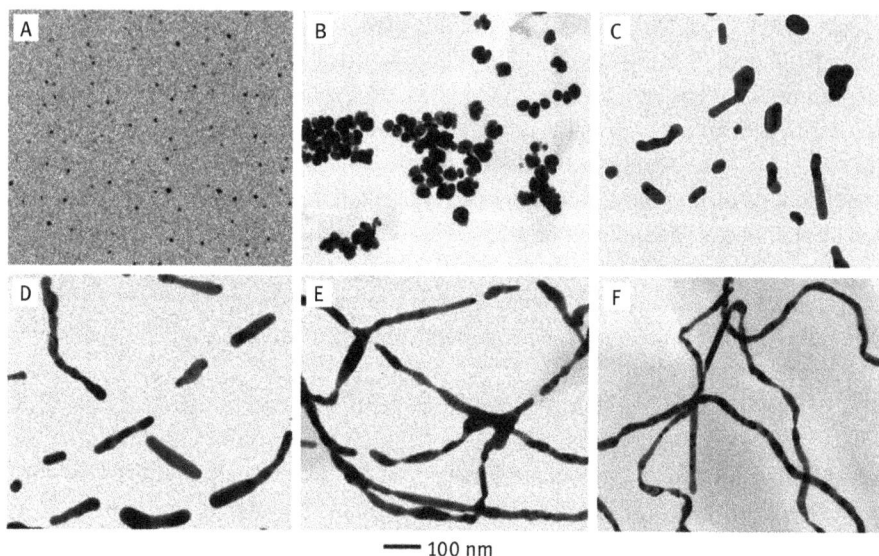

——— 100 nm

Fig. 3.37: Wavy gold nanowires. Electron microscopy images showing time evolution of Au nanowires. Images were recorded for the reaction mixture from 5 minutes (**A**) to 90 minutes (**F**), and reveal the "welding" of individual fragments into elongated wavy Au nanowires. Reprinted with permission from Zhu et al., *JACS* **134** (2012), 20234–20237, © 2012, American Chemical Society.

Variations of the basic gold nucleation-reduction-growth methods for Au nanowire synthesis have been reported. Figure 3.37, for example, depicts electron microscopy images of *wavy* Au nanowires and their intermediate species. The scheme for generating these nanowires, developed by Y. Xia and colleagues at the Georgia Institute of Technology, was surprisingly simple, involving a single step gold reduction and fusion or "welding" of shorter rod-like fragments, ultimately producing the long "wavy" structures. Interestingly, the CTAB surfactant capping the nanowires adopted a monolayer structure (rather than the bilayer organization usually encountered as coating layers of Au nanorod structures), consequently rendering the Au nanowires *hydrophobic* – a property that could be exploited to create self-assembled structures. The formation of Au nanowires instead of the more common Au nanorods which are also synthesized in the presence of surfactants such as CTAB has been ascribed to the fact that the reaction scheme was not based on a seed-mediated procedure. Rather, a slower reduction process of the gold precursor produced the gold fragments which were subsequently merged, generating the nanowire morphology.

Au nanowires of distinct lengths and aspect ratios were produced via a unique synthetic scheme which did not require the presence of reducing agents or capping substances (Fig. 3.38). The generic approach, developed by R. Jelinek and colleagues at Ben Gurion University, Israel, was based on dissolution of a sparingly-soluble gold complex – $Au(SCN)_4^-$ – in solvent mixtures comprising of water and organic solvents.

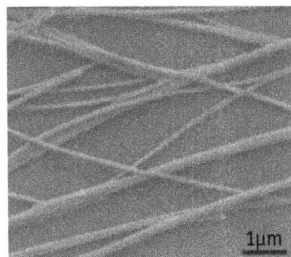

Fig. 3.38: Gold nanowires produced by spontaneous crystalliza-
tion/reduction of Au(SCN)$_4^-$ in solvent mixtures. Production of
Au nanowires does not require addition of reducing agents or
surfactants. The Au nanowires form transparent, conductive film
networks on surfaces.

The length, thickness, and abundance of the Au nanowires formed using this approach were dependent on the nature of the organic solvent (polar/apolar) and solvent ratios. Interestingly, the Au nanowires spontaneously formed interspersed film
networks which were electrically conductive and also allowed transmission of light,
thereby potentially to be used as *transparent conductive electrodes* (TCE), a basic component of many electro-optic and photonic devices.

   An interesting phenomenon associated with Au nanowires is the generation of
propagating surface plasmon waves. This feature essentially integrates the localized
surface plasmon resonance in Au nanoparticles and the highly anisotropic morphology of the nanowires to generate plasmon waves moving along the nanowire axis (similar to waves propagating on the surface of the ocean). As such, Au nanowires can be
employed as "waveguides" in hybrid photonic/electronic circuits and devices. A recurring deficiency in nanowire-based plasmonic waveguides, however, has been the
significant energy loss incurred by the propagating waves, thereby limiting their practical applications.

3.1.8 Other morphologies of gold nanoparticles

While nanospheres, nanorods, and to some extent nanowires, have been by far the
most prominent and widely-reported Au NP morphologies, other Au nanostructures
have been produced. Impetus to develop new Au NP configurations is the fact that important physical properties, particularly localized surface plasmon resonance (LSPR)
and surface-enhanced Raman scattering (SERS), are intimately linked to the particles'
shapes and their surface features. Efforts to synthesize Au nanostructures in non-
spherical and noncylindrical shapes also stem from the desire to use faceted nanoparticles such as nanocubes, nanoprisms, and the like as building blocks for more
complex "superstructures".

   The development of synthetic routes yielding nonspherical, faceted Au NPs in a
controlled manner has been a central activity in this field. Indeed, the close dependence of Au NP structural properties on synthetic parameters has been a primary reason for the striking variety of Au NP morphologies reported. In practical terms, slight
variations in reaction conditions are known to result in considerable changes to the fi-

Fig. 3.39: Gold nanoprisms. Reprinted with permission from Ah et al., *Chem. Mater.* **17** (2005), 5558–5561, copyright (2005), American Chemical Society.

nal NP products and their structural features. Several synthetic schemes have utilized Au NP/polymer hybrids as a platform for tunable NP morphologies. Such mixed systems usually prevent *isotropic* growth of Au crystallites, thereby blocking formation of spherical or rod-like particle symmetries.

Figure 3.39, for example, shows crystalline Au nanoplatelets and nanoprisms synthesized by W. S. Yun and colleagues at the Korea Research Institute of Standards and Science through incubation of the gold NP precursor (chlorauric acid) with a reducing agent and a polymer, poly(vinyl pyrrolidone), or PVP, acting as a structure-shaping constituent. Au nanoprisms such as those depicted in Figure 3.39 exhibit unique optical properties closely linked to their configurations. Specifically, unlike spherical Au NPs, the crystalline planes (i.e. surfaces) of the nanoprisms give rise to higher-order plasmon resonance absorbances ("dipole" plasmon resonances and "multipole" plasmon resonances) reflected in resonances appearing in higher wavelengths compared to the plasmon peaks observed in spherical Au NPs. These spectral features (which endow distinct color properties to Au nanoprism suspensions) could be utilized in electro-optical and sensing devices. The challenges, however, of the practical utilization of Au nanoprisms primarily focus on devising effective synthetic routes to attain sufficient control over shape, size, and uniformity of these unique Au NPs.

Gold nanoprisms have been used as precursors for more complex-shaped nanoparticles which further utilized DNA as a template agent (Fig. 3.40). In that study, Y. Lu and colleagues at the University of Illinois, Urbana-Champaign, demonstrated that oligonucleotides comprised of 20–30 single bases (i.e. A_{20}, T_{30}, $A_{10}G_{20}$, etc.) produced distinct "star-like" NP morphologies when incubated with the nanoprism seeds. In a sense, the Au NP shapes were "encoded" by the oligonucleotide sequences containing each of the four bases (A, T, C, G). This remarkable phenomenon is likely related to different affinities of the four bases to the crystal facets of the Au nanoprisms, consequently giving rise to the distinct growth dynamics and structural features of the resultant nanoparticles.

Au *nanocubes* and *nano-octahedra* (Fig. 3.41) belong to an interesting class of multifaceted Au NP morphologies. Au nanocubes (as well as Au *octahedra* and other polyhedral structures) have been particularly useful as substrates for SERS-based sensing; it is hypothesized that the sharp edges and corners in such polyhedral structures generate pronounced localized electric fields responsible for the enhanced Ra-

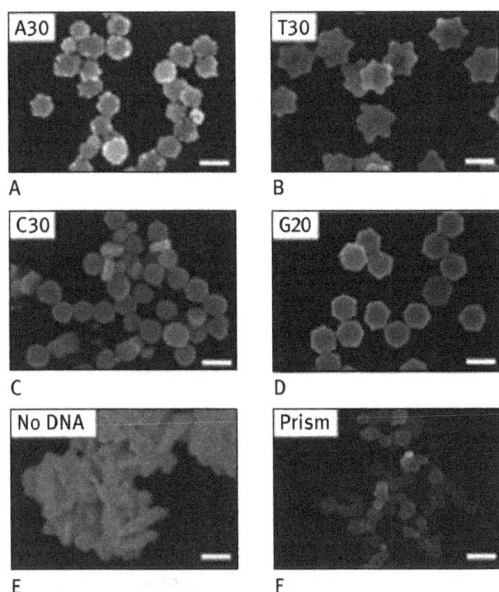

Fig. 3.40: Oligonucleotide-directed gold nanoparticles. The different nanoparticle shapes were produced by adding the Au nanoprisms (shown in panel **F**) to solutions containing the indicated single-base oligonucleotides. Scale bars correspond to 200 nm. Reprinted with permission from Wang et al., *Angew. Chem. Int. Ed.* **51** (2012), 9078 –9082, © 2012, John Wiley & Sons.

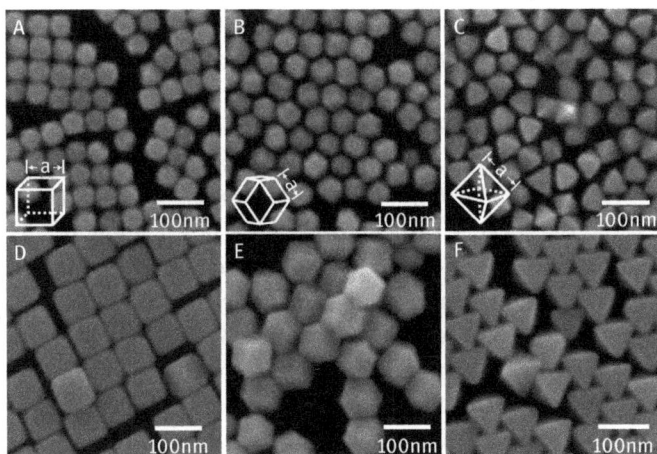

Fig. 3.41: Polyhedral gold nanoparticles. The different-shaped Au nanoparticles are prepared by varying the ratios of the reaction constituents: chlorauric acid, surfactant (cetyltrimethylammonium-chloride, CTAC), and salt (NaBr). Reprinted with permission from Wu et al., *Inorg. Chem.* **50** (2011), 8106–8111, © 2011, American Chemical Society.

man signals. In addition, in some cases polyhedral Au NPs were shown to be catalytically active, which is ascribed to the crystalline facets themselves. The main synthetic breakthroughs in realizing symmetrical, uniform cubic and polyhedral NP morphologies have been the development of reaction pathways controlling the interplay between the Au seeds initiating NP growth, the types of surfactant capping agents and the counter *halide ions* of the surfactants (mainly bromide and chloride). The highly-uniform Au polyhedral nanoparticles prepared in the laboratory of M. H. Huang and colleagues in National Tsing Hua University, Taiwan (Fig. 3.41), were synthesized using seed-mediated growth of crystalline Au nuclei, co-incubated with the surfactant cetyltrimethylammonium-chloride (CTAC), and small amounts of NaBr. The concentrations of the NaBr salt, in particular, had a pronounced effect on the ultimate morphologies of the nanoparticles.

Photoluminescence is another interesting physical property associated with the sharp-edged structure of Au nanocubes. Specifically, Au nanocubes at certain dimensions produce intense luminescence presumed to arise from the plasmon bands which develop around the edges of the nanocubes. In practical terms, the high photoluminescence of Au nanocubes can be exploited for bioimaging applications. Similar to their spherical NP counterparts, Au nanocubes can be internalized by cells, providing a useful probe for microscopic imaging.

While many studies demonstrated formation of Au polyhedral nanoparticles with a relatively low number of facets (such as cubes, octahedrons), a more formidable synthetic challenge has been the construction of "high-index faceted" Au NPs – i.e. complex polyhedrons with large numbers of facets (> 20). A major impetus for progress in this field has been the observation that such multifaceted particles are excellent catalysts. Figure 3.42 presents a fine example of a high-index faceted Au nanocrystal – evidence for the advances in synthetic acumen enabling construction of sophisticated nanostructures in controlled and uniform dimensionalities. In this study, J. Y. Lee and colleagues at the National University of Singapore employed successive seed-mediated growth steps using CTAC as the nanoparticle capping agent. It should be noted that while the *concave trisoctahedra* produced were highly uniform, the exact crystallization pathways are not fully understood and more research is needed to develop routes for controlled fabrication of multifaceted NPs.

Branched or "multipod" Au NPs such as the "nanocrosses" and nanohexapods shown in Figure 3.43 have attracted interest because of their unique shapes and optical properties. Similar to Au nanorods, multipod Au NPs generally exhibit two plasmon resonances, one associated with the particles' cores, while the other corresponds to the sharp tips. Furthermore, the SPR positions (i.e. color) exhibit pronounced sensitivity to the length of the multipod "arms" and tip morphologies. This feature is important in a therapeutic context, since the plasmon resonances can be shifted to the near IR (NIR) spectral region (i.e. the "water window", see discussion of Au NRs above) making multipod NPs into possible photothermal agents.

An important upshot of the NIR plasmon resonance of branched Au NPs has been the observation of "two-photon luminescence" – fluorescence generated by absorption of two light photons in the IR range. This phenomenon, mostly observed for Au nanostars, stands in contrast to the prevalent fluorescence quenching which occurs in most Au NPs. This intense two-photon fluorescence might enable tracking of the Au nanostars inside the body, thus making the NPs an effective vehicle for imaging.

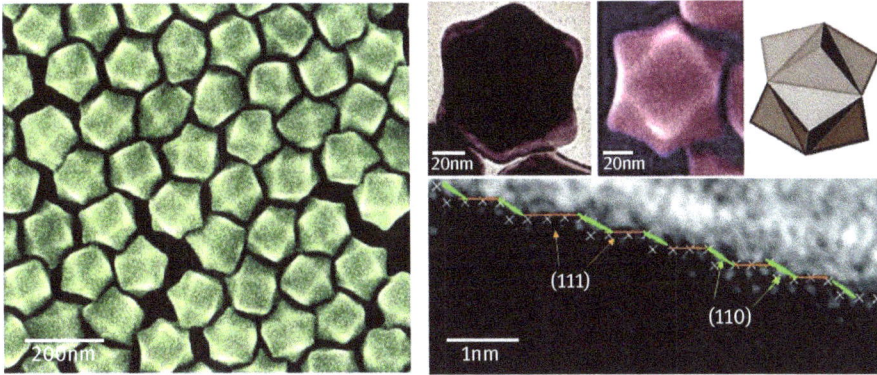

Fig. 3.42: High-index faceted gold nanoparticles. Uniform concave trisoctahedra gold nanoparticles produced via a stepwise seed-mediated synthesis procedure. Reprinted with permission from Yu et al., *J. Phys. Chem. C.* **14** (2010), 11119–11124, © 2010, American Chemical Society.

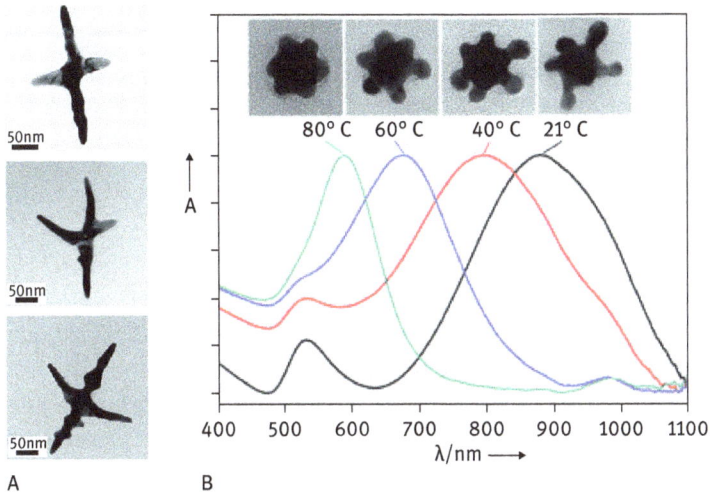

Fig. 3.43: Branched gold nanoparticles. **A:** Au "nanocrosses". Reprinted with permission from Ye et al., JACS **2011**, *133*, 8506–8509, copyright (2011), American Chemical Society. **B:** Au nanohexapods (approximately 20 nm in size) with different arm lengths produced by incubation of the reaction mixture in the indicated temperatures. The corresponding shifts in plasmon resonance are shown in the spectrum below the images. Reprinted with permission from Kim et al., *Angew. Chem. Int. Ed.* **50** (2011), 6328 – 6331, © 2011, John Wiley & Sons.

Similar to other NP species such as spherical Au NPs and Au NRs, Au nanostars can be functionalized with recognition elements designed to deliver the particles to specific physiological targets such as cancer cells. The cells/tissues can then be visualized or destroyed via the photothermal effect. Aiding this route of biological applications

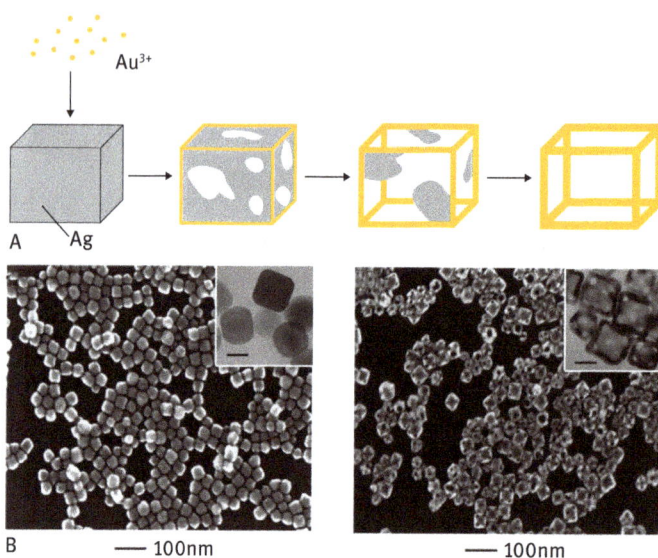

Fig. 3.44: Hollow gold nanocages. **A:** Synthesis scheme: a silver nanocube serves as a "sacrificial template"; Au^{3+} ions oxidize the silver and consequently metallic Au^0 substitutes the silver template. **B:** Electron microscopy images showing the silver nanocube templates (left) and hollow Au nanocubes (right). Reprinted with permission from Chen et al., *Nano Lett.* **5** (2005),, 473 – 478, © 2005, American Chemical Society.

has been the observation that *cell uptake* of Au nanostars and similar multipod NPs is more pronounced in many instances than uptake of Au NRs or spherical NPs.

Au "nanoshells" are another interesting family of Au NPs exploiting NIR irradiation for biological applications. Au nanoshells can either be hollow or surround dielectric cores (the latter NPs are discussed in Chapter 6). Hollow Au "nanocages" were prepared by Y. Xia and colleagues at Washington University via a "galvanic replacement" process (Fig. 3.44): silver nanocubes were initially prepared; subsequent addition of Au^{3+} ions resulted in electron transfer from the metallic silver atoms at the nanocube surface to the gold ions, eventually producing an Au^0 shell. Note that since each Au^{3+} ion consumes three electrons, the galvanic replacement process results in oxidation (and dissolution) of three silver atoms, eventually etching the entire Ag nanocube (appropriately termed a "sacrificial template"), leaving behind a hollow Au nanocage.

The nanometer dimensions of the gold shells translate into tunable LSPR, determined by both the inner and outer diameters of the shell. For example, the Au nanoshell size can be tuned to achieve SPR absorbance in the NIR spectral region, making the particles a potentially useful conduit for photothermal therapy. An elegant application utilizing the photothermal effect of Au nanocages is illustrated in Figure 3.45. Truncated hollow Au nanocages, prepared by Y. Xia and colleagues via the galvanic replacement method, further encapsulated drug molecules. The "nanocon-

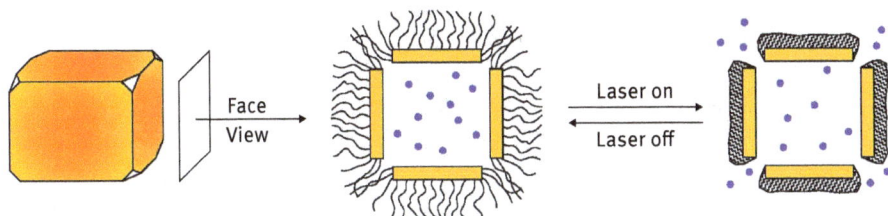

Fig. 3.45: Drug delivery using functionalized gold nanocages. Truncated hollow Au nanocages encapsulated drug molecules are prepared, further coated with a temperature-sensitive polymer "brush". At lower temperatures the polymer brush prevents leakage of the drug molecules; however when the temperature is elevated the polymer network collapses and the embedded molecular cargo is released. Figure based on Yavuz et al., *Nature Mater.* **8** (2009), 935–938.

tainers" were additionally coated with a dense layer of a heat-sensitive polymer brush which prevented release of the loaded cargo into the surrounding solution. However, following irradiation with NIR light, localized heating at the nanoshell surface resulted in collapse of the polymer layer and consequent release of the encapsulated molecules.

Surfaces have often served as powerful mediums for the generation of Au nanostructures with diverse morphologies. Indeed, research in the field of surface-induced nanostructures is highly active, exhibiting vast commercial potential in light of the prevalent use of surfaces, films, and related structures in numerous applications. In a sense, nanostructures grown on surfaces might not be considered purely "nanoparticles" in the context of this book, since in many instances the interest and utilization of nanostructure-on-surface systems stem from their overall cooperative properties, rather than the properties of the individual nanoparticles produced. However, for the sake of completeness an example of surface-generated Au NPs is presented below.

Surfaces usually dictate *anisotropic* growth of nanoparticles. This is both due to the space restrictions of particle growth, and also related to the physico-chemical properties of the surface itself – the functional groups present, chemical reactivity, smoothness, and crystallinity. Indeed, the nature of the surface upon which NPs are formed is a major parameter affecting the structures and properties of the nanostructures grown. Figure 3.46 depicts Au "nanothorns" formed spontaneously upon incubation of chloroauric acid ($HAuCl_4$) with polyaniline (PANI) surfaces, in the presence of a polymeric capping agent. In this study, P. Xu and colleagues at Los Alamos National Laboratory, US, demonstrated that aniline moieties contributed the reducing electrons to Au ions attracted to the surface, while the capping agent promoted growth of surface-protruding NPs. Similar to other Au NP systems, the surface-grown Au NPs shown in Figure 3.46 exhibit SERS activity and could be used as sensitive sensors.

Fig. 3.46: Gold nanoparticles grown on a polymer surface. Au nanoparticles formed spontaneously by attachment and reduction of Au^{3+} ions on polyaniline (PANI). Reprinted with permission from Li et al., *ACS Appl. Mater. Interfaces* **51** (2013), 49–54, © 2013, American Chemical Society.

3.2 Silver nanoparticles

In many respects, silver NPs are similar to gold NPs, and this resemblance is apparent in their properties and applications. The variety of Ag NP structures, in fact, is even greater than that of Au NPs. Numerous reports featuring Ag nanospheres, nanorods, nanowires, and other types of NPs in different dimensions and exhibiting varied functions have been reported. The diverse *synthesis routes* of Ag NPs have also contributed to the variety of Ag NP morphologies. Most synthetic schemes are based on mixing silver ion precursors (such as $AgNO_3$) with soluble reducing agents required to generate elemental silver (*borate* and *citrate* have been popular reducing agents in both silver and gold NP synthesis processes). Usually, surfactants or polymers are also added to the reaction mixtures, to stabilize the nanoparticles and prevent aggregation. Such stabilizers also serve, in many instances, as pivotal factors determining the eventual size and shape of the NP products. Figure 3.47 depicts a generic strategy implemented by Y. Xia and colleagues at Washington University for the construction of Ag NPs in a remarkably broad range of shapes and sizes. In this approach, Ag^+ ions are reduced by ethylene glycol (EG), which also serves as the *solvent* for the reaction. The resultant silver nuclei (also referred to as "fluctuating nanostructures") are intrinsically unstable and further assemble into varied nanoparticle morphologies. Careful modulation of the reaction conditions, such as temperature, reagent concentrations, and reaction duration, facilitated control of the Ag NP structural features.

Other Ag NP synthesis procedures have been reported. *Seed mediated NP growth* has been one of the more popular approaches. This versatile technique involves "seeding" the precursor solution with nanocrystals (which, in most cases, do not comprise of silver but rather other crystalline species). These nucleation units catalyze crystal-

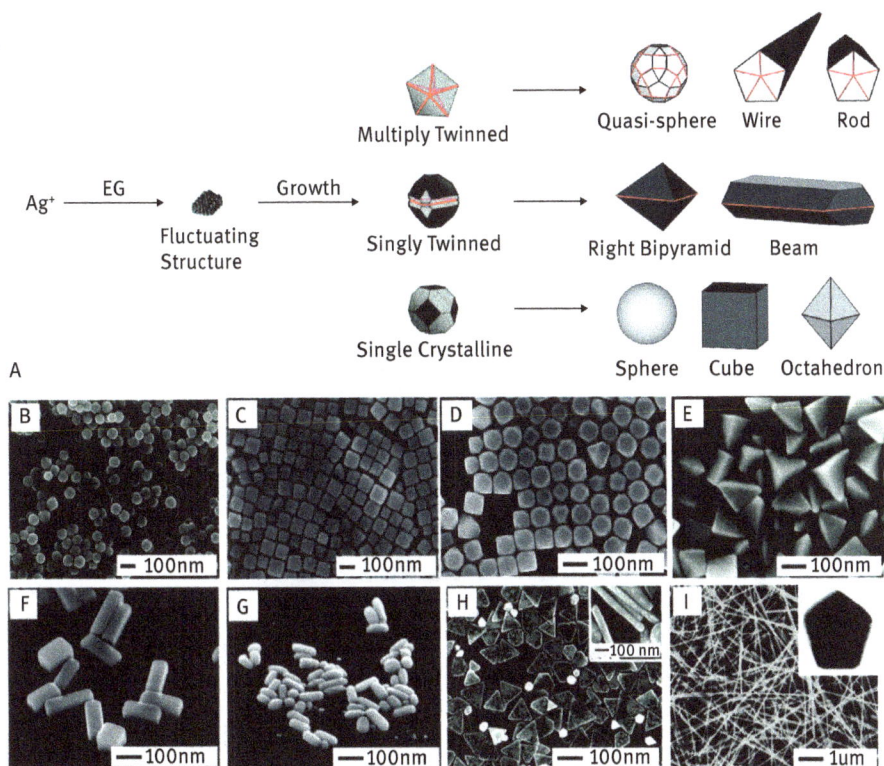

Fig. 3.47: Structural diversity of silver nanoparticles. **A:** Generic synthesis scheme depicting Ag$^+$ reduction by ethylene glycol (EG) and formation of silver seeds, which further aggregate, yielding varied nanoparticle morphologies. **B–I:** Electron microscopy images of diverse Ag nanoparticles produced via the EG-reduction strategy. Reprinted with permission from Rycenga et al., *Chem. Rev.* **111** (2011), 3669–3772, © 2011, American Chemical Society.

lization and growth of Ag NPs. In fact, separating the NP nucleation and growth stages provides better control of NP structural properties. In particular, the extent of *crystal lattice mismatch* between the nanocrystal seed and the crystalline Ag layers deposited is considered a fundamental parameter affecting the morphology and physical properties of the synthesized Ag NPs.

Template-directed synthesis of Ag NPs has also been demonstrated, employing different molecular "scaffolds" for assembly of the silver nanostructures. Figure 3.48 presents an example of this approach. The template, developed by J. P. Rabe and colleagues at Humboldt Universitaet zu Berlin, Germany, consisted of a tubular structure formed upon assembly of an amphiphilic organic molecule (a cyanine-based dye) in a methanol/water mixture. Importantly, the cyanine dye also acted as a reducing agent, transforming silver ions into metallic Ag nanowires inside the hollow nanotube. As shown in Figure 3.48B, the resultant Ag nanowires exhibit a well-defined and uni-

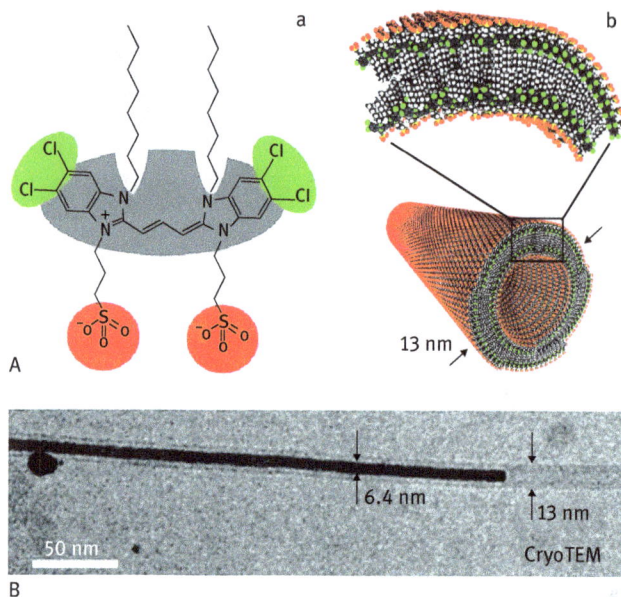

Fig. 3.48: Template-directed synthesis of silver nanowires. **A:** Chemical structure of the amphiphilic cyanine dye forming the hollow nanotube; the dye also functions as the reducing agent, thereby producing the Ag nanowires inside the tubes. **B:** Electron microscopy image showing a uniform Ag nanowire inside the hollow tube. Reprinted with permission from Eisele et al., *JACS* **132** (2010), 2104–2105, © 2010, American Chemical Society.

form diameter, determined by the dimensionality of the tubular template. Numerous variations of template-directed synthesis have been reported, including hollow nanotubes assembled from inorganic, organic, and biological materials, natural and synthetic porous matrixes, patterned films acting as surface masks, and others. Indeed, the choice of template constitutes a powerful vehicle for the creation of diverse Ag nanoparticle structures.

A different kind of template is portrayed in Figure 3.49, in which chemically-modified peptides were employed as scaffolds for Ag NP synthesis. Peptide template-generated Ag NPs, reported by H. Wennemers and colleagues at the University of Basel, Switzerland, were synthesized on covalently attaching aldehyde moieties – acting as reducing agents for Ag^+ – in specific positions within *polyproline* peptide sequences. Polyprolines adopt helical structures, of which the helix length is determined by the number of proline residues in the sequence. The aldehydes were displayed in each third position within the polyproline sequence, i.e. in the same spatial orientation within the helical structure (Fig. 3.49). Accordingly, the length of the polyproline scaffolds determined the dimensions of the nanoparticles. Indeed, using this strategy the researchers generated Ag NPs with particularly narrow size distributions.

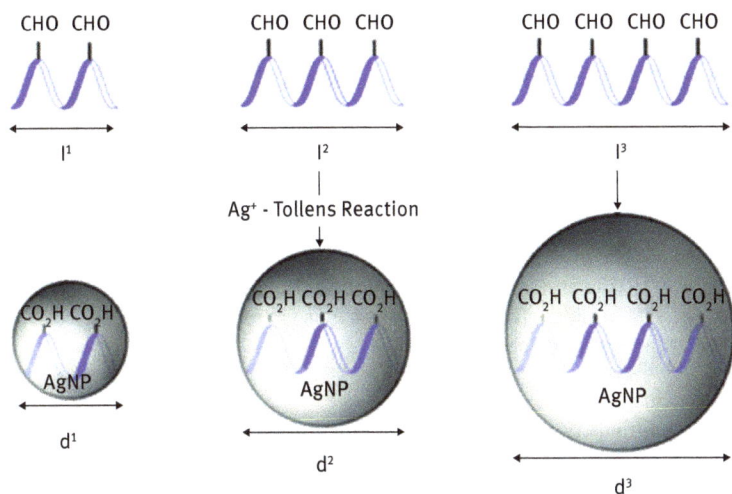

Ag$^+$ - Tollens Reaction

A

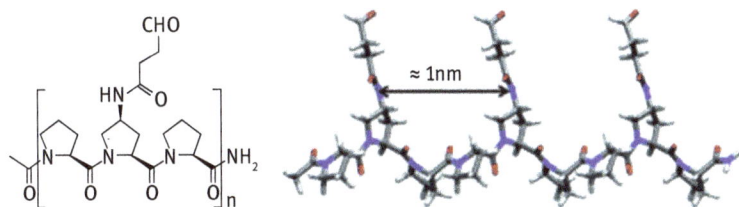

B

Fig. 3.49: Silver nanoparticles synthesized on peptide-aldehyde templates. **A:** Diagram showing the aligned aldehyde residues on the polyproline helical peptide; the length of the peptide determines the size of the Ag nanoparticle. **B:** Chemical structure of the peptide chain and the single covalently-attached aldehyde in a three-residue motif. Reprinted with permission from Upert et al., *Angew. Chem. Int. Ed.* **51** (2012), 4231 – 4234, © 2012, John Wiley & Sons.

Considering the significant research focused on Ag NP systems, it might seem somewhat surprising that no consensus has emerged yet as to the *mechanism* of Ag NP formation in solution. This might reflect the many variables affecting NP growth and dimensions, including reduction kinetics, the propensity to aggregate vs. stabilization of individual NPs, surface reactivity of the evolving crystal facets of NPs, and other parameters. Figure 3.50 illustrates a proposed mechanism for Ag NP growth, outlined by J. Polte and colleagues at the Technical University Berlin, Germany. According to the model, NP assembly and ultimate size are determined by "coalescence steps" in which both the initial small Ag crystallites formed upon Ag$^+$ reduction, as well as the larger aggregates assembled via the first coalescence step are "fused together", eventually forming stable Ag NPs. The main driving force for coalescence and NP formation according to this model is the competition between nanoparticle stability and aggregation – the final NP size is thus determined by the colloidal stability which prevents further aggregation. In this context, surfactants and other agents contribute to NP stabil-

1. Reduction   2. Coalescence   3. Metastable state   4. Coalescence   Final nanoparticles

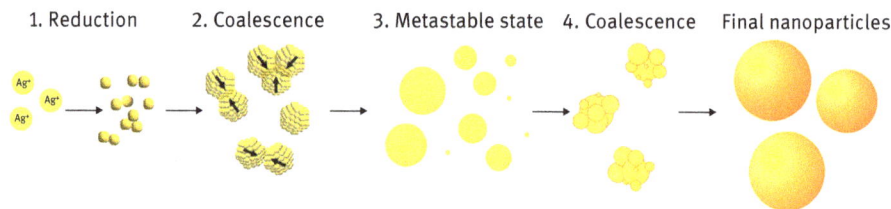

Fig. 3.50: Mechanism of silver nanoparticle formation. Small metallic Ag aggregates form following consecutive coalescence steps. The final state of the nanoparticles is determined by colloidal stability minimizing further aggregation. Reprinted with permission from Polte et al., *ACS Nano* **6** (2012), 5791–5802, © 2012, American Chemical Society.

ity throughout the assembly process. Interestingly, the main difference between silver NP and gold NP growth according to the proposed mechanism concerns the *duration* and thus relative contributions of the coalescence steps, leading to greater homogeneity of Au NP size compared to their Ag NP counterparts.

Besides the conventional chemical deposition and solution-phase synthetic routes of Ag NPs, intriguing methods affecting *shape transformations* of the nanoparticles have been reported. Figure 3.51 illustrates a simple technique, demonstrated by C. Mirkin and colleagues at Northwestern University, by which *spherical* Ag NPs can be converted to triangular Ag "nanoprisms" simply by irradiation with ultraviolet (uv) light. The conversion process is presumed to occur via "photo-fragmentation" of the Ag nanospheres, with subsequent re-assembly of crystalline nanoprisms. The light-induced nanosphere-nanoprism conversion has important implications, as Ag nanoprisms have been shown to exhibit remarkable optical and electronic properties which are significantly different to spherical Ag NP configurations.

Surface plasmon resonance (SPR) or localized SPR (LSPR), which is more specifically associated with nanoparticles, have both been widely observed in Ag NPs. As discussed above for Au NPs, SPR arises from strong interactions between incident light and surface electrons of metallic nanostructures. This "coupling" between electromagnetic radiation (light) and collective motion (oscillations) of surface electrons is translated into resonances – light emitted in specific wavelengths – which in the realm of nanotechnology exhibit significant dependence upon size, shape, and sur-

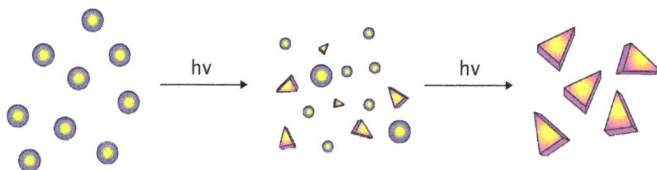

Fig. 3.51: Light-induced transformation of silver spherical nanoparticles into nanoprisms. Figure inspired by Rongchao et al., *Science* **294** (2001), 1901–1903.

face properties of the nanostructures examined. SPR phenomena have been observed in many metal and metal-oxides, however gold and silver are by far the most important and widely-used materials for SPR applications. While gold nanostructures exhibit attractive features as SPR platforms in terms of sensitivity, resonance wavelengths in convenient spectral range, and diverse biological applications, silver offers advantages as it exhibits very high efficiency in generating SPR (in fact silver has the highest surface plasmon "quality factor" of all metals). Moreover, the significantly lower cost of silver compared to gold is a favorable factor for Ag NPs as a conduit for SPR applications.

As the sizes and shapes of Ag NPs have been shown to exert significant effects on LSPR, many studies have attempted to characterize the relationship between this optical phenomenon and the structural features of Ag NPs, i.e. isotropic vs anisotropic NP morphologies, areas of exposed crystal planes, and abundance of edges and vortices. Ag nanocubes have been useful models for analysis of LSPR properties. Figure 3.52 presents electron microscopy images of Ag nanocubes of different sizes and the corresponding LSPR spectra. In this carefully-conducted study, Y. Xia and colleagues at Washington University demonstrated that nanocube size (particularly *edge length*) intimately affected both the color (e.g. SPR peak position) and resonance width of the LSPR generated.

An elegant example of *shape-dependent* LSPR in Ag NPs is given in Figure 3.53. In this study, S. W. Han and colleagues at KAIST, Korea, showed that fine tuning

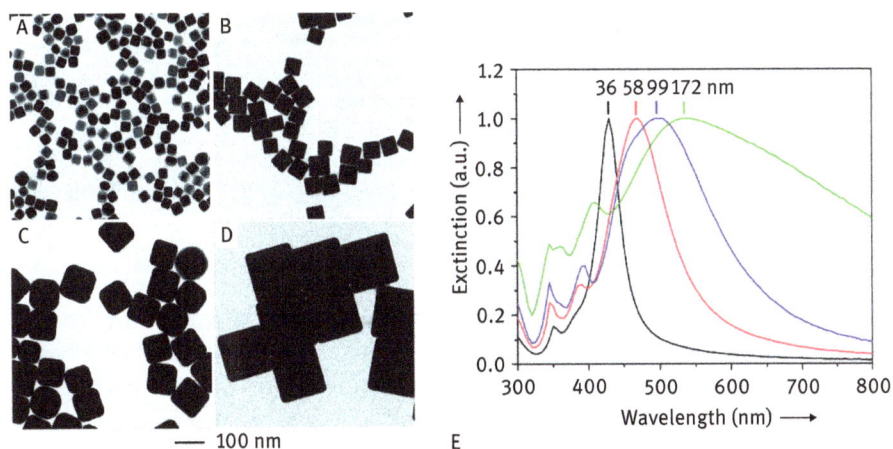

Fig. 3.52: Silver nanocubes of different sizes generate distinct shifts and widths of localized surface plasmon resonances. **A–D:** Electron microscopy images of the silver nanocubes, underscoring the uniform size distribution; **E:** Localized surface plasmon resonances associated with Ag nanocubes of different edge lengths. Both the resonance shifts and widths are affected by nanocube size. Reprinted with permission from Rycenga et al., *Chem. Rev.* **111** (2011), 3669–3712, © 2011, American Chemical Society.

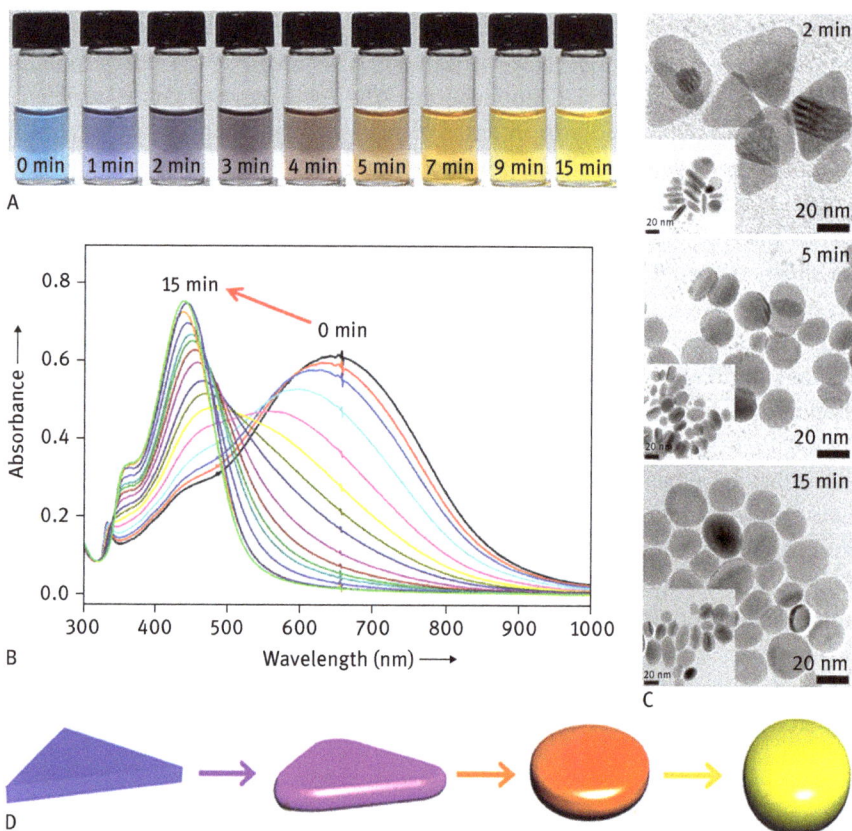

Fig. 3.53: Shape-dependent surface plasmon resonance of silver nanoparticles. Color changes of silver nanoparticle solutions (**A–B**) affected by different incubation times of silver nanoprisms with bromide ions. The Br⁻ ions etch the corners of the nanoprisms, eventually producing spherical silver nanoparticles (**C–D**). Reprinted with permission from Kim et al., *ACS Appl. Mater. Interfaces* **4** (2012), 5038–5043, © 2012, American Chemical Society.

nanoparticle *shape* and *morphology* resulted in dramatic color changes of NP suspensions (corresponding to shifts in SPR peak positions). The experiments were based on incubating Ag nanoprisms with Br⁻ ions for varying lengths of time. The bromide ions gradually *etched* the nanoprisms (i.e. reacted and removed silver ions), however as illustrated in Figure 3.53D, the etching occurred primarily at the corners of the Ag nanoprisms, resulting in changes in shape. Accordingly, by controlling the Ag NP incubation time with the bromide ions, the researchers effectively tuned the particle shapes (from nanoprism into spheroid) and demonstrated that the morphological transformation was accompanied by dramatic color changes.

Heterogeneous catalysis (e.g. catalysis occurring at solid-solution interfaces) is a broad field in which Ag NPs and Au NPs have often been bundled and discussed

together (and compared in terms of their catalytic properties). Like their Au NP counterparts, Ag NPs exhibit high surface-to-volume ratios which are conducive to adsorption of target molecules (particularly organic molecules, which have been widely
investigated in the context of heterogeneous catalysis). Furthermore, the abundance
of partly-free electrons at the Ag NP surface is believed to play a major role in catalytic reactions; these electrons can be excited by light (visible or ultraviolet) and
subsequently enhance reactions involving molecules adsorbed onto the nanoparticle
surface.

The effectiveness of heterogeneous catalysis is directly linked to the properties of
the catalyst's active sites, particularly the organization and electronic properties of the
atoms close to the surface-adsorbed target molecules. Accordingly, in the case of Ag
NPs, the catalytic properties are intimately affected by the shapes, surface areas, and
crystalline organization of the NP facets. Indeed, the substantial economic benefits of
better catalytic profiles, for example in the petroleum and plastics industries, provide
impetus for research activities in this field. Major efforts have been directed towards
"fine tuning" the structural parameters of Ag NPs, particularly as related to the crystalline organization of their surfaces, with the goal of generating better catalysts.

As an example relevant to the above observations, Figure 3.54 presents three crystalline Ag NP structures tested for their catalytic action upon the oxidation of *styrene* to

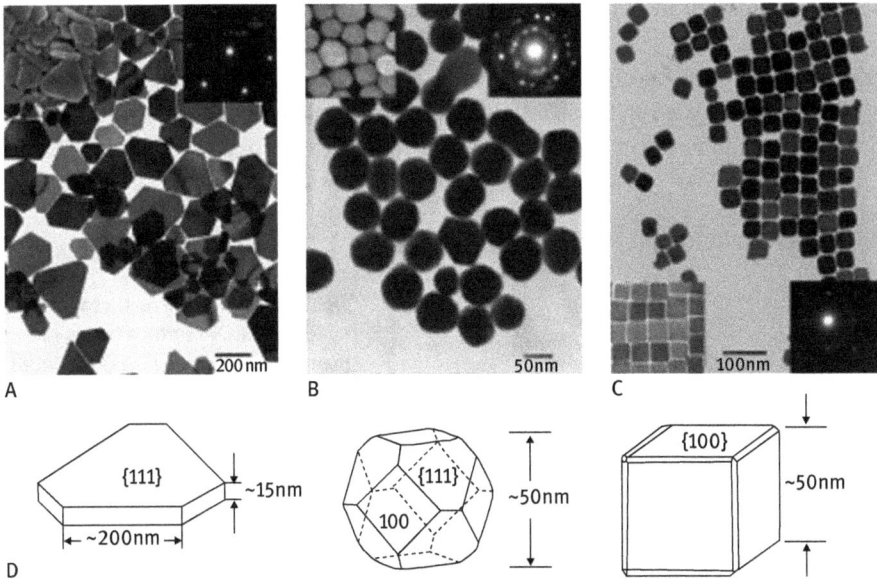

Fig. 3.54: Catalytic silver nanoparticles. The most pronounced catalytic activity was recorded in the
cubic Ag nanoparticles (**c**), ascribed to the greater area of the 100 crystal planes in the nanocubes.
Reprinted with permission from Xu et al., *Chem. Asian J.* **1** (2006), 888–893, © 2006, John Wiley &
Sons.

produce an *epoxide* molecule. The experiments, carried out by Y. Li and colleagues at Tsinghua University, China, recorded the most pronounced catalytic action in the case of Ag *nanocubes*. This result signifies two aspects of Ag NP catalysis – the greater contribution to the catalytic action of *planar facets* within the NPs as compared to *edges and corners*, and the distinct catalytic effects of the *crystalline organization* of the NP surface. Indeed, several studies have proposed that the relatively lower surface stability and greater exposure of silver atoms in the (100) crystal planes of the nanocubes in comparison with the (111) planes in the semi-spherical and prism-like particles is the likely driving force for enhanced adsorption of the target molecules (styrene in this case) and their reactivity.

Silver in general, and silver *colloids* in particular, have long been known to be powerful *antimicrobial agents*. In fact, *Ag NP*-based antimicrobials are probably the most widely used commercial nanomaterial. The origin of bactericide action by Ag NPs is still debated. While it is generally accepted that the oxidative capacity of Ag^+ ions released from silver-containing materials is an insidious toxic factor, some proposals have focused on the nanoparticles themselves as exerting bacterial toxicity. Lending support to the latter hypothesis are studies reporting on a dependence of toxicity on Ag NP size, shape, and surface chemistry. However, as Ag NPs are readily oxidized by air in aqueous solutions, thereby producing Ag^+, silver ions might exert the underlying toxic effect. Figure 3.55 illustrates a proposed mechanism for Ag NP-induced cell lysis. The model, outlined by P. J. J. Alvarez and colleagues at Rice University, portrays Ag NPs as vehicles for delivery of Ag^+ ions (produced by oxidation of Ag NPs) into the cytoplasm and membranes of bacterial cells, resulting in cell destruction. Indeed, when Ag NPs were incubated with bacterial cells in strict anaerobic conditions (in which silver oxidation does not occur), minimal antibacterial activity was observed.

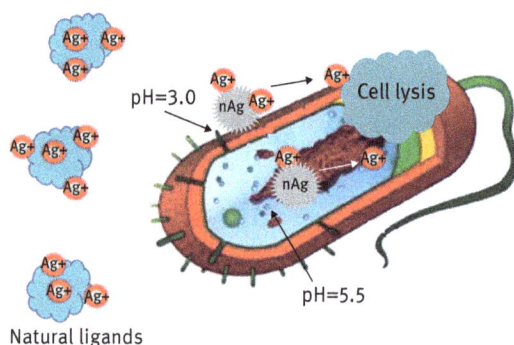

Fig. 3.55: Bactericide by silver nanoparticles. According to this proposed mechanism, Ag^+ ions produced by oxidation of Ag nanoparticles exert the toxic effect upon interaction with and penetration of the bacterial cell. Reprinted with permission from Xiu et al., *Nano Lett.* 12 (2012), 4271–4275, © 2012, American Chemical Society.

While numerous studies have addressed the crucial issue of whether, and to what extent, Ag NPs pose health risks, their effects upon human cells and tissues have not yet been fully determined. The concern is that the biological consequences of Ag NP usage have not been adequately assessed, particularly in light of their increased commercial applications. Ag NPs have large surface areas potentially available for adsorption of biological molecules (which might consequently oxidize the silver atoms). Furthermore, their nanometer-scale might lead to cell penetration and interference in physiological processes. Overall, the need for further research in this field is widely recognized.

3.3 Other noble metal and transition metal nanoparticles

This section discusses NPs comprising of noble metals other than Ag and Au, as well as *transition metal* NPs. Most applications of noble and transition metal NPs center on *catalysis* (more specifically, *heterogeneous* catalysis), reflecting the well-known contributions of transition metals as catalysts. Similar to other NPs exhibiting catalytic activities, the size, shape, and crystalline organization of metal NPs intimately affect their catalytic action, furthermore distinguishing the activity of nanoparticles from bulk materials. Indeed, *shape diversity* is one of the most striking properties of metal NPs, and constitutes a powerful tool for modulation of their functionalities.

Palladium (Pd) is a broadly-used catalyst in numerous industrial and commercial processes. Pd nanoparticles (also referred to as Pd nanocrystals) have been examined as substrates for catalysis and synthetic techniques for fine-tuning the shape and crystal organization of Pd NPs have contributed to expanded applications in this field. In particular, studies demonstrated that modification of surface areas and orientations of the exposed Pd NP facets had profound effects upon the catalytic action. These observations have been partly ascribed to the relationship between adsorption and surface orientation of target molecules upon the crystalline organization of Pd NPs.

Different to Au and Ag NPs, which have mostly been studied in *spherical* or *cylindrical* organization, Pd NPs (and other transition metal NPs) appear in many instances as crystalline *polyhedral* particles. Pd NP synthesis has been systematically investigated, yielding important mechanistic insights most likely pertinent to other metallic NPs. Figure 3.56 portrays proposed mechanisms for assembly of truncated Pd nanorods and nanobars. The study, conducted by Y. Xia and colleagues at Washington University, underscored the significance of both *kinetic* aspects (i.e. leading to formation of transient and/or meta-stable intermediate species) and *thermodynamic* parameters (responsible for assembly of energetically-favored, stable particles) in determining the ultimate NP structures and properties. Specifically, the researchers postulated that initial *reduction* and *nucleation* steps are kinetic processes which generate the core NP morphologies (nanocubes, nanobars, or nanorods), while aging processes, which are thermodynamic in nature, dictate formation of nanorods and nanocubes with truncated crystal facets (Fig. 3.56).

Figure 3.57 presents concave Pd NPs in which the high number of crystalline facets (e.g. "high index facets") enhanced the catalytic activity. The concave Pd NPs synthesized by Y. Xia and colleagues were produced by surprisingly simple synthetic routes modulating the kinetics (e.g. rates) of the different steps in NP assembly (seeding, nu-

Nucleation and growth: Kinetic control

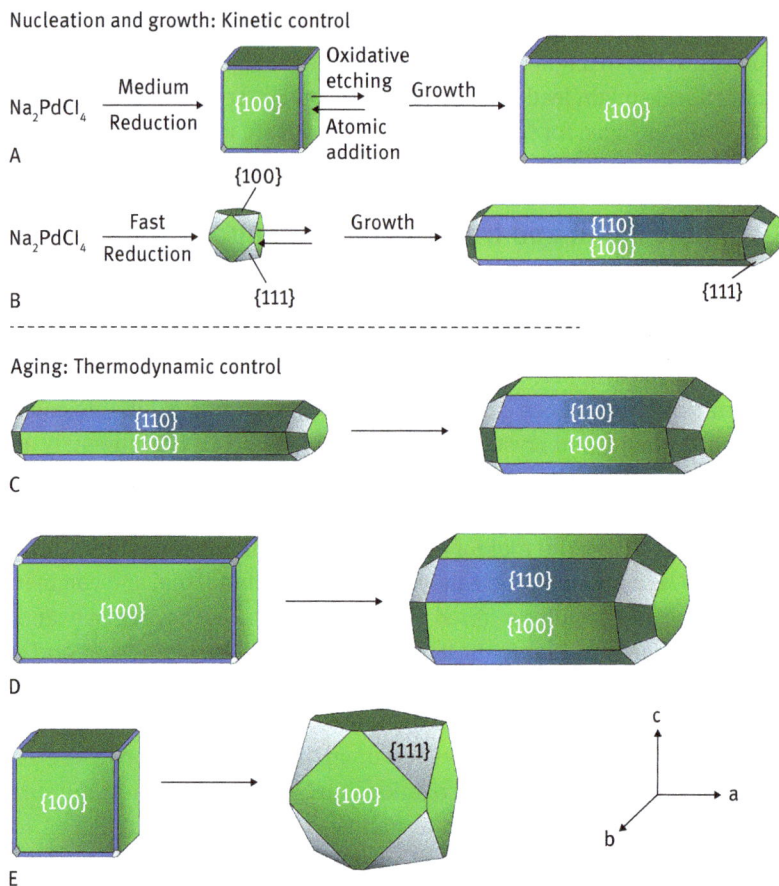

Aging: Thermodynamic control

Fig. 3.56: Formation mechanisms of truncated palladium nanorods and nanocubes. The proposed mechanisms distinguish between kinetically-controlled steps generating the nucleating seeds and intermediate nanoparticle morphologies, and the thermodynamic processes producing the final truncated structures. Reprinted with permission from Xiong et al., *JACS* **129** (2007) *129*, 3665–3675, © 2007, American Chemical Society.

cleation, crystal growth, and maturation), primarily via alteration of reagent concentrations. The researchers reported enhanced catalysis of organic molecule decomposition when using the *concave* structures as catalysts. Indeed, the catalytic activity of the concave Pd NPs was greater than *planar-faceted* nanocubes, presumably due to their steep edges and "kinks" providing higher densities of exposed Pd atoms, onto which greater adsorption of target molecules and surface reactivity of the Pd atoms likely occurred.

Several studies systematically investigated the relationship between the catalytic activity and crystal organization of the exposed facets of Pd NPs. Figure 3.58 depicts a

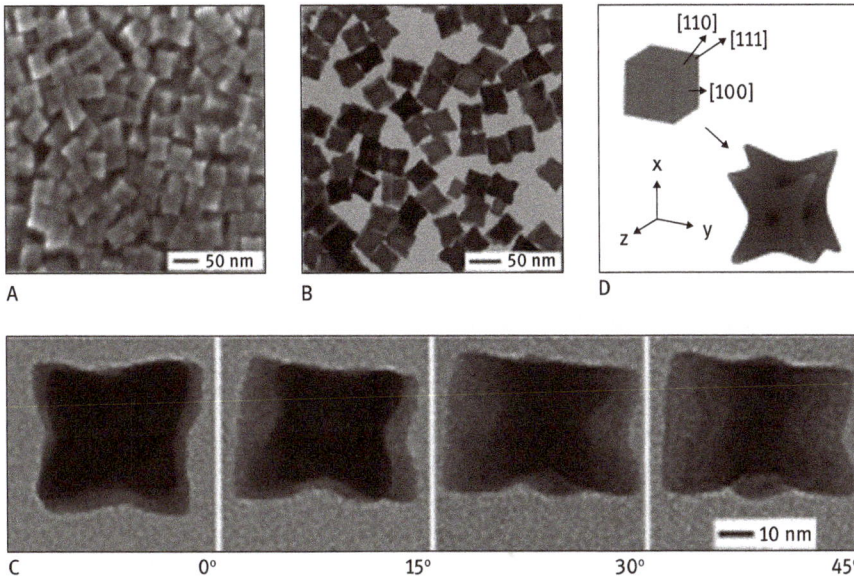

Fig. 3.57: Concave palladium nanocubes. **A–C:** Electron microscopy images of concave nanocubes. **D:** Scheme showing formation of the concave structure through specific growth of the corners and edges of the nanocube. Reprinted with permission from Jin et al., *Angew. Chem. Int. Ed.* **50** (2011), 7850–7854, © 2011, John Wiley & Sons.

series of Pd NP configurations containing different ratios of the *{100}* and *{111}*crystal facets, ubiquitous in crystalline materials. Notably, Y. Xia and colleagues found that the catalytic performance of the NPs towards oxidation of a small organic molecule (formic acid) clearly depended upon surface organization – enhanced catalytic activity was recorded in cases when the Pd NPs had more abundant {100}surfaces. This effect might be ascribed to stronger binding of the organic substances onto the {100}facets, and/or greater mobility of electrons (presumed to play fundamental roles in promoting catalysis) on the {100}surface.

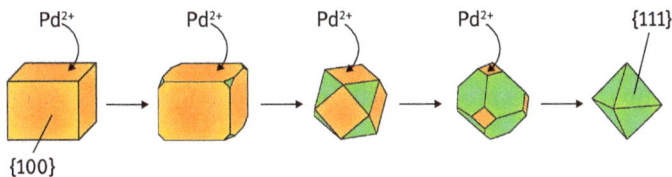

Fig. 3.58: Controlled transformation of palladium nanocube bearing {100} crystal facets and octahedron with exposed {111} planes. Greater catalytic activity was observed for the nanocubes, ascribed to the {100} crystalline organization of the nanoparticle's facets. Figure inspired by Jin et al., *Energy Environ. Sci.* **5** (2012), 6352.

Fig. 3.59: Hollow palladium nanocubes . Reprinted with permission from Mahmoud et al., *Nano Lett.* **10** (2010), 3764–3769, © 2010, American Chemical Society.

Hollow Pd nanocubes (Fig. 3.59) have also garnered interest in light of their excellent catalytic activity. Synthesis of hollow Pd nanocubes has been achieved via "galvanic replacement", similar to the process described in Chapter 3.2 in the case of Ag NPs. In this process metal atoms within a "sacrificial template" are oxidized by Pd^{2+} ions which transforms into Pd^0, subsequently forming a complementary framework tracing the initial template. Figure 3.59 displays hollow Pd nanocubes prepared by M. A. El-Sayed and colleagues at the Georgia Institute of Technology by use of a silver sacrificial template. Incubation of Pd^{2+} with Ag cubic nanoparticles resulted in oxidation of the metallic silver atoms and their substitution by Pd, generating hollow nanocubes. The Pd nanocages were effective catalysts, exhibiting better activity than the solid Pd NP counterparts. Indeed, *cavity-confinement effects* (also referred to as "cage effects") ascribed to adsorption of reagents *inside* the nanocages were postulated to be primary factors contributing to the enhanced catalytic properties of the hollow nanocubes.

Platinum (Pt) is another noble metal playing a prominent role in catalysis and analogous to palladium. Pt NPs (or Pt *nanocrystals)* have attracted interest for their potential as new and effective catalysts. Functional investigations of Pt NPs have progressed hand in hand with the development of sophisticated synthesis schemes yielding a variety of NP structures. Figure 3.60, for example, illustrates Pt NPs in uniform and diverse morphologies. The synthetic routes developed by C. B. Murray and colleagues at the University of Pennsylvania were based on the use of reaction mixtures comprising of Pt^{2+} ions, capping agents (oleic acid), and carbonyl (CO) as a structure-directing and reducing ligand. Significantly, shape control of the Pt NPs was achieved by modifying concentration and types of solvent of the carbonyl ligands, temperature, and reaction times.

Figure 3.61 presents other interesting Pt NP structures and their synthesis route. This study, carried out by R. D. Tilley and colleagues at Victoria University of Wellington, New Zealand, has aided in pinpointing important parameters affecting crystal growth of Pt NPs (and other noble metal NPs for that matter). In particular, the researchers observed that formation of "branched" Pt nanocrystals was associated with high concentrations of the Pt precursor molecules. This relationship most likely indi-

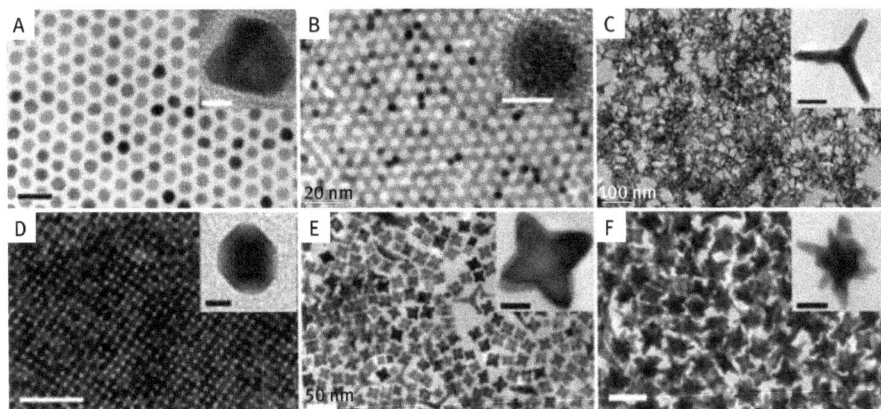

Fig. 3.60: Controlled synthesis of platinum nanoparticles with different morphologies. The distinct nanoparticle shapes were produced by addition of metal carbonyl in different concentrations. Reprinted with permission from Kang et al., *ACS Nano* **7** (2013), 645–653, © 2013, American Chemical Society.

cates that the NP branches were formed via a fast kinetic process compared to Pt NP morphologies displaying more planar facets which assemble in low precursor concentration via slow thermodynamic deposition processes. This model is consistent with the assumption that the high concentration regime promotes reactivity of the precursors and consequent selective deposition of Pt on the growing branches. The mechanism outlined in Figure 3.61B further suggests that growth of Pt NP branches is accompanied by gradual etching of the NP surfaces in between the growing branches.

Branched or "multipod" nanoparticle structures have been generated in other metal NP systems and this structural motif has been linked to interesting functionalities. Figure 3.62 depicts branched *rhodium (Rh)* NPs, synthesized by Y. Xia and colleagues at Washington University through selection of specific Rh precursor molecules and careful optimization of experimental conditions. Notably, these multipod Rh NPs exhibited surface plasmon resonance (SPR); moreover, the SPR peak was blue-shifted in comparison to spherical Rh NPs, which was ascribed to modulation of the localized electromagnetic field at the nanocrystal branches. This feature, which echoes SPR shifts recorded in other branched NP systems (particularly silver and gold), might also explain a dramatic surface-enhanced Raman scattering (SERS) effect which could make the branched Rh NPs a convenient sensing platform.

Nickel (Ni) NPs have attracted interest because of their catalytic and magnetic properties (nickel being a well-known magnetic element). Being NPs of a "classic" transition metal, Ni NPs have been investigated as a model for the differences and similarities in structural and functional properties in comparison to more widely studied noble metal families, such as Au NPs. In fact, Ni NPs display many structural features encountered in other metal NP species, underscoring the *universality* of the growth

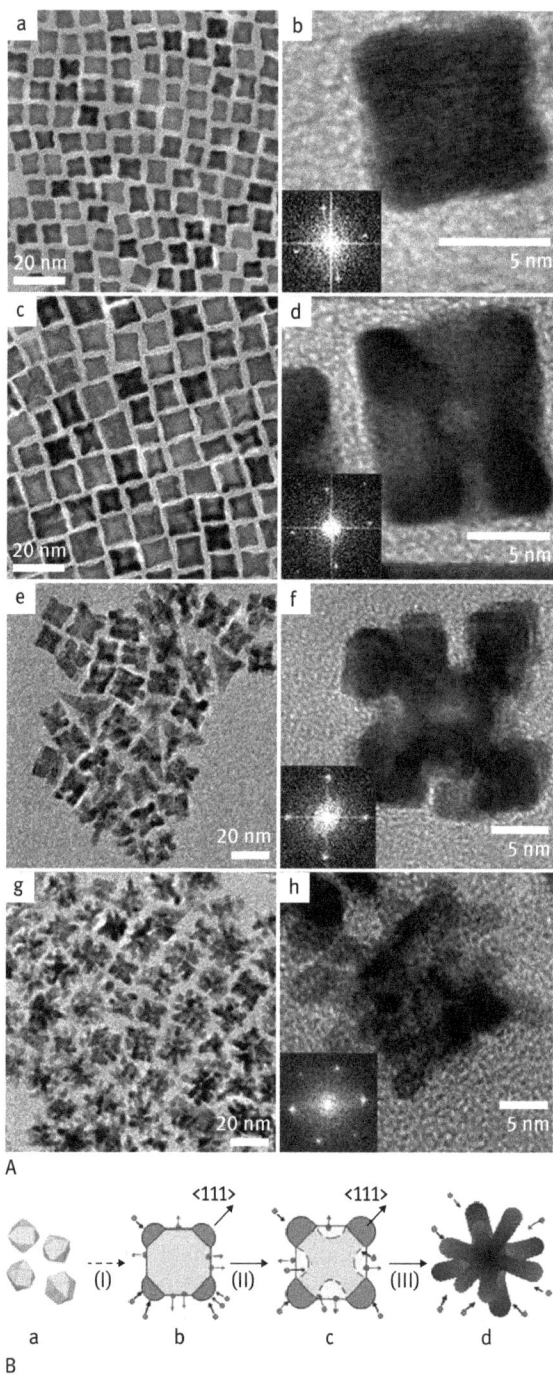

Fig. 3.61: Branched platinum nanoparticles. **A:** Electron microscopy images of Pt nanoparticles with branches of different lengths. Branch lengths were dependent upon duration of the growth reaction in a high-concentration precursor solution (from 75 min in A–B to 500 min in G–H). **B:** Proposed model for the growth of the branched Pt nanoparticles: a rapid kinetic process of deposition of Pt on the branches, occurring simultaneously with etching of the surface planes between the branches. Reprinted with permission from Cheong et al., *JACS* **131** (2009), 14590–14595, © 2009, American Chemical Society.

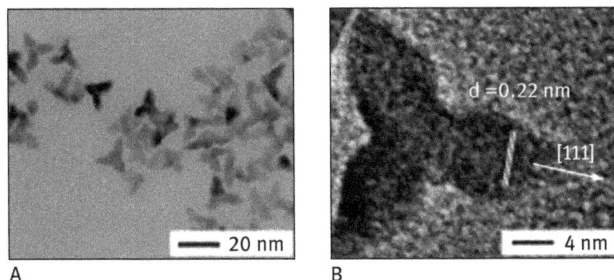

Fig. 3.62: Branched rhodium nanoparticles. The branched morphology gives rise to shifts in the surface plasmon resonance (SPR), and also significant surface-enhanced Raman scattering (SERS) properties. Reprinted with permission from Zettsu et al., *Angew. Chem. Int. Ed.* **45** (2006), 1288–1292, © 2006, John Wiley & Sons.

mechanism aspects of metal NPs. Figure 3.63 shows polymorphic Ni NPs and a model depicting the growth process. Specifically, R. D. Tilley and colleagues at Victoria University of Wellington, New Zealand, proposed that, similar to other metal NP systems discussed above, branched (i.e. multipod) Ni NPs assemble via a two-stage process. The initial nucleating site consists of a single crystalline *truncated octahedron* core. Subsequent growth of the branches occurs via anisotropic stacking of crystal planes, initiated on specific facets. While the model outlined in Figure 3.63 does not actually predict the number of branches – i.e. the exact multipod structure – it provides important insights into the molecular factors affecting the structural features of Ni NPs (and by extension other metal NP species).

Similar to Nickel, *Cobalt (Co)* also belongs to the group of highly magnetic transition metals and Co NPs have been investigated as potential constituents of magnetic sensing and magnetic storage applications. Like other metal NPs discussed in this section, Co NPs have been prepared in diverse morphologies. Intriguing *hollow* Co NP structures (or "nanoskeletons"), synthesized by J. Yao and colleagues at the Chinese Academy of Sciences, are shown in Figure 3.64. The hollow Co nanocubes displayed magnetic properties that were different to the solid NP counterparts. Synthesis of the hollow Co nanocubes was not carried out via the "galvanic replacement" strategy discussed above in the case of Pd or Ag nanocubes. Rather, as depicted in Figure 3.64, the hollow nanocubes were formed via spontaneous dissolution of the crystalline nanocube interior (a process termed "Ostwald ripening"), followed by active etching of Co through reaction with halides such as F^- ions which were added to the reaction mixture. The researchers demonstrated that tuning the reaction parameters, particularly the "time window" in which the reaction was terminated, enabled production of relatively uniform Co nanoskeletons.

Copper (Cu) NPs are considered to be promising candidates for electronic applications, since copper exhibits high conductivity and is also an abundant and inexpensive element. Cu NPs of different sizes have been examined, for example, as

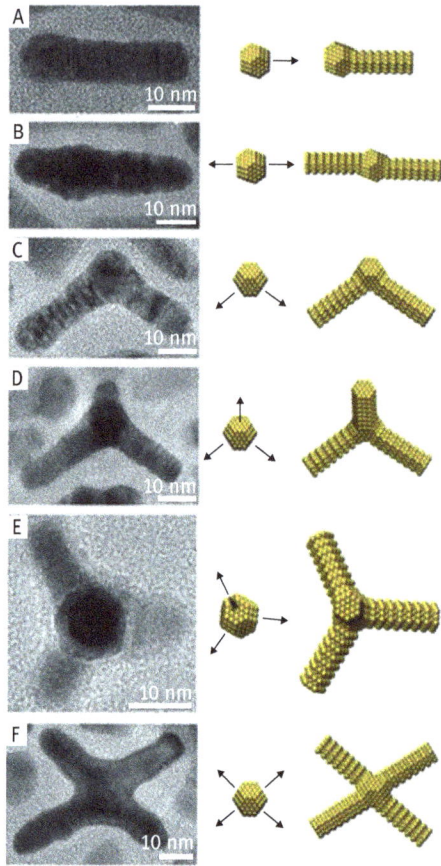

Fig. 3.63: Branched nickel nanoparticles. Electron microscopy images (*left*) and schematic growth models (*right*) of multipod Ni nanoparticles. In all scenarios nanoparticle growth starts at the truncated octahedron core, while the crystal planes of the branches grow in anisotropic directions. Reprinted with permission from LaGrow et al., *Adv. Mater.* **25** (2013), 1552–1556, © 2013, John Wiley & Sons.

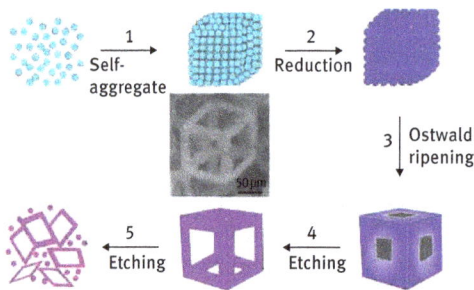

Fig. 3.64: Hollow cubic cobalt nanoparticles. Scheme depicting the synthesis steps, and an electron microscopy image of a hollow nanocube. Reprinted with permission from Wang et al., *Adv. Mater.* **21** (2009), 1636–1640, © 2009, John Wiley & Sons.

substances for electronic ink, since they can be sintered (i.e. fused), forming a continuous conductive layer in relatively low temperatures (below 150 °C). Similarly, Cu nanowires exhibit useful functions as a conductive media. A primary challenge in this field, however, is synthesis of sufficiently long NWs which could be further assembled into macro-scale configurations used in practical applications – for example in large-area films employed as transparent conductive electrodes. Figure 3.65 outlines

A

B

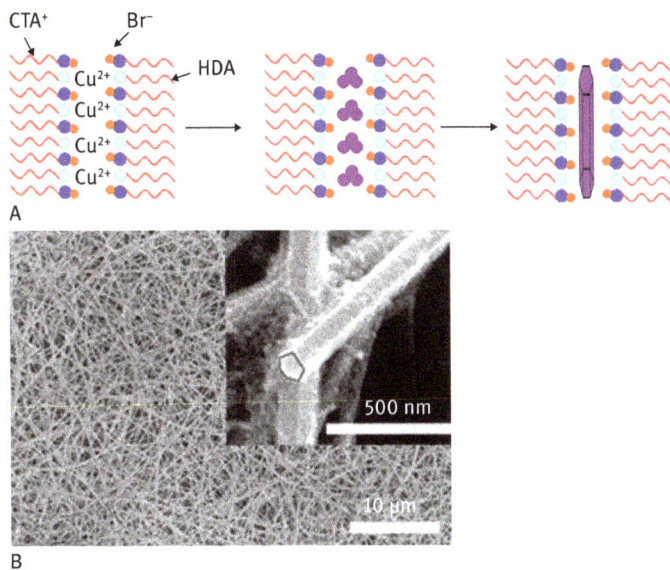

Fig. 3.65: Copper nanowires. **A:** Schematic drawing of the fabrication procedure: Cu^{2+} ions are aligned in the channels of a liquid crystalline medium. Subsequent reduction yields copper nanowires. **B:** Electron microscopy images showing the Cu nanowire mesh and cross section of a single nanowire. Reprinted with permission from Zhang et al., *JACS* **134** (2012), 14283–14286, © 2012, American Chemical Society.

an elegant synthesis scheme based on liquid crystalline medium as a conduit for Cu nanowire formation. The synthesis strategy, developed by Y. Lu and colleagues at the University of California, Los Angeles, relies on the intrinsic molecular alignment of liquid crystals as scaffolding for organization of the copper ions and subsequent reduction and nanowire growth. This synthetic approach provided additional benefits as the Cu NWs were easily dispersed and did not aggregate – a problem frequently encountered in metal NW fabrication methods.

3.4 Hybrid metal nanoparticles

As the title implies, hybrid metal nanoparticles (also referred to as *bimetallic* or *hetero-structured metal* NPs) consist of more than a single metal constituent. Most hybrid metal NPs comprise of two metal elements, but some NPs containing three or more metal components have been investigated. In many instances, the hybrid nature of the NPs gives rise to unique structures and functionalities emanating from both complementarity and synergy between the two (or more) metal components. Hybrid metal NPs can be divided into three main groups relating to the organization of the atomic constituents: *alloys*, in which the different metal species are randomly or semi-randomly interspersed within the NP; *Janus* NPs, which constitute non-symmetrical (i.e. anisotropic) NPs; and *core-shell* NPs, in which the core comprises of one metal element and the shell/s contain one or more other metal constituents. While the two latter NP types also encompass *non-metallic* particles (hybrid NPs which contain, for example, organic, polymeric, or semiconductor building blocks, discussed in Chapter 6), this section focuses on hybrid metallic species (also including metal/metal oxide species), illuminating their structural and molecular properties and applications.

3.4.1 Alloy nanoparticles

The presence of more than a single metal element in alloy NPs endows these NPs with interesting characteristics, primarily modified (and/or enhanced) chemical and physical functionalities in catalysis, optical, and sensing applications. Notably, in many cases the physico-chemical properties of alloy NPs are distinct from comparable NPs comprising of pure metal constituents. Synthesis of alloy NPs is conceptually similar to schemes developed for production of pure metal NPs. Generally, alloy NPs are synthesized by co-incubation of metal precursors with reducing and stabilizing agents (Fig. 3.66A). An alternative approach is based on the galvanic replacement reaction scheme: first creating pure metal nanoparticles, subsequently replacing some of the atoms with a different metal through a reduction-oxidation reaction (Fig. 3.66B); alloy NPs are formed when the substitution reaction results in a *random* distribution of the two metal constituents (some galvanic replacement processes yield core-shell configurations; see below). Other synthesis routes for alloy NPs have been demonstrated using vapor deposition and template-assisted schemes.

The effect of alloy formation on the color of NPs has been among the most useful phenomena in these NP species (this feature is not in fact limited to nanoparticles – "white gold" common in jewelry, for example, is essentially an alloy of gold and other

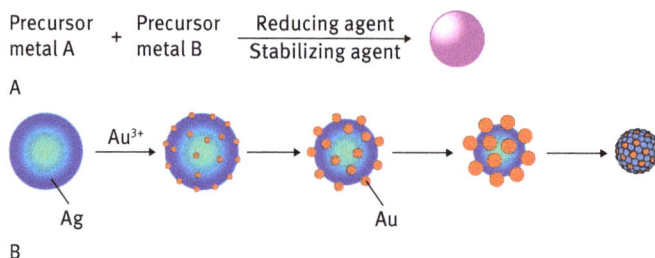

Fig. 3.66: Generic synthesis routes for alloy hybrid metal nanoparticles. **A:** Co-incubation of precursors for different metal nanoparticles. **B:** Galvanic replacement process (demonstrated for creation of AuAg alloy nanoparticle) – initial formation of single-metal nanoparticles and partial substitution with another metal species via electron transfer.

metals, resulting in "masking" of the typical yellow color of pure gold). Tunable color range – reflecting shifts in the surface plasmon resonance (SPR) wavelengths – is a hallmark of alloy NPs, although it should be emphasized that color tuning is achieved in other configurations of hybrid metal NPs; see below. Indeed, since the electronic oscillations giving rise to SPR are intimately associated with the composition and organization of surface metal atoms, the presence of a metal mixture in an alloy NP could result in significant modulation of the SPR wavelengths. Figure 3.67, for example, portrays the composition-dependent color transitions of Ag-Au alloy NPs. The NPs were prepared by S. Barcikowski and colleagues at the University of Duisburg-Essen, Germany, through a galvanic replacement process, starting with Ag NPs which were then mixed with gold ions (at high temperature), yielding alloy NPs according to the chemical reaction:

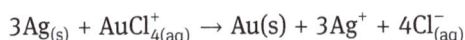

$$3\,Ag_{(s)} + AuCl_{4(aq)}^{+} \rightarrow Au(s) + 3\,Ag^{+} + 4\,Cl_{(aq)}^{-}$$

The photograph in Figure 3.67 manifests the compositional dependence of the Ag-Au alloy NP colors. Specifically, while pure Ag NPs and Au NPs exhibit yellow and red

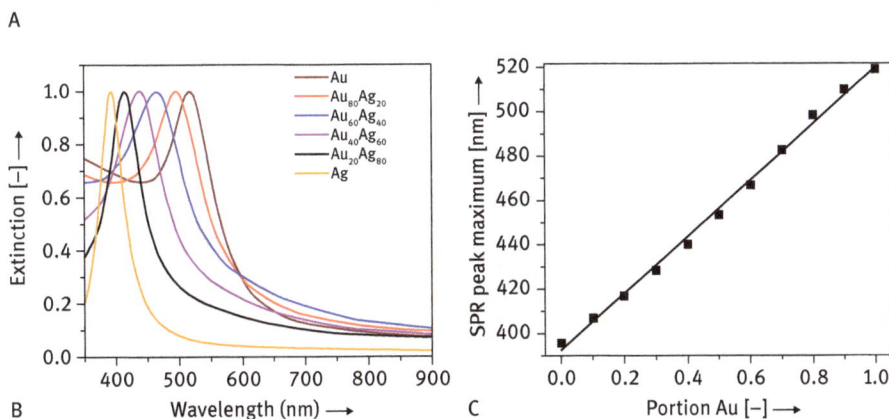

Fig. 3.67: Composition-dependent color of gold-silver alloy nanoparticles. Changes in visible color (*top*) and position of surface plasmon resonance peak (bottom spectra) in Au-Ag alloy nanoparticles produced through galvanic replacement. Reprinted from Rebock et al., *Beilstein J. Nanotech.* **5** (2014) 1523–1541.

colors, respectively (reflecting their distinct SPR absorbance), alloy NPs produce intermediate colors which depend on the mole ratios between the two metal components.

In addition to spherical configurations, other alloy NP morphologies have been reported, echoing the diversity of single-metal nanoparticle structures. Figure 3.68 depicts *polyhedral* alloy PtNi$_2$ NPs, synthesized through co-reduction of the respective metal precursors (i.e. synthetic route in Fig. 3.66A). Importantly, the reaction mixture also contained crystal growth inhibitors (phenyl-displaying molecules such as benzoic acid and aniline). While the composition of the alloy NPs was determined by the ratio between the metal precursors, the crystal inhibitors were responsible for

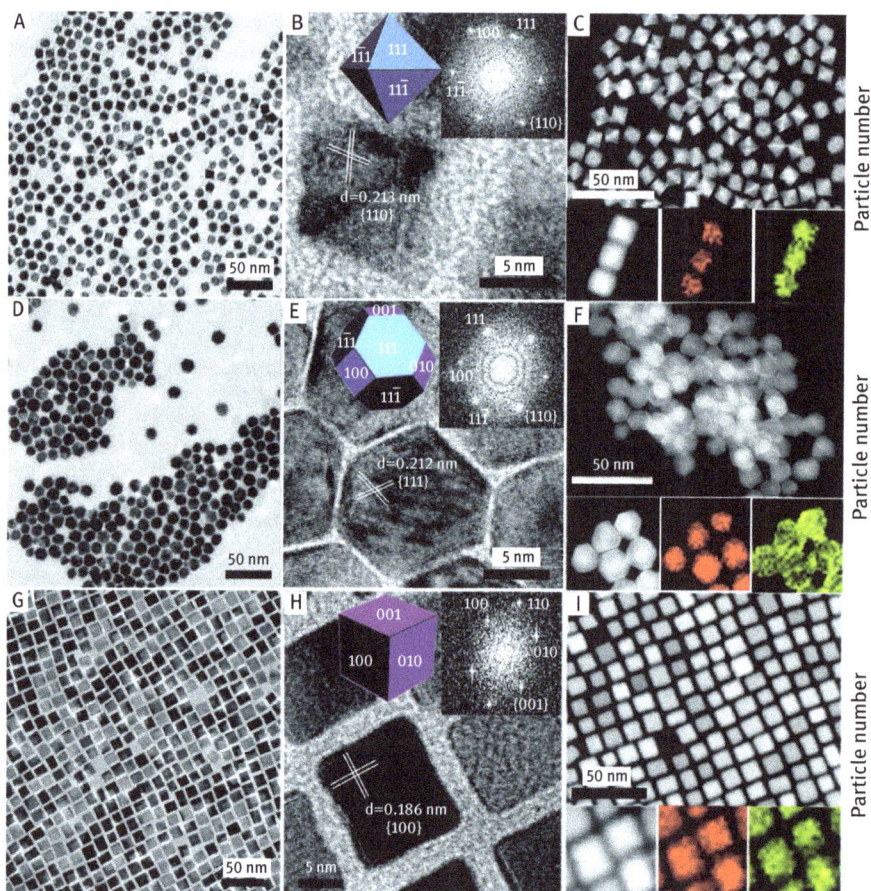

Fig. 3.68: Polyhedral alloy PtNi$_2$ nanoparticles. The electron microscopy images show the distinct polyhedral structures synthesized by changing the type of crystal growth inhibitor co-added to the reaction mixture. The elemental color maps in the microscopy images in panels C, F, and I confirm the homogeneous distribution of both platinum (red) and nickel (green) in the alloy nanoparticles. Reprinted with permission from Wu et al., *JACS* **134** (2012), 8975–8981, © 2012, American Chemical Society.

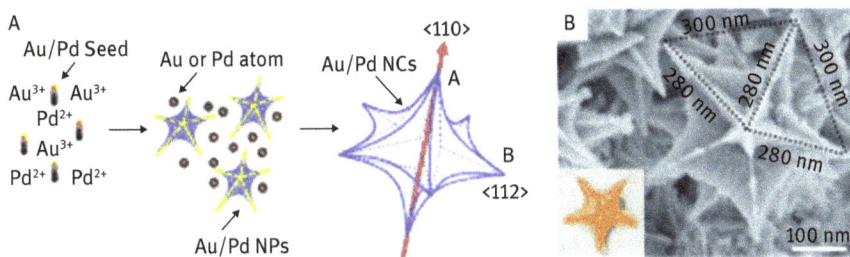

Fig. 3.69: Gold/palladium concave alloy nanostars. **A:** Synthetic procedure: the metal ions are incubated with Au/Pd nanorods acting as nucleating seeds; gradual growth of the nanoparticles results in nanostar formation. **B:** Electron microscopy image and a model of the Au/Pd nanostars. Reprinted with permission from Zhang et al., *Angew. Chem. Int. Ed.* **52** (2013), 645–649, © 2013, John Wiley & Sons.

non-isotropic growth and eventual polyhedral morphologies. Indeed, optimization of the type and concentration of the inhibitors enabled Y. Li and colleagues at Tsinghua University, China, to fine-tune the crystalline parameters and NP configurations. Like other hybrid metal NPs, the polyhedral alloy NPs highlighted in Figure 3.68 were used as *heterogeneous catalysts*, in which the enhanced catalytic activity has been attributed to the planar crystalline facets serving as adsorption sites for the target substrates.

Other intriguing alloy NP shapes were reported. Figure 3.69 presents Au-Pd "nanostars" – multi-branched concave NP structures – produced through co-nucleation of Au^{3+} and Pd^{2+} ions in the presence of a reducing agent. The synthetic strategy, devised by A. W. Xu and colleagues at the University of Science and Technology of China, utilized in addition to the two metal precursors a mixture of surfactant molecules and pre-synthesized rod-shaped "seeds" which were added to the reaction vessel. Remarkably, the researchers induced formation of concave alloy NP structures by tuning the mole ratio between the two metal precursors. Similar to the polyhedral NPs discussed above, the star-shaped alloy NPs displayed useful functionalities, particularly surface enhanced Raman scattering (SERS) and enhanced catalytic activity. Both properties have been ascribed to the extended surface areas of the nanostar facets and abundant sharp edges presumed to result in high concentration of surface electrons.

Catalysis has been a primary application of alloy NPs (and hybrid metal NPs in general). Figure 3.70 illustrates a proposed mechanism for the catalytic action of PtPdTe alloy nanowires (NWs) towards *methanol oxidation* (converting methanol to CO_2). The nanowires were synthesized by S. H. Yu and colleagues at the University of Science and Technology of China by first constructing *tellurium (Te)* NWs, which were subsequently used as sacrificial framework and a source of reducing electrons – generating Pt and Pt in specific mole ratios. $Pt_{23}Pd_{21}Te_{56}$ NWs, in particular, were shown to catalyze methanol oxidation to a higher degree than PtPdTe NWs with different mole

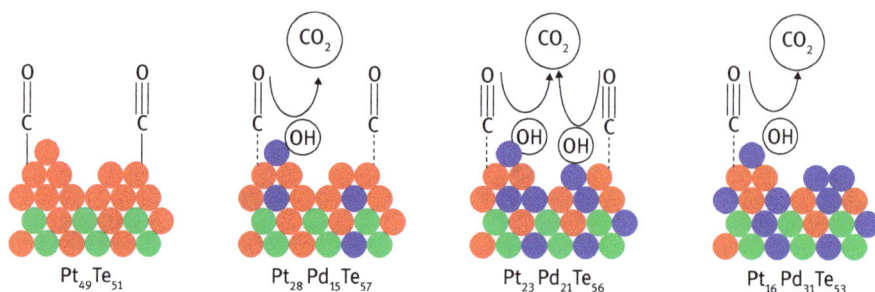

Fig. 3.70: Catalytic mechanism of methanol oxidation on the surface of PtPdTe alloy nanowires. The most effective catalysis occurs on $Pt_{23}Pd_{21}Te_{56}$ nanowires, which promote both adsorption of OH molecules and weakening of the surface binding of CO (an intermediate in methanol decomposition). Reprinted with permission from Li et al., *Angew. Chem. Int. Ed.* **52** (2013), 7472–7476, © 2013, John Wiley & Sons.

ratios between the three components or commercially available catalysts. According to the proposed catalytic mechanism in Figure 3.70, palladium atoms on the nanowire surface are believed to contribute to both immobilization of OH residues (produced through dissociation of adsorbed water molecules) as well as weakening of the bond between the NP surface and the CO molecules which are intermediates in the methanol oxidation reaction. Together, the two effects were most pronounced in $Pt_{23}Pd_{21}Te_{56}$ NWs, resulting in maximal catalytic activity as compared to NWs comprising only of Pt and Te, or NWs exhibiting different mole ratios between the three atoms.

Pronounced catalytic activities were reported for other alloy NP structures. Figure 3.71 shows Pd-Pt hollow *nanocubes* (or "nanocages"), found to catalyze carbon monoxide oxidation, an environmentally important process mostly associated with catalytic converters in cars. The nanocages were prepared by Y. Xia and colleagues at Washington University via a galvanic replacement process, albeit with a crucial modification. Synthesis started with generation of solid Pd nanocubes serving as the template. The Pd nanocubes (capped by Br^- ions which were found by the researchers to promote the galvanic replacement reaction) were then placed in a solution of the Pt^{2+} precursor ($PtCl_4^{2-}$), which started to "leach" the metallic Pd through oxidation by Pt^{2+}. Importantly, at this stage of the process a powerful reducing agent (citric acid) was added to the reaction mixture, giving rise to co-reduction of both Pd and Pt on the nancube template, resulting in the hollow nanocube structures. The PdPt nanocages displayed enhanced catalysis of CO oxidation, ascribed both to the presence of the two elements in the NP surface, as well as the hollow (and likely porous) morphology, which together promoted CO adsorption and decomposition.

Underscoring the remarkable variety of alloy NP morphologies, Figure 3.72 shows electron microscopy images of PdPt "nanodendrites". These NPs were synthesized by Y. Yamauchi and colleagues at the National Institute for Materials Science, Japan, through a simple one-step process of co-incubating Pd^{2+} and Pt^{2+} precursors in the

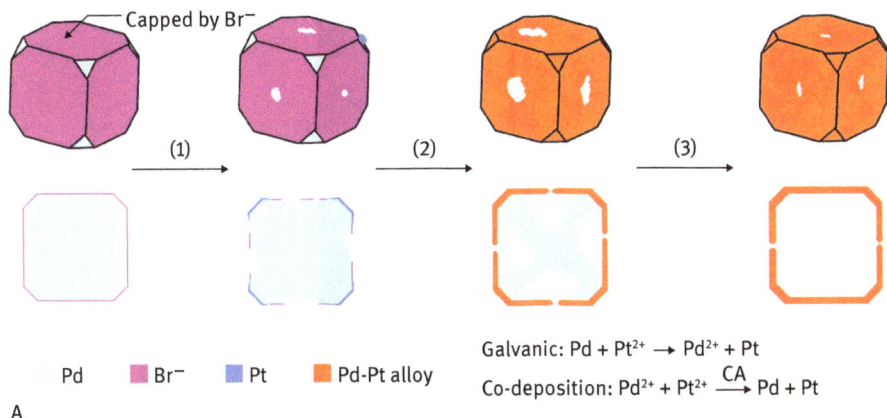

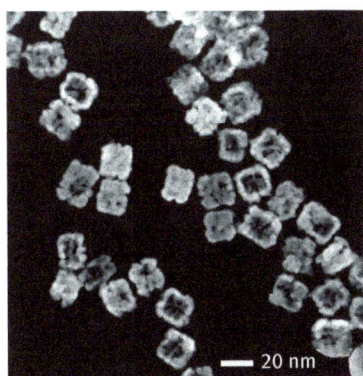

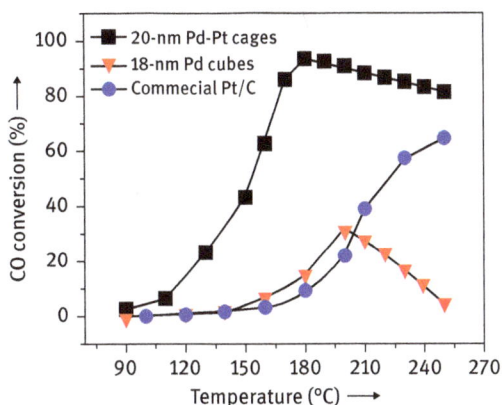

Fig. 3.71: Platinum-palladium alloy nanocages as effective catalysts. **A:** Preparation scheme: galvanic replacement of solid Pd nanocubes by Pt^{2+} ions (1); co-deposition of Pd and Pt on the Br^--stabilized nanocube facets (2); complete decomposition of the Pd inner nanocube, leaving the Pd-Pt hollow nanocage. **B:** Electron microscopy image of the nanocages (left) and graph showing the enhanced catalytic activity of the Pt-Pd nanocages for carbon monoxide oxidation (black squares) compared with solid Pd nanocubes (blue circles) and a commercial catalyst (red triangles). Reprinted with permission from Zhang et al., *ACS Nano* **5** (2011), 8212–8222, © 2011, American Chemical Society.

presence of a co-polymer stabilizer and a reducing agent. The presence of the polymer additive (pluronic F127, a well-known polymer additive which stabilizes nanoparticle structures in various organic and inorganic systems) was critical for attaining the nanodendrite morphologies and their uniform distributions. Furthermore, size and shape of the nanodendrites were found to be intimately dependent on the ratio between the two metal constituents. The nanodendrite particle structure is particularly attractive for catalysis since it provides a very high surface area accessible for atomic adsorption and permeability.

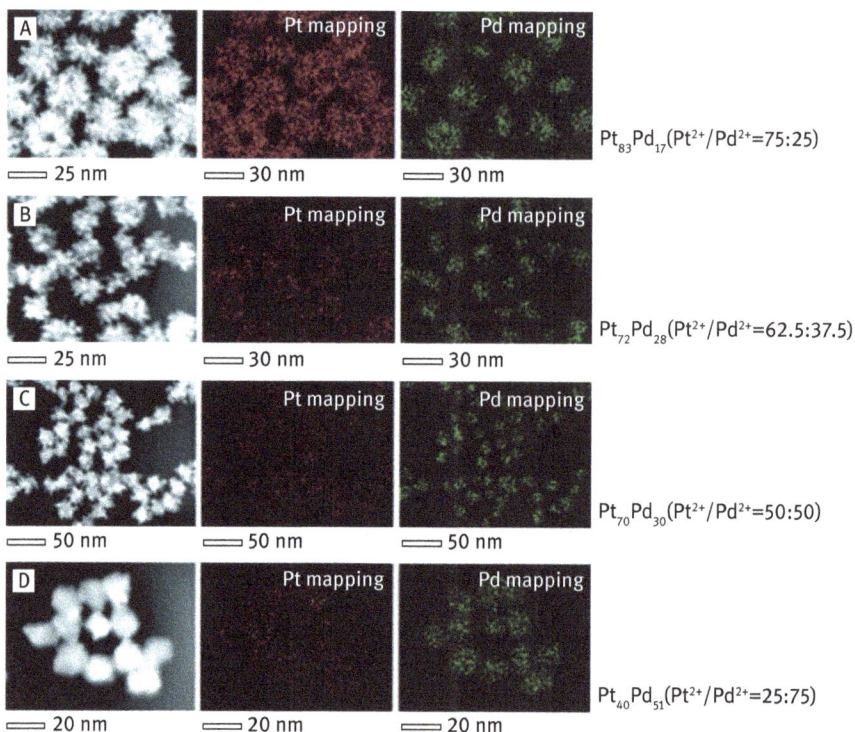

Fig. 3.72: Palladium-platinum alloy nanodendrites. Electron microscopy images (left column) and elemental maps of the alloy nanodendrites. Nanodendrite size and shape depends on the mole ratio of the two metals in the alloy nanoparticles. Reprinted with permission from Wang and Yamauchi, *Chem. Asian J.* **5** (2010), 2493–2498, © 2010, John Wiley & Sons.

3.4.2 Core-shell metal NPs

Core-shell NPs (also referred to as core-shell nanocrystals) have attracted interest as a means for expanding the universe of metallic NPs, in particular since their physico-chemical properties have been found in many instances to be different to those of NPs comprising of pure, individual metal components. Figure 3.73A depicts several core-shell configurations; while spherical, concentric core-shell NPs have been the most widely-studied, other core-shell morphologies have been reported and are discussed below.

Figure 3.73B outlines the generic "seeded growth" approach for synthesis of core-shell metal NPs. Essentially, the core is synthesized first (metal NP synthetic processes are discussed in more detail in preceding chapters devoted to single-element NPs). The NP cores subsequently serve as "seeds" for further growth of the shell through placing in solutions containing precursors for the shell material. The seed-mediated process offers distinct advantages as it enables control of the core properties, and particularly

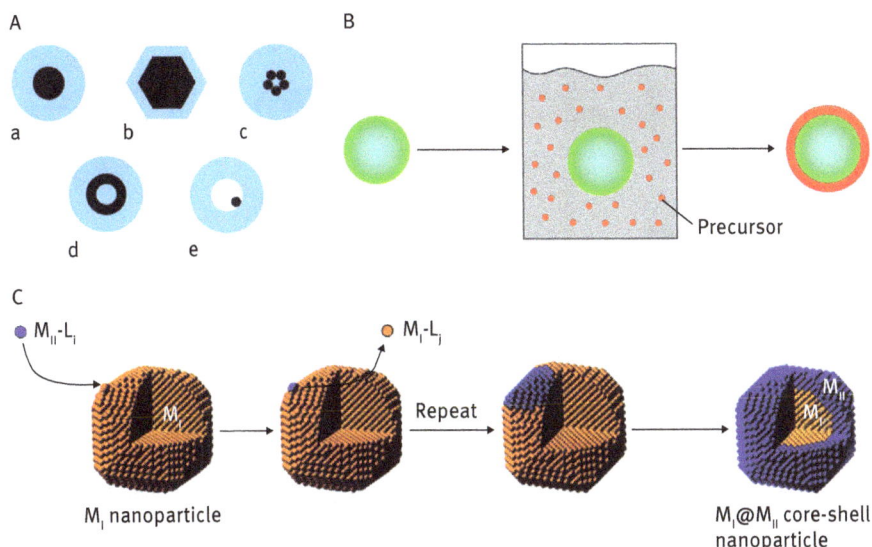

Fig. 3.73: Bimetallic core-shell nanoparticle configurations and synthesis. **A:** Distinct morphologies of hybrid metallic nanoparticles: concentric (**a**); polyhedral (**b**); multicore (**c**); multishell (**d**); semi-hollow core (**e**). Synthesis processes: **B:** Seed-mediated growth; the cores are synthesized separately and then added to a solution containing the precursors for the metal shell. **C:** Core-shell nanoparticle formed via a galvanic replacement process: the surface layer of the metal nanoparticle core is oxidized and substituted by another metal. Reprinted with permission from Lee et al., *JACS* **127** (2005), 16090–16097, © 2005, American Chemical Society.

shell dimensions (i.e. diameters) – providing the means for tuning desired properties of the resultant core-shell NPs. Several examples illuminate this important aspect of core-shell NP technologies.

Galvanic replacement (Fig. 3.73C), also employed in alloy NP fabrication (see discussion in preceding section), is conceptually similar to the seed-mediated approach. The process begins with synthesis of the NP core and the shell is subsequently assembled through gradual reduction and deposition of the second metal constituent, accomplished via electron transfer from surface atoms of the core. In principle, the process of metal substitution (e.g. replacement) is dependent on the difference between the redox potentials of the core and shell metal atoms and thus permits flexibility in the choice of metallic elements used in the cores and shells. Furthermore, the process is simple to carry out as it neither requires the presence of precursor molecules of the shell metal, nor co-addition of reducing agents; in essence, only metal ions need to be added to the reaction mixture. Indeed, one of the recurring obstacles in bimetallic core-shell NP synthesis, avoided by use of the galvanic replacement technique outlined in Figure 3.73B, is the formation of metal aggregates comprising the second metal constituent in solution, leading to inhomogeneous and distorted shells.

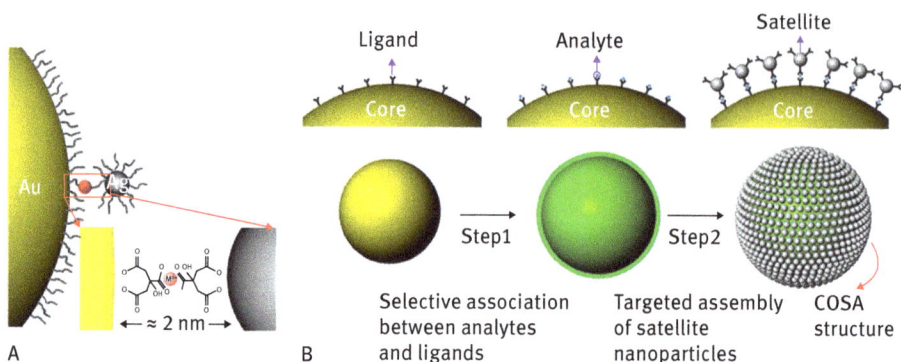

Fig. 3.74: Core-shell bimetallic nanoparticles with non-continuous shells. **A:** Diagram showing the attachment of small Ag nanoparticles to the surface of the Au core through ligand complementarity. **B:** The assembly process: Au nanoparticles serving as the cores are coated with a layer of ligands; analytes specifically recognized by the ligands are then bound to the Au nanoparticle surface; formation of the core-satellite (COSA) structure through binding of Ag nanoparticles coated with the same ligands as the core. Reprinted with permission from Choi et al., *JACS* **134** (2012), 12083–12090, © 2012, American Chemical Society.

Interesting bimetallic NPs comprising *non-continuous* shells were also reported. Figure 3.74 depicts an elegant example of such NPs, termed "core-shell satellite" (COSA) NPs, in which the Au core was coated with a layer of smaller Ag NPs. The assembly procedure, developed by J. Yi and colleagues at Seoul National University, Korea, started with formation of the nanoparticle core, coated with a layer of ligands exhibiting recognition elements. Following binding of the target analytes recognized by the ligands, the "satellite" Ag NPs, coated with the same ligands as the core, were added and consequently formed a uniform layer on the Au core through ligand-analyte recognition. The ligand-analyte-ligand "sandwich" configuration gave rise to stable hybrid NPs, and furthermore provided a powerful means for sensing (for example when the analytes were metal ions recognized by the ligands) – since the plasmon resonance wavelength of the Au NP core shifted in response to coating by the Ag NP satellite NPs.

Similar to the alloy metal NPs discussed above, bimetallic core-shell NPs exhibit enhanced catalytic activity in comparison to single-element NPs. This feature is believed to reflect synergistic effects between the core and shell metals, and is ascribed to both structural and electronic factors. Specifically, it has been hypothesized that the crystalline core alters the organization of metal atoms at the shell surface, increasing the concentration of binding sites for target substances. Moreover, electron migration from the core atoms towards the shell surface is also presumed to contribute to a more pronounced catalytic action of the core-shell NPs. Both factors are highly dependent on the structural features of the NPs; accordingly, "core-shell NP engineering" is important for attaining optimal catalytic properties.

An example of enhanced catalysis in core-shell NPs is presented in Figure 3.75. In this study, S. Sun and colleagues at Brown University synthesized NPs comprising a Fe-Cu-Pt alloy core and a Pt shell. Importantly, the researchers created "strains" in the platinum shell by modulating the crystal "mismatch" between core and shell, apparent as the periodic Pt peaks in the cross-sectional electron microscopy image in Figure 3.75. Crucially, the Pt shell strain resulted in enhanced catalysis (of an oxygen reduction reaction), attributed to more pronounced adsorption of the oxygen-containing substrates onto the NP surface.

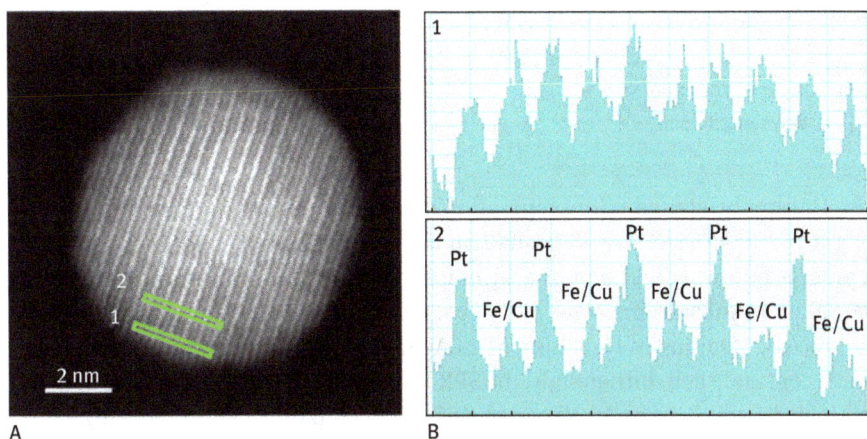

Fig. 3.75: "Corrugated" core-shell bimetallic nanoparticles. Core-shell FeCuPt@Pt nanoparticles in which a crystal mismatch between the core and shell creates periodic strains apparent in the electron microscopy image (left) and element mapping profile (right. The line profile in **2** is a normalized cross-section taking the distinct metal species into account). Reprinted with permission from Choi et al., *JACS* **136** (2014), 7734–7739, © 2014, American Chemical Society.

The surface plasmon resonance (SPR) shift is an important physical property which can be modulated and tuned in bimetallic core-shell NPs. In principle, the spectral position of the plasmon resonance in core-shell NPs is determined by both metallic species, as the collective motion of the surface electrons of the shell is affected by the electronic properties of the core material. Accordingly, varying the mole ratio between the core and shell atoms, or modulating the thickness of the shell, both effect the SPR and corresponding color of the NP suspensions. Figure 3.76 highlights this optical effect. The core-shell Ag@Au NPs, designed by C. Narayana and colleagues at the Jawaharlal Nehru Centre for Advanced Scientific Research, India, contained different mole ratios between the Au and Ag constituents. As is clearly apparent, tuning the mole ratio gave rise to dramatic shifts of plasmon resonance.

SPR tuning was also accomplished in a core-shell "nanorattle" design, outlined in Figure 3.77. The synthetic procedure, developed by Y. Xia and colleagues at Wash-

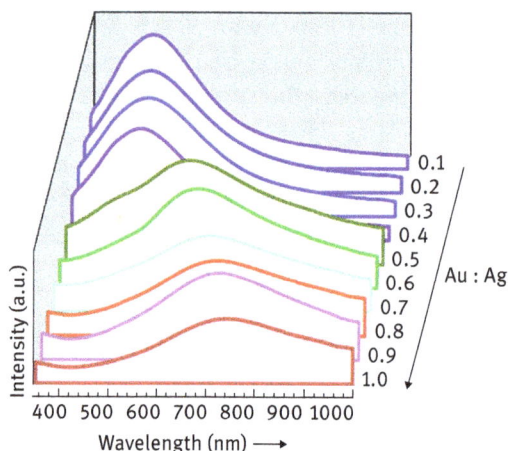

Fig. 3.76: Surface plasmon resonance shifts in core-shell Ag@Au nanoparticles depend on the mole ratio between the two metals. Reprinted with permission from Pavan Kumar et al., *J. Phys. Chem. C* **111** (2007), 4388–4392, © 2007, American Chemical Society.

ington University, started with preparation of the Ag-Au alloy NP core. In a second step, a metallic Ag layer was deposited on the Ag-Au core by a conventional reduction process following addition of a silver ion precursor. Lastly, Au ions were added to the solution, substituting some of the Ag atoms in the shell via a galvanic replacement process, resulting in formation of an Ag-Au shell encapsulating a "void space" between core and shell. Intriguingly, the SPR response of the nanorattles was both different to pure (solid) Ag or Au NPs, and shifted in comparison to hollow Au/Ag NPs. Furthermore, the color of the NPs (i.e. position of the SPR peak) could be tuned by modifying the core and shell diameters.

While the bimetallic core-shell NPs thus far reported mostly exhibit spherical (or, more accurately, concentric) morphologies, non-spherical core-shell NPs have also been demonstrated. In general, however, reproducible synthesis of non-spherical core-shell NPs is more challenging due to the impact of even slight variations in experimental conditions. Despite the difficulties, dramatic progress has been achieved in designing novel non-spherical core-shell bimetallic NPs and fine-tuning their struc-

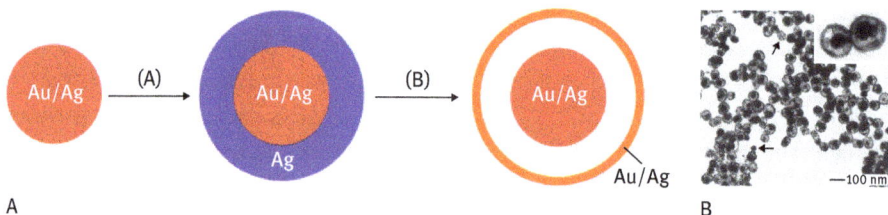

Fig. 3.77: Silver/gold "nanorattles". **A:** Synthesis procedure: alloy Au-Ag nanoparticle core is coated with Ag shell; galvanic replacement by Au^{3+} results in partial removal of the shell creating a "rattle" structure. **B:** Electron microscopy image of the nanorattles. Reprinted with permission from Sun et al., *JACS* **126** (2004), 9399–9406, ©(2004), American Chemical Society.

tural properties. Figure 3.78 depicts remarkable multi-faceted (e.g. polyhedral) Au@Ag NP morphologies. These core-shell NPs, synthesized by Z. L. Wang and colleagues at the National Center for Nanoscience and Technology, China, required careful modula-

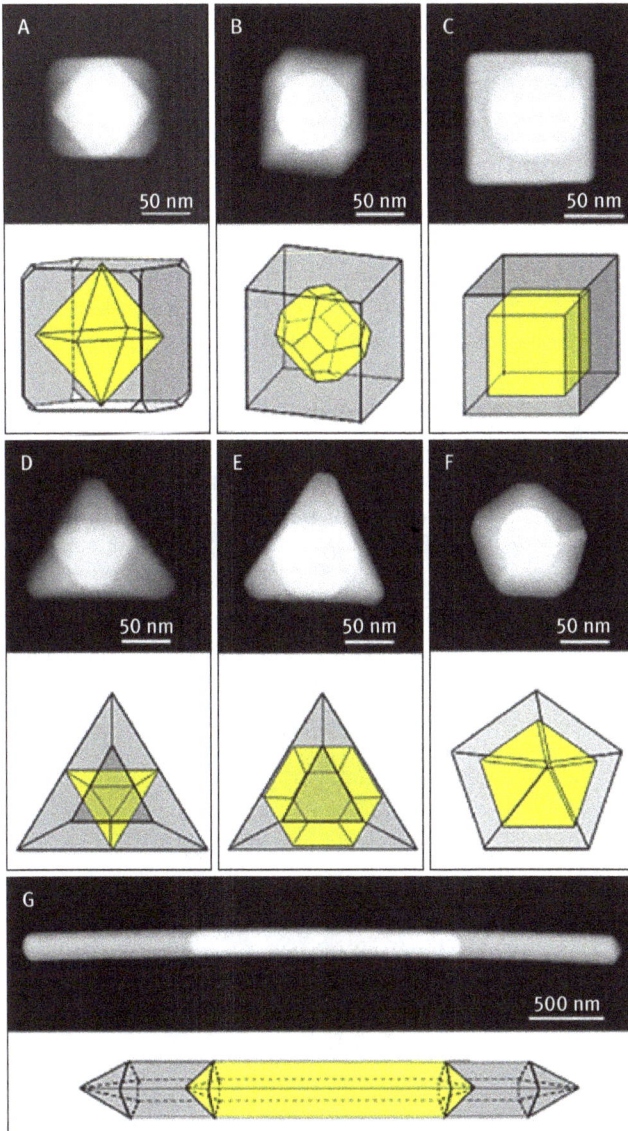

Fig. 3.78: Polyhedral core-shell nanoparticles. High angle annular dark field (HAADF) electron microscopy images and corresponding models of core-shell Au@Ag nanoparticles. The shapes of the Au cores are clearly resolved. Reprinted with permission from Wu et al., *Nanotechnology* **20** (2009), 305602, © 2009, IOP Publishing.

Tetrahexahedra   Concave octahedra   Octahedra

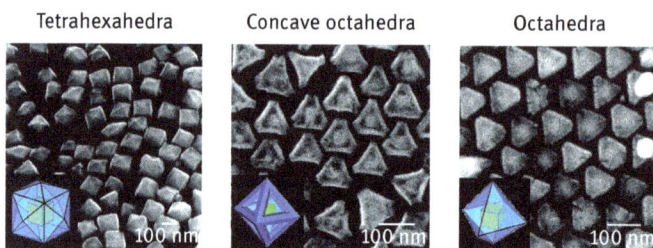

Fig. 3.79: Concave octahedral core-shell Au@Pd nanoparticles. Electron microscopy images and corresponding models of the core-shell nanoparticles viewed from different crystal planes. Reprinted with permission from Sun et al., *JACS* **126** (2004), 9399–9406, © 2004, American Chemical Society.

tion of the type of solvent, reaction temperature, and concentration of NP stabilizing agents. As apparent in Figure 3.78, the researchers have accomplished independent control of both the structural features of the Au seeds (constituting the NP cores) and the Ag shells.

Figure 3.79 presents another example of non-spherical bimetallic core-shell NPs. The unusual polyhedral Au@Pd core-shell NPs were synthesized by M. H. Huang and colleagues at National Tsing Hua University, Taiwan, using *Au* nanocubes as seeds. The researchers found that such complex polyhedral core-shell NPs could be generated by careful tuning of the reaction parameters, including temperature, reaction time, types of surfactants added to the Pd^{2+} precursors, and the presence of oxidant species such as oxygen. The unique concave core-shell NPs functioned as good catalysts, presumably related to the high number of crystalline facets (e.g. "high index number") serving as binding sites for target organic molecules.

Core-shell *nanoprisms* were constructed via an unusual *plasmon-induced* process (Figure 3.80). Intriguingly, C. Mirkin and colleagues at Northwestern University

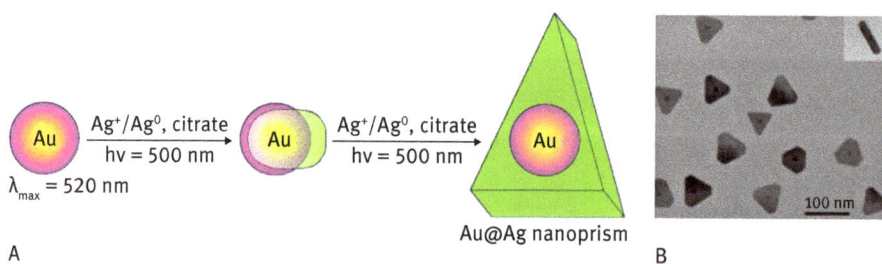

Au@Ag nanoprism

A   B

Fig. 3.80: Plasmon-induced core-shell Au@Ag nanoprisms. **A:** Synthesis scheme: Au seeds are immersed in Ag^+ ion solution (the Ag^+ ions are released from small metallic Ag^0 particles). Light irradiation at the approximate Au plasmon resonance initiates deposition of Ag layer on the Au seeds; extended irradiation results in formation of the core-shell nanoprisms. **B:** Electron microscopy image showing monodispersed nanoprisms. Reprinted with permission from Xue et al., *Angew. Chem. Int. Ed.* **46** (2007), 8436–8439, © 2007, John Wiley & Sons.

observed that anisotropic Ag nanoprism shells could be grown around Au NP cores through extended light irradiation at a wavelength close to the SPR of the Au NPs. This process was attributed to excitation of the surface electrons of the Au NP cores which initiated deposition of the Ag atoms, albeit in an anisotropic manner, producing the prism-like organization.

Other bimetallic core-shell NP architectures have included *nanowheels, nanoplates*, and *nanostars*. Such NP morphologies might be perceived as somewhat "exotic"; however they have opened new routes for various applications and could serve as conduits for developing synthesis procedures leading to novel nanostructures. Figure 3.81, for example, depicts Pd@Ag core-shell nanoplates. Synthesis of these hybrid NPs, demonstrated by N. Zheng and colleagues at Xiamen University, China, followed the generalized route outlined in Figure 3.73B – first creating Pd nanoplate seeds, which were subsequently placed in a solution of Ag^+ ions in the presence of a reducing agent. The choice of this particular hybrid composition, however, was not random, since palladium provided tunable SPR emission peaks extended into the important infrared region of the electromagnetic spectrum. Indeed, different colors of the hybrid NP suspensions could be produced through modifying the mole ratio between the palladium and silver constituents (Figure 3.81B).

In many instances, non-spherical core-shell NPs containing *dielectric* cores and *metallic* shells exhibit pronounced enhancement and tunability of the SPR response of the shell material, particularly in comparison with similar spherical configurations. These observations were often recorded in the case of rod-like structures, and were attributed to both the effect of the nanorod geometry on the shell-associated SPR (specifically the longitudinal plasmon resonance associated with the long rod axis), and the contribution of the interface between the metallic shell and dielectric core. Figure 3.82 presents Fe_3O_4@Au core-shell NPs denoted "nanorice" by N. J. Halas and colleagues at Rice University. These core-shell nanoparticles were synthesized in a step-wise manner through initial coating of the Fe_3O_4 nanorods with a silane-bearing layer which attracted Au colloids; these particles, in turn, served as nucleation seeds for further reduction and deposition of Au^0, resulting in formation of the gold shell (Figure 3.82A). The optical properties – e.g. SPR wavelengths – of these intriguing core-shell NPs could be tuned both by changing the length of the iron oxide seeds, and also by varying the shell thickness. The close dependence of the optical response of the Fe_3O_4@Au NPs on shell properties makes these particles promising candidates for sensing applications, as the shell surface can be functionalized with molecular recognition elements.

An unusual, visually striking Ag@Au core-shell NP structure is shown in Figure 3.83. This NP morphology, comprising of Ag core and Au-Ag alloy shell, was produced by J. Y. Lee and colleagues at the National University of Singapore via a galvanic replacement process. A truncated Ag octahedron NP was first synthesized. Galvanic replacement of the silver ions took place on introduction of Au^{3+} ions. However, substitution of the Ag atoms did not occur randomly but rather pref-

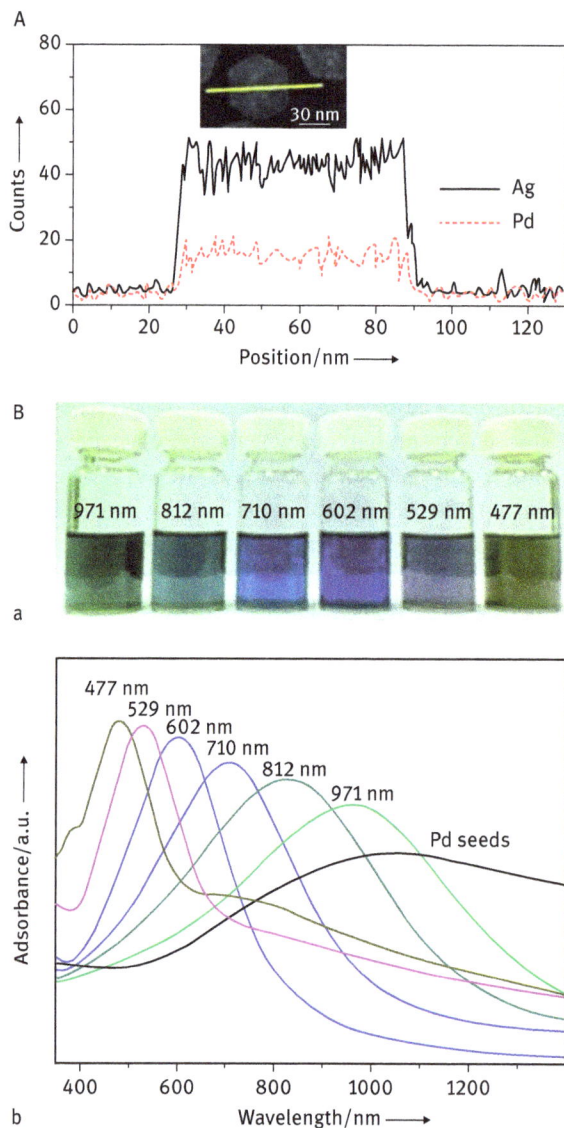

Fig. 3.81: Core-shell Pd@Ag nanoplates. A: Electron microscopy image of the nanoplate and elemental map showing the Ag shell deposited on the Pd core. B: Photographs (a) and corresponding plasmon resonance spectra (b) underscoring the spectral shift induced by varying the Ag:Pd mole ratio in the nanoplates. Reprinted with permission from Huang et al., *Adv. Mater.* **23** (2011), 3420–3425, © 2011, John Wiley & Sons.

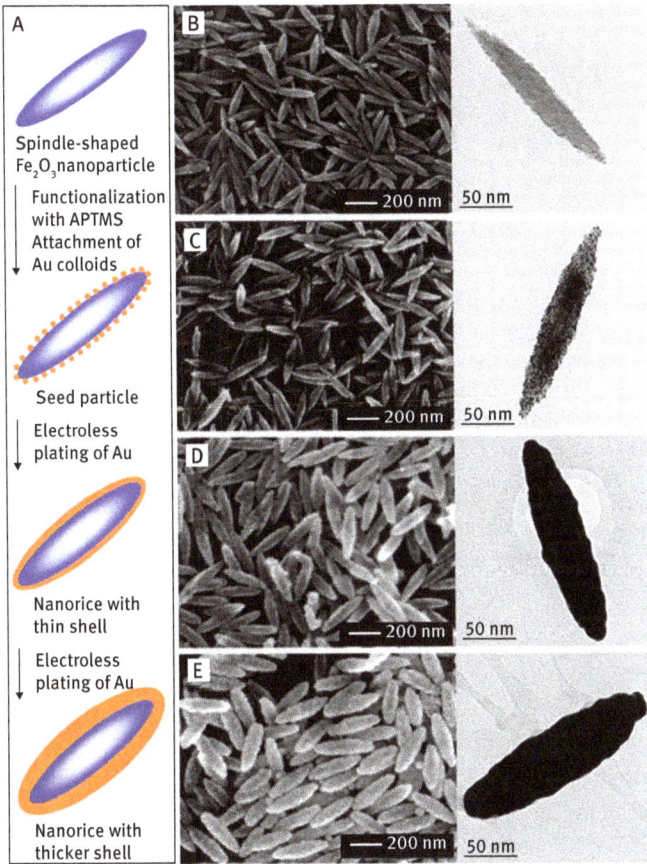

Fig. 3.82: Core-shell Fe_3O_4@Au "nanorice". **A:** Synthesis procedure: gold seeds are attached to the iron oxide nanorod surface through coating with a silane-displaying polymer. The Au^0 seeds promote reduction and deposition of the Au shell. **B–E:** Electron microscopy images depicting the nanoparticles at different stages of synthesis. Reprinted with permission from Wang et al., *Nano Lett.* **6** (2006), 827–832, © 2006, American Chemical Society.

erentially on distinct crystalline facets, yielding core-shell NPs exhibiting localized "protrusions".

Other sophisticated bimetallic core-shell NP synthesis schemes have been demonstrated. Figure 3.84 presents trimetallic nanoparticles comprising octahedral Au cores and Pd-Pt alloy shells. The Au@PdPt NPs were synthesized by S. W. Han and colleagues at KAIST, Korea, using a "one-pot" strategy – essentially mixing the metal precursors and respective reducing agents for each metal ion. Notably, through careful tuning of the concentrations of precursors and reducing agents, the researchers obtained highly uniform core-shell NPs exhibiting *dendritic* alloy shells (Fig. 3.84). These core-shell NPs were found to possess excellent catalytic activity, likely ascribed

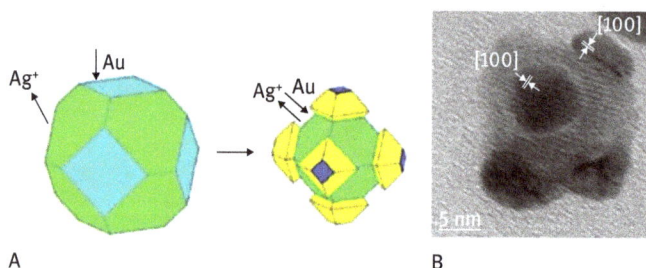

Fig. 3.83: Branched core-shell Ag@Au nanoparticles. **A:** Assembly route: truncated Ag octahedron serving as the core undergoes galvanic replacement, occurring preferably on the truncated crystal planes. The green and cyan facets represent the Ag {111} and {100} crystal planes, respectively, while the yellow and blue facets correspond to the crystal planes of Au. **B:** Electron microscopy image of the core-shell nanoparticle. Reprinted with permission from Zhang et al., *Small* **4** (2008), 1067–1071, © 2008, John Wiley & Sons.

to the specific metal composition and dendritic shell morphology providing high surface area for adsorption of target analytes.

3.4.3 Janus nanoparticles

Janus NPs have anisotropic structures due to the non-symmetrical distribution of the atomic constituents within the particles ("Janus" evokes the mythological Roman god with two adjoined heads facing in opposite directions. The discussion here encompasses asymmetric bimetallic NPs in general, however, not only particles made up of "two halves" each comprising of different elements). Colloidal asymmetry is a major factor affecting the properties of Janus NPs, and can usually be modulated by the synthesis parameters, primarily the ratio between the (usually two) atomic components. Most Janus NPs reported thus far comprise *metal/non-metal* compositions and are discussed in Chapter 6. Some bimetallic Janus NPs have been designed as well and are highlighted below.

Synthesis of bimetallic Janus NPs is generally carried out via "two-step" processes, in which one of the metallic nanoparticle components constitutes a nucleation and growth site for the second metal. Importantly, the growth of the second metal component is anisotropic, thus generating a Janus nanoparticle configuration (Fig. 3.85A). It should be noted, however, that in some bimetallic NP systems the formation of Janus vs. core-shell structures depends on fine-tuning of the experimental parameters. Figure 3.85B, for example, schematically shows cubic Janus Ag-Pd configurations produced by Y. Xia and colleagues at Washington University via tuning of the *injection rate* of the Ag^+ precursors into the solution containing the Pd seeds; in certain experimental conditions a core-shell Pd@Ag NP organization actually prevailed.

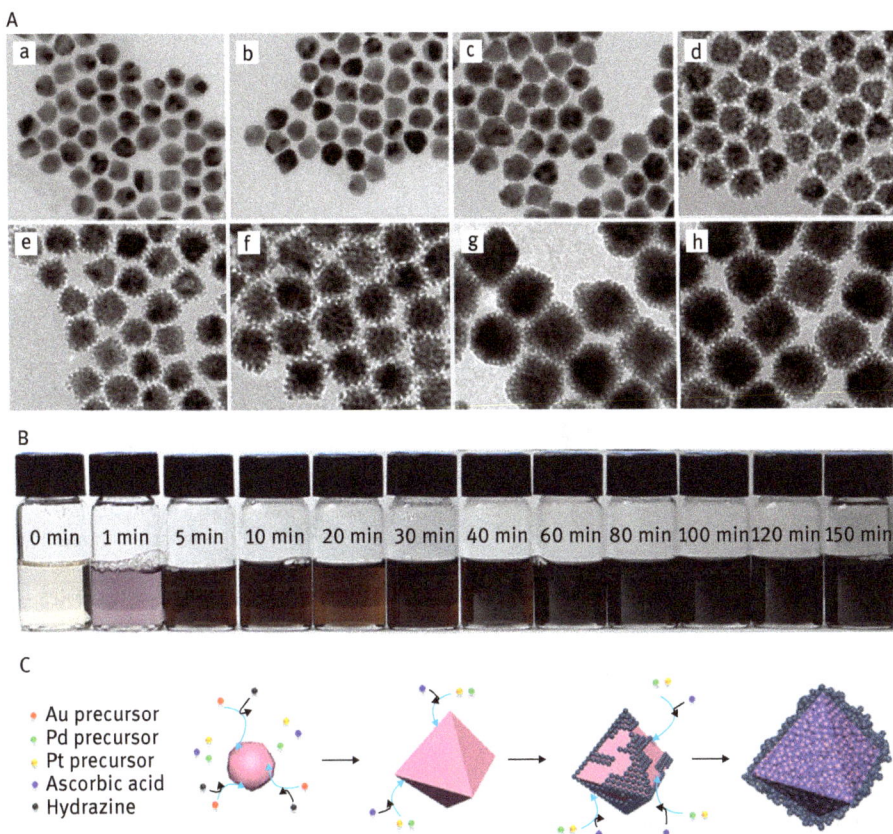

Fig. 3.84: Trimetallic Au@PdPt core-shell nanoparticles. **A:** Electron microscopy images of the nanoparticles throughout the synthesis process (different reaction times). Scale bars correspond to 50 nm. **B:** Colors of the nanoparticle solution as the nanoparticle assembly progresses; **C:** Schematic model for assembly of the core-shell nanoparticles: formation of Au octahedral seeds and deposition of the Pd-Pt alloy shell adopting dendritic structure. Reprinted with permission from Kang et al., *ACS Nano* **7** (2013), 7945–7955, © 2013, American Chemical Society.

Figure 3.86 portrays growth mechanisms of bimetallic cubic-based Janus Pd-Ag NPs. Specifically, Y. Xia and colleagues at Washington University showed that the rate of injection of Ag^+ ions into a solution containing Pd seeds affected the extent of growth of crystalline Ag on the cubic Pd NPs. The researchers refined their synthesis procedures, achieving a remarkable control over NP morphology and atomic distribution within the bimetallic NPs, growing Ag layers on specific facets of the cubic Pd seeds. Importantly, the shift of the Ag plasmon resonance was dependent on the configuration of the Janus NPs (Fig. 3.86B), likely reflecting transfer of electrons from palladium to silver atoms through the interface between the two metallic species. The SPR shifts

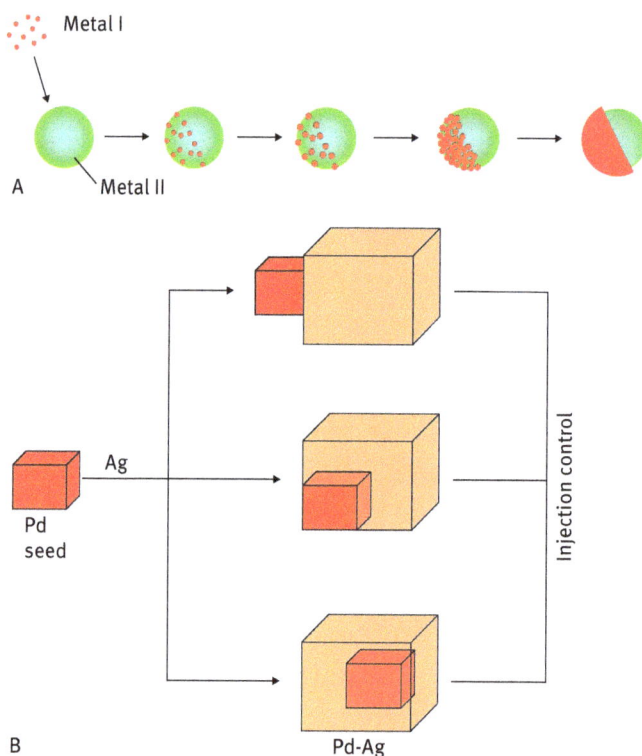

Fig. 3.85: Synthesis of bimetallic Janus nanoparticles. A: Generic synthesis scheme showing anisotropic growth of a metal component on a nanoparticle comprising a different metallic species. B: Generating Janus or core-shell nanoparticle structures by control of experimental conditions.

highlighted in Figure 3.86B reflect the modulation of *optical properties* in Janus NPs in comparison with NPs comprising of the individual elements.

Janus NPs have not been limited to two component systems. Incorporation of three and even more atomic species resulted in fabrication of "multi-modal" NPs in which each element had a specific functional role. Figure 3.87 presents an example of such a Janus NP designed by J. Cheon and colleagues at Yonsei University, Korea, for possible biomedical applications. As shown in Figure 3.87, one side of the hetero-structured NP comprised of an iron-platinum alloy. This component had two important functions: serving as a contrast agent in magnetic resonance imaging applications (through the ferromagnetism of Fe; see more detailed discussion in Chapter 4), and constituting a catalytic platform (through the Pt metal surface) for nucleation and growth of the Au NP, which was the second component of the Janus NP. Gold bestows additional important functionalities on the NPs, specifically easy conjugation with hydrophilic residues designed to achieve water solubility and biological targeting. Furthermore, the Au component can be also used as a vehicle for sensing applications through the SPR properties.

Asymmetric hybrid metal NPs have sometimes (quite fittingly...) been referred to as "dumbbell" NPs, and studied extensively in light of their unique structural fea-

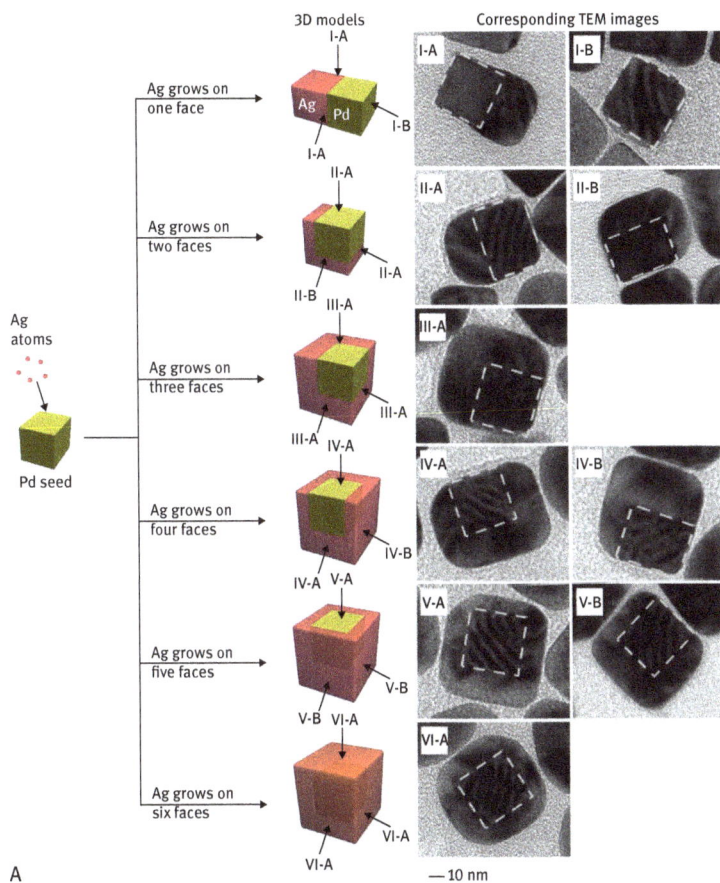

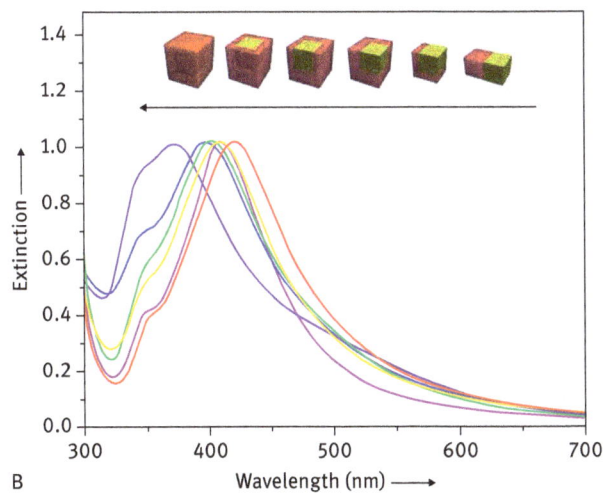

Fig. 3.86: Controlled formation of Ag-Pd Janus nanoparticles. **A:** Structural models and electron microscopy images of the Ag-Pd nanoparticles exhibiting different Ag growth profiles upon the Pd seed. **B:** Shifts of the surface plasmon resonance correlating with the Janus nanoparticle configurations. Reprinted with permission from Zhu et al., *JACS* **134** (2012), 15822–15834, © 2012, American Chemical Society.

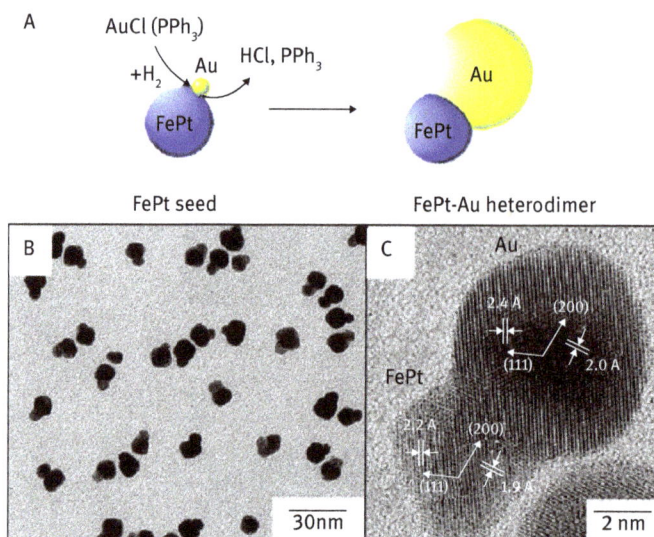

Fig. 3.87: "Multi-modal" trimetallic Janus nanoparticles. **A:** Synthesis scheme: initial formation of the Fe-Pt alloy seeds is followed by growth of the Au constituent, catalyzed by platinum. **B–C:** Electron microscopy images of the Janus nanoparticles. Reprinted with permission from Choi et al., *JACS* **128** (2006), 15982–15983, © 2006, American Chemical Society.

tures and potential applications in sensing, catalysis, and more. Figure 3.88 illustrates a synthesis strategy in which Au NPs (as well as NPs of other noble metals such as Ag and Pt) were utilized as growth platform for metal oxide constituents. Essentially, iron carbonyl precursor molecules were adsorbed on the surface of the Au NPs, undergoing decomposition and oxidation, yielding a crystalline iron oxide aggregate. The Au-Fe_3O_4 dumbbell NPs produced by this approach by S. Sun and colleagues at Brown University were quite homogeneous and could be employed as conduits for heterogeneous catalysis – specifically decomposition of toxic vapors, such as CO gas. The catalytic effects of these hybrid NPs were ascribed to immobilization of the gas molecules onto higher-reactivity mixed-atom sites at the NP surface combined with transfer of electrons from the iron oxide onto the Au NP surface.

An alternative synthesis procedure for fabrication of metal dumbbell NPs utilized core-shell NPs as starting materials (Fig. 3.89). In this study, reported by L. Manna and colleagues at the Instituto Italiano di Tecnologia, Italy, core-shell NPs comprising Au-Pd alloy (core) and iron oxide (shell) were prepared. High-temperature annealing of the core-shell NPs resulted in gradual distortion of the shell, ultimately creating dumbbell morphology. Similar to the above system, these hybrid NPs exhibited notable catalytic activity. Intriguingly, the dumbbell NPs could be further treated with strong oxidizing agents (such as I_2) leading to dissolution (etching) of the metallic gold and producing asymmetric iron oxide "nanocontainers".

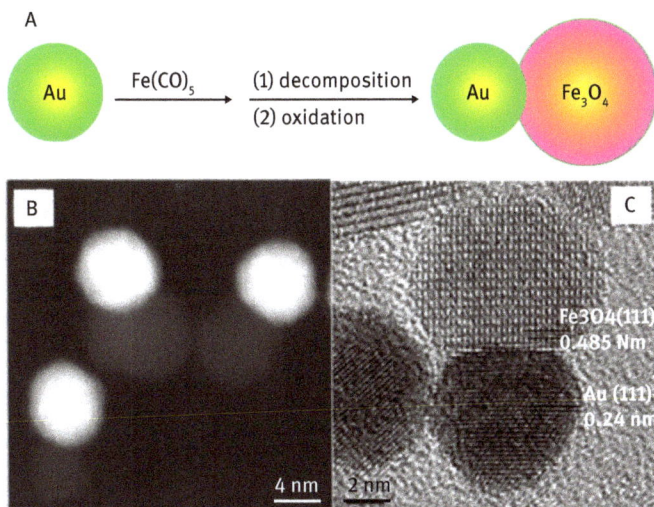

Fig. 3.88: "Dumbbell" Au-Fe$_3$O$_4$ nanoparticles. **A:** Formation of the dumbbell nanoparticles through adsorption of iron carbonyl onto Au nanoparticles followed by decomposition and oxidation of the iron precursor yielding Fe$_3$O$_4$. HAADF electron microscopy (**B**) and high-resolution electron microscopy (**C**) images underscoring the distinct crystalline components of the dumbbell nanoparticles. Reprinted with permission from Yu et al., *Nano Lett.* **5** (2005), 379–382, © 2005, American Chemical Society.

Metal-metal oxide dumbbell NPs display interesting surface plasmon resonance properties, ascribed to the strong localized electron fields at the interface between the two components, and the overall asymmetry of the NPs. Figure 3.90 presents an experimental and theoretical analysis of Au-TiO$_2$ nanodumbbells. Localized SPR (LSPR) experiments, carried out by M. Y. Han and colleagues at A*STAR, Singapore, revealed that the Au-TiO$_2$ dumbbell NPs produced LSPR at a wavelength different to that of core-shell NPs (and also different to that of *bare* Au NPs). Computer simulations indeed indicated a dramatic enhancement of the electric field close to the NP surface in the asymmetric Au-TiO$_2$ NP configuration (Fig. 3.90B). Importantly, the enhanced plasmon fields on the surfaces of the dumbbell Au-TiO$_2$ NPs were probably responsible for the greater photocatalytic activity (towards generation of hydrogen from water-dissolved alcohol) recorded for these NPs compared to the symmetric core-shell or bare Au NPs, as the plasmon fields at the Au-TiO$_2$ interface are believed to aid formation of electron-hole pairs which play a prominent role in TiO$_2$-based photocatalysis.

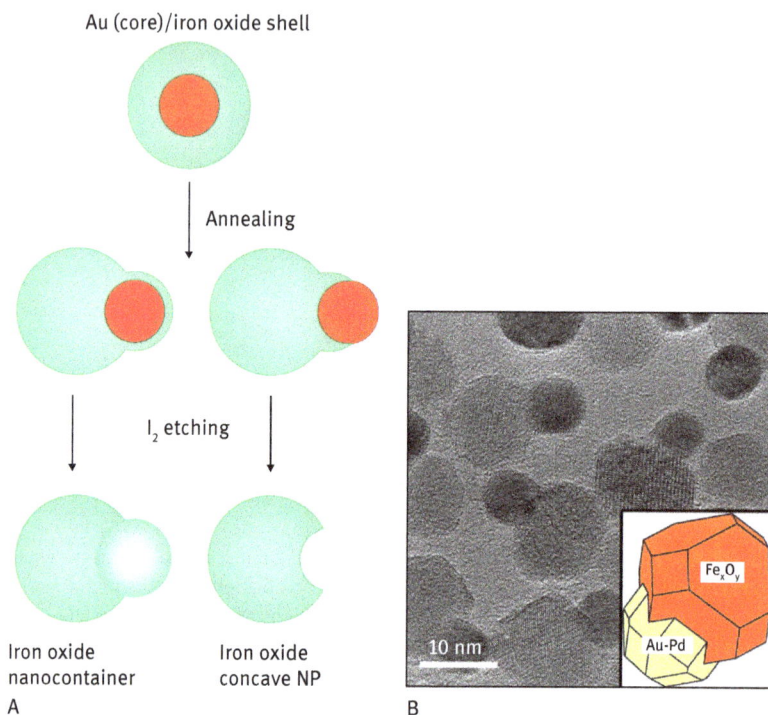

Fig. 3.89: Hybrid bimetallic nanodumbbells. **A:** Generic synthesis scheme: initially-prepared core-shell Au@Fe$_3$O$_4$ nanoparticle is annealed, inducing distortion of the iron oxide shell to produce a Janus nanoparticle. The nanodumbbell can be further treated with a strong oxidizer such as I$_2$, resulting in etching of the Au core and formation of a "nanocontainer". **B:** Electron microscopy image and corresponding model of an AuPd/iron oxide nanodumbbell. Reprinted with permission from George et al., *Nano Lett.* **13** (2013), 752–757, © 2013, American Chemical Society.

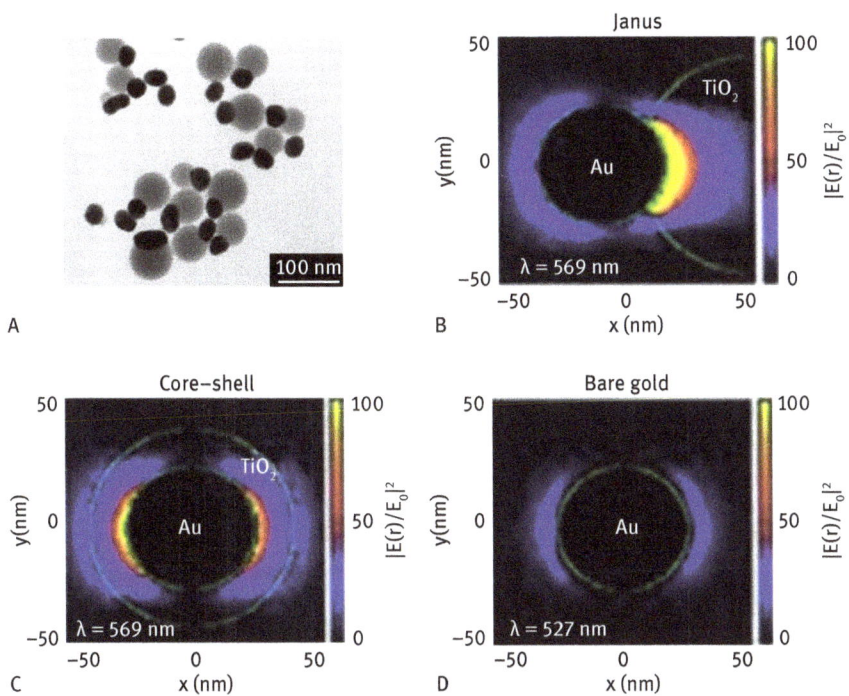

Fig. 3.90: Au-TiO$_2$ Janus nanoparticles. A: Electron microscopy image of the Janus nanoparticles. B–D: Computer simulations depicting the plasmonic maps of a Janus Au-TiO$_2$ nanoparticle, Au@TiO$_2$ core-shell nanoparticle, and Au nanoparticle, respectively. A significant plasmon enhancement is apparent in the Janus Au-TiO$_2$ nanoparticle, likely responsible for their pronounced photocatalytic properties. Reprinted with permission from Seh et al., *Adv. Mater.* **24** (2012), 2310–2314, © 2012, John Wiley & Sons.

4 Metal oxide nanoparticles

Metal oxides are a broad class of materials, both naturally-occurring and synthetic, exhibiting diverse properties and applications. Metal oxide nanoparticles (NPs) have likewise attracted significant interest. This chapter discusses widely-studied oxide NP systems such as *iron oxide NPs, silicon oxide NPs, titanium oxide NPs, zinc oxide NPs*, and *rare earth oxide NPs*.

4.1 Iron oxide nanoparticles

Iron oxide (Fe_3O_4 familiarly known as *magnetite*, and its oxidized species *maghemite* Fe_2O_3) forms highly *magnetic* nanoparticles, as iron is one of the most ferromagnetic elements in the periodic table. The sizable magnetic moment of iron oxide NPs (IONPs) – i.e. *superparamagnetism* – is directly related to the nanoscale dimensions of the particles, specifically the *single crystalline* domains they comprise of. In a somewhat simplistic picture, the electron spins in such single-domain IONPs are all oriented in the same direction, producing a strong local magnetic field (Fig. 4.1). IONPs are thus commonly referred to as "magnetic NPs" or MNPs. Furthermore, below a certain size threshold (typically around 20 nm for IONPs), the magnetic properties of the NPs become *size-dependent* due to the enhanced contribution of magnetic and energy properties of the nanoparticle *surface* as compared to the *core* (or "bulk" component).

While a detailed discussion of the fundamental physical properties of IONPs, particularly the size-dependent magnetic properties, is beyond the scope of this book, it is safe to say that the pronounced *magnetism* of IONPs (and other magnetic NPs for

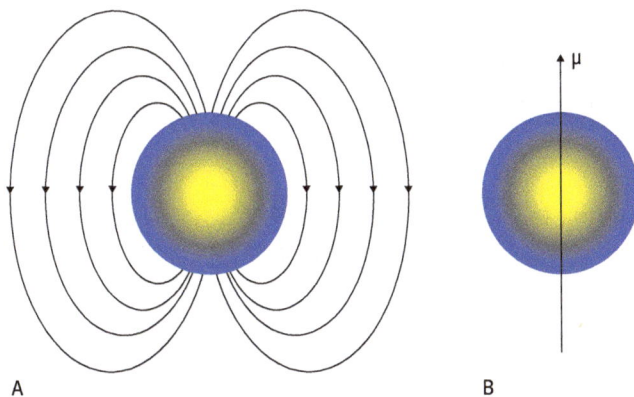

Fig. 4.1: Iron oxide nanoparticles. **A:** The magnetic field around the nanoparticle. **B:** Direction of magnetic moment.

that matter) is the defining scientific and technological feature of these NP species. Indeed, the interactions of IONPs with magnetic fields are the primary attraction of these systems for biological imaging and therapeutics, and representative examples of this vast research area are presented below.

IONPs are amenable to biological applications since iron is an integral participant in many physiological processes. Primary among these is iron's role in oxygen and carbon dioxide transport in the bloodstream (as part of the *hemoglobin* in red blood cells). Accordingly, IONPs are presumed to exhibit minimal toxicity, and could be recycled and/or discarded via natural biodegradation pathways in the human body. Indeed, intensive R&D efforts in this field have already resulted in the commercialization of several pharmaceutical formulations containing IONPs.

Many iron oxide NP *synthesis schemes* have been developed. Synthetic routes for production of magnetic NPs are not complex, however they vary significantly in terms of particle homogeneity, stability of the NP products, and their biocompatibility. Solution-based synthesis has been widely used to generate IONPs. Most methods start with iron salts or iron complexes, which are subsequently converted to IONPs on reaction with oxygen donors at elevated temperatures. Solution-based syntheses can be divided into *aqueous phase* and *organic phase* processes. *Water-based* synthetic methods are easy to implement and usually produce large quantities of biocompatible particles. The downside of such techniques is low particle uniformity, i.e. less monodispersity. *Organic phase* magnetic NP synthesis generally yields more homogeneous reaction products, albeit particle coating with molecules displaying hydrophilic moieties is then required for further use in biological applications (or for other applications carried out in aqueous environments).

Figure 4.2A illustrates a widely used method for synthesizing crystalline and monodisperse γ-Fe_2O_3 NPs (e.g. the *maghemite* sub-family of iron oxides). Iron pentacarbonyl, $Fe(CO)_5$, is treated with oleic acid at 100°C, forming an iron-oleic acid metal complex. Upon heating at a higher temperature of 300°C, the iron-oleic acid complex decomposes to form Fe nanoparticles which can be converted to γ-Fe_2O_3 nanoparticles via controlled oxidation with a mild oxidant, trimethylamine oxide, $(CH_3)_3NO$. Figure 4.2B depicts an alternative method, in which co-precipitation of

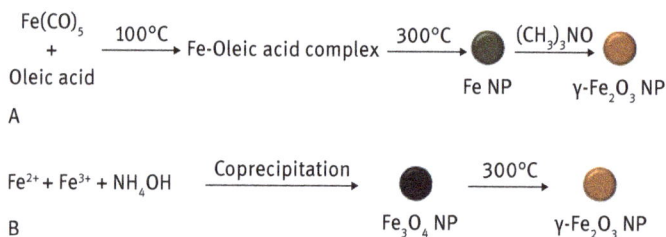

Fig. 4.2: Common synthetic processes for production of iron oxide nanoparticles.

Fe^{2+} and Fe^{3+} ions in the presence of ammonium hydroxide (NH_4OH) in aqueous medium results in the formation of *magnetite* (Fe_3O_4) nanoparticles, which can be converted to Fe_2O_3 NPs through oxidation in air at 300°C.

Developments of IONP synthesis methods have led to intriguing discoveries, such as the creation of *hollow* IONPs. A. P. Alivisatos and colleagues at the Lawrence Berkeley Laboratory, for example, found that oxidation of metallic iron NPs at between 200°C–300°C led to formation of hollow and core-shell IONPs (Fig. 4.3). This interesting phenomenon was ascribed to rapid diffusion of the oxidized iron ions from the metallic core outwards to the iron oxide shell of the NP, leaving behind "vacancies" which eventually coalesced into larger empty spaces inside the nanoparticles.

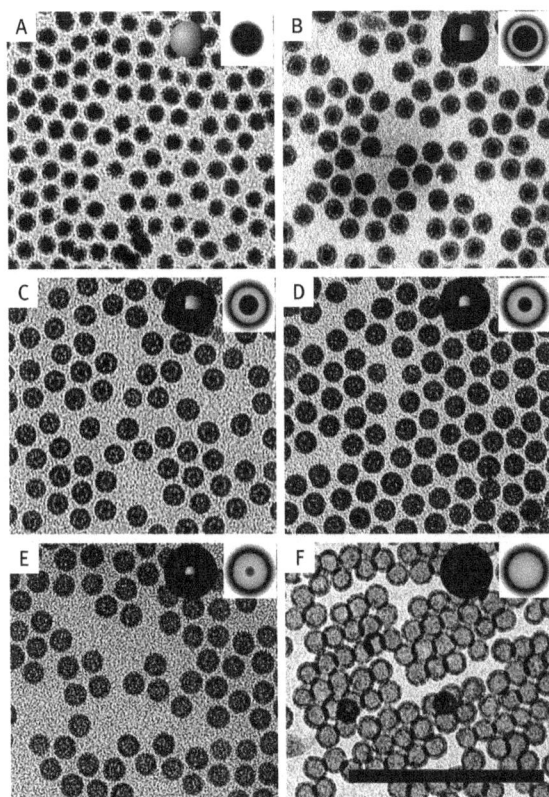

Fig. 4.3: Hollow iron oxide nanoparticles. Electron microscopy images of iron nanoparticles exposed to oxygen gas for increasing lengths of time. Nanoparticles with different configurations of iron oxide shell/iron core/voids are formed. Schematic depictions of nanoparticle structures and cross-section projections are shown in the top right of each image. Scale bar corresponds to 100 nm. Reprinted with permission from Cabot et al., *JACS* **129** (2007), 10358–10360, © 2007 American Chemical Society.

As with the other types of NPs discussed in this book, chemical functionalization of IONPs is a key feature of their wide applicability, particularly in biology and biomedicine. Numerous synthetic routes have been introduced for covalent display of chemical and biological recognition elements, enzymes, peptides, oligonucleotides, and other molecular constituents on IONP surfaces. *Noncovalent* association of hydrophobic/hydrophilic layers on IONPs has also been widely carried out. Such shells generally prevent NP aggregation, but also play critical roles in many instances, including NP biocompatibility, minimizing immune response, and enabling NP passage through physiological barriers such as the blood-brain barrier (BBB) and cell membranes. Furthermore, the molecular layers encapsulating the iron oxide core can act as "protective shields", preventing enzymatic degradation in the bloodstream and extracellular space.

Figure 4.4 depicts a strategy for construction of functionalized IONPs with molecular recognition properties via a combined noncovalent/covalent approach. The single-step synthetic procedure, developed by W. Tan and colleagues at the University of Florida, enclosed IONPs within a hydrophobic layer comprising of phospholipid residues covalently attached to *hydrophilic* recognition elements. The approach is generic, as different residues can be attached to the phospholipids. Specifically, the researchers demonstrated the concept by conjugating *oligonucleotides* (DNA fragments) to the phospholipids. Binding the functionalized IONPs was achieved through *hybridization* between the NP-displayed DNA and its complementary oligonucleotide target. Alternatively, *specific protein binding* could be accomplished through the use of *aptamers* – short DNA fragments designed to recognize and bind protein targets. Modified IONPs such as those shown in Figure 4.4 can be functionalized with multiple, different ligands, and the hydrophilic residues extending from the NP surface had a further role in stabilizing the particles in aqueous solutions.

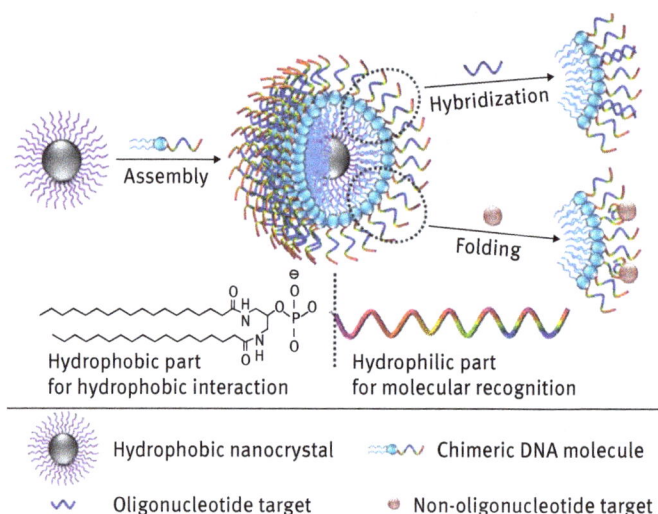

Fig. 4.4: Iron oxide nanoparticles displaying biomolecular recognition elements. The nanoparticles are coated with a layer of phospholipid molecules functionalized with hydrophilic residues (DNA sequences). Target recognition is achieved either through DNA hybridization (top route) or peptide-binding and consequent folding (bottom route). Reprinted with permission from Chen et al., *JACS* **134** (2012), 13164–13167, © 2012 American Chemical Society.

Another example of IONP functionalization through noncovalent interactions is presented in Figure 4.5. In this simple yet powerful system, developed by D. G. Anderson and colleagues at MIT, the IONPs were interspersed with *cationic lipids* which prevented aggregation of the NPs in water solutions. The NPs were subsequently loaded with oligonucleotides (either DNA, or "small-interfering RNA", siRNA; widely employed as a tool for "gene silencing" as it interferes with gene expression pathways) which were bound to the cationic residues through electrostatic attraction. The assemblies generated were quite uniform in size, resilient in physiological solutions, and could efficiently deliver their oligonucleotide cargoes to cell targets.

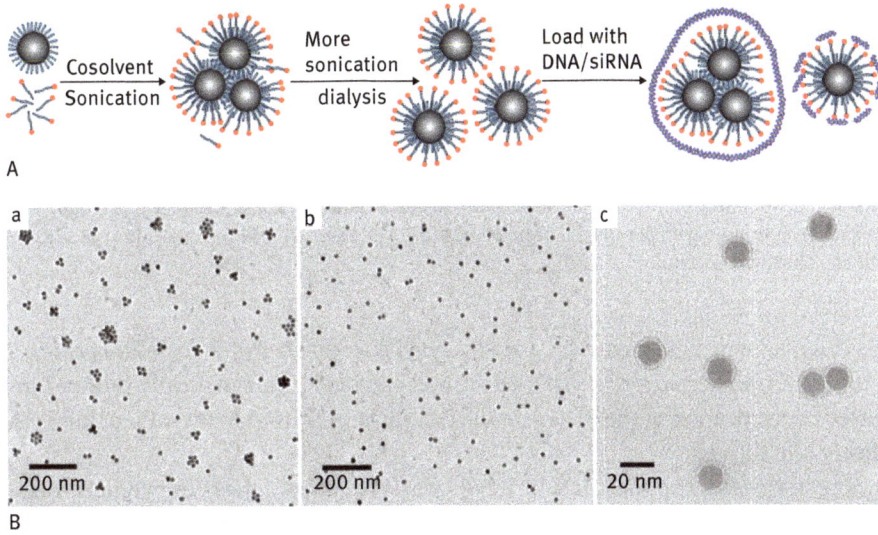

Fig. 4.5: DNA delivery through lipid coating of iron oxide nanoparticles. **A:** Experimental scheme: the iron oxide nanoparticles are interspersed and coated with cationic lipids. Following dispersion through sonication and dialysis the lipid-coated nanoparticles are mixed with DNA or RNA which associate with the nanoparticles through electrostatic attraction. **B–D:** Electron microscopy images showing different stages in the process: **B** cationic lipid/iron oxide nanoparticle aggregates; **C** dispersed lipid-coated iron oxide nanoparticles; **D** DNA-associated lipid-coated iron oxide nanoparticles. Reprinted with permission from Jiang et al., *Nano Lett.* **13** (2013), 1059–1064, © 2013 American Chemical Society.

One of the major advantages of using IONPs such as those shown in Figures 4.4 and 4.5 for biomolecular targeting is the possibility of using an externally-applied *magnetic field* for physical separation and spatial manipulation of the NPs. Magnetically-guided targeting has been successfully demonstrated, for example, for the DNA/RNA-loaded IONPs prepared according to the scheme in Figure 4.5. In the experiment (summarized in Fig. 4.6) the researchers loaded the IONPs with DNA encoding for green fluorescent protein (GFP), a fluorescence-emitting protein. As shown in Figure 4.6, placing the

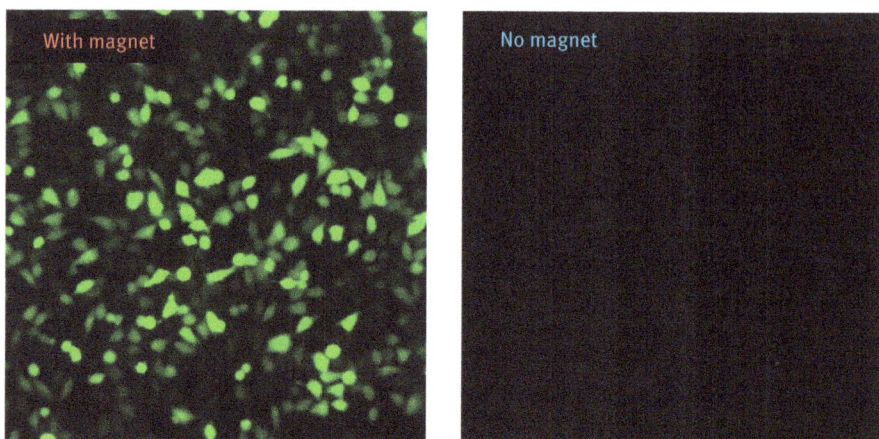

Fig. 4.6: Magnetic targeting of bio-functionalized iron oxide nanoparticles. Fluorescence microscopy images showing the effect of placing suspensions of cells and DNA/lipid-coated iron oxide nanoparticles in a magnetic field. The DNA is encoded for the green fluorescent protein (GFP), resulting in enhanced fluorescence of the cells incubated with the nanoparticles in the presence of a magnet (top image). Reprinted with permission from Jiang et al., *Nano Lett.* **13** (2013), 1059–1064, © 2013 American Chemical Society.

cells in a magnetic field gave rise to pronounced fluorescence (due to GFP expression by the DNA-transfected cells). This effect was ascribed to magnetically-induced increased concentration of the IONPs on the cell surface, consequently enhancing DNA delivery into the cell.

One of the early applications of IONPs was as *contrast agents* in magnetic resonance imaging (MRI). MRI is a powerful, noninvasive imaging technology based on the magnetic resonance signals of water protons' nuclei. The magnetic signals are highly sensitive to their chemical/biological environments, thereby providing detailed spatial information on organs and tissues. The significant role of IONPs in MRI stems from the shifts and in particular the attenuation of magnetic resonance signals of atoms in the vicinity of ferromagnetic atoms such as iron. Accordingly, accumulation of IONPs in cells and tissues generally results in substantially lower image brightness – e.g. higher contrasts with surrounding tissue areas not containing IONPs.

Numerous examples of IONP use in bio-imaging have been reported. Figure 4.7 depicts an MRI experiment utilizing IONPs coated with oleic acid (to prevent aggregation), which were further encapsulated within *oil micro-emulsions* (essentially microscale oil drops). When these NP assemblies were incubated with human cells the IONPs were taken up by the cells, thus allowing visualization of these cell populations thanks to enhanced contrast in the MRI images (Fig. 4.7B). While the oil-based formulations clearly promoted cell entry of IONPs, G. A. F. van Tilborg and colleagues at the University of Utrecht, Netherlands, found that the oil emulsions had an overall toxic effect, thereby limiting the practical impact of the strategy. Nevertheless, the

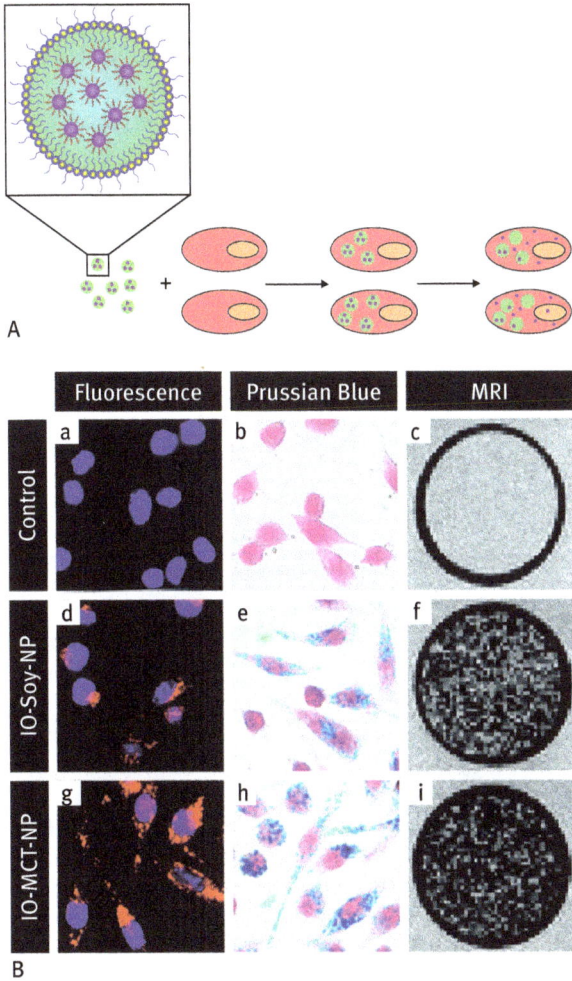

Fig. 4.7: Oil-encapsulated iron oxide nanoparticles for magnetic resonance imaging (MRI). **A:** Experimental scheme: the iron oxide nanoparticles are encapsulated within oil drops (i.e. micro-emulsions); following incubation with cells, the oil drops containing the iron oxide nanoparticles insert into the cells and subsequently release the nanoparticles in the cell interior. **B:** Results obtained for different formulations. **a–c:** control (no iron oxide nanoparticles; **d–f:** iron oxide nanoparticles embedded within soy oil micro-emulsion; **g–i:** iron oxide nanoparticles embedded within medium chain triglyceride oil (MCT) micro-emulsion. The fluorescence microscopy images (left column) depict the nuclei in blue, while the red fluorescent areas correspond to fluorescently-labeled nanoparticles taken up by the cells; Prussian blue (middle column) stains the iron nanoparticles within the cells; the MRI images (right column) show cell populations in which the iron oxide nanoparticles attenuated the magnetic resonance signals (i.e. darker areas). Reprinted with permission from van Tilborg et al., *Bioconjugate Chem.* **23** (2012), 941–950, © 2012 American Chemical Society.

experiments summarized in Figure 4.7 underscore the fact that IONPs can be transported into cells through diverse chemical avenues and successfully employed in MRI applications.

Localized heating of the NP environment through application of alternating magnetic fields (also referred to as *magnetic hyperthermia*) is one of the most important and therapeutically-promising phenomena associated with IONPs. This well-known effect stems from heat dissipation when the direction of the iron's electron spins is continuously inverted by an alternating magnetic field, and it provides a powerful therapeutic tool. For example, IONPs can be targeted to tumors by chemical means (surface-display of tumor recognition elements) or physical strategies (i.e. an external magnet directing the nanoparticle to the tumor area). An alternating magnetic field can then be applied to destroy the tumor via localized magnetic hyperthermia (Fig. 4.8).

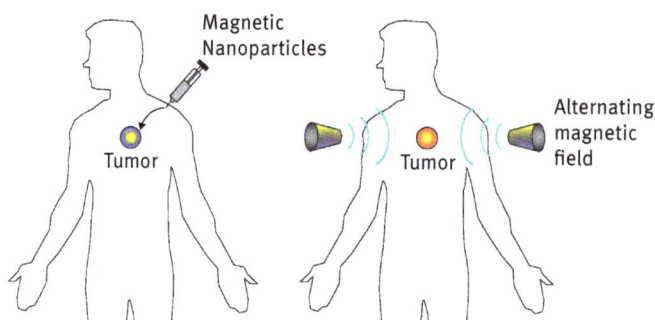

Fig. 4.8: Anti-tumor therapy using magnetic hyperthermia. Iron oxide nanoparticles are injected into the body and accumulate in a tumor area by chemical or physical means. An externally-applied alternating magnetic field subsequently induces localized heating of the nanoparticles at the tumor site through magnetic hyperthermia, destroying the tumor.

As expected, IONP-assisted cancer therapy has attracted intense interest in recent years. Several central issues in this field are illustrated in Figure 4.9. A major impediment of IONP delivery to their site of action, outlined by P. Couvreur and colleagues at Universite Paris-Sud, France, concerns inhibited transport within blood vessels (Fig. 4.9A). Upon injection into the bloodstream, IONPs, like other foreign objects, are quickly coated with a layer of *opsonins* – a class of blood-plasma proteins. Opsonin coating triggers a prominent physiological clearance mechanism, relying on capture by opsonin-binding antibodies present on the surface of macrophage cells ("Kupffer cells") within epithelial tissues (Fig. 4.9A). To avoid this clearance route and extend blood circulation times, IONPs can be coated with *polyethylene glycol (PEG)*, a commonly used biologically-inert polymer endowing "stealth" properties to the IONPs (Fig. 4.9B). Subsequent accumulation of IONPs in tumors can then be achieved ei-

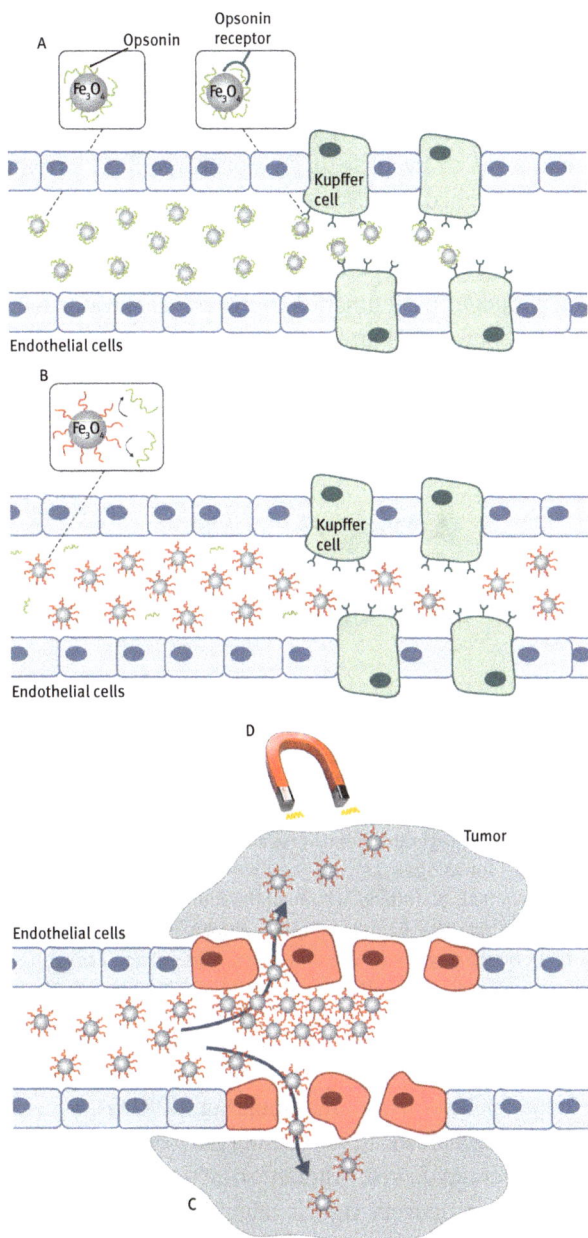

Fig. 4.9: Iron oxide nanoparticles in cancer therapy. **A:** The nanoparticles are coated with opsonins and consequently captured by anti-opsonin antibodies on epithelial cells. This process inhibits nanoparticle transport in blood vessels. **B:** Coating the nanoparticles with polyethylene glycol makes them biologically-inert. **C:** The nanoparticles can reach the tumor due to the high leakage of blood vessels around tumors; **D:** the iron oxide nanoparticles can be targeted via application of an external magnetic field. Reprinted with permission from van Reddy et al., *Chem. Rev.* **112** (2012), 5818–5878, © 2012 American Chemical Society.

ther through the intrinsic permeability and leakage of blood vessels around tumors (Fig. 4.9C), or by physical targeting via an externally-applied magnetic field (Fig. 4.9D).

Other biological applications of magnetic hyperthermia have been proposed. Figure 4.10 portrays an experiment in which localized heating induced by IONPs in an alternating magnetic field was employed as a vehicle for "remote controlled" enzyme activation. The experimental setup, designed by S. Daunert and colleagues at the University of Miami, comprised of enzyme molecules and IONPs embedded within a rigid hydrogel (transparent water-containing silica network). The enzyme selected by the researchers (dehalogenase) is *thermophilic*, i.e. it functions optimally at elevated temperatures (around $70\,^\circ$C). Indeed, as shown in Figure 4.10, no enzymatic activity was observed at room temperature; however, on placing the sample in an alternating magnetic field enzyme activity was triggered by the increased temperature generated locally within the hydrogel by the encapsulated IONPs.

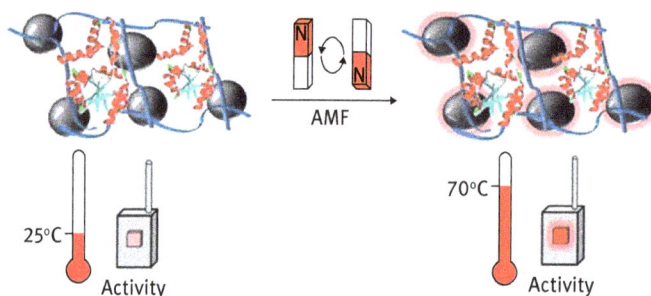

Fig. 4.10: "Remote control" enzyme catalysis via magnetic hyperthermia of iron oxide nanoparticles. Experimental scheme: the enzyme molecules (dehalogenase) and iron oxide nanoparticles are incorporated within a hydrogel matrix. At room temperature no activity of the thermophilic enzyme is observed. Upon application of an alternating magnetic field, localized heating is induced by the iron oxide nanoparticles (hyperthermia effect), increasing the temperature within the gel and activating the enzyme. Reprinted with permission from Knecht et al., *ACS Nano* **6** (2012), 9079–9086, © 2012 American Chemical Society.

The use of external magnetic fields for both spatial manipulation of IONPs as well as for induction of physical processes such as localized heating have opened up dramatic possibilities for IONPs in the emerging field of "theranostics" – combining both *therapeutics* and *diagnostics* in the same delivery vehicle. Figure 4.11 outlines typical modulations of IONPs for theranostic applications. The magnetic iron oxide core serves as the vehicle for spatial manipulations (via application of an external magnetic field), conduit for magnetic hyperthermia, and contrast agent in MR imaging. The molecular coating surrounding the magnetic core exhibits complementary functionalities, including delivery of therapeutic cargoes, display of recognition elements for cell- and tissue-targeting, and providing NP biocompatibility.

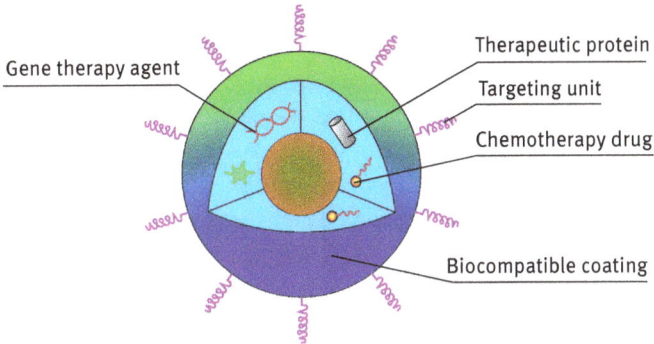

Fig. 4.11: Bio-functionalized iron oxide nanoparticles for theranostic applications. The generic scheme highlights different functional constituents of iron oxide nanoparticle assemblies exhibiting therapeutic and diagnostic applications.

Figure 4.12 presents an example of functionalized IONPs used for theranostics. The IONP system, designed by X. Chen and colleagues at the National Institutes of Health, US, enabled both delivery of a cancer therapeutic agent as well as imaging of the tumor area by MRI. Specifically, the IONPs were first treated with *dopamine*, rendering them hydrophilic (Fig. 4.12A). The NPs were then coated with *human serum albumin (HSA)* – an abundant human protein functioning here both as a vehicle for attaining biocom-

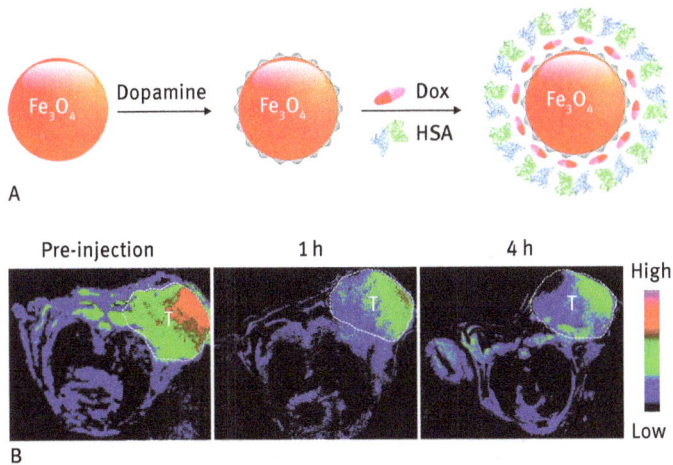

Fig. 4.12: Cancer theranostics with iron oxide nanoparticles. **A:** System design: iron oxide nanoparticles first coated with hydrophilic moieties (dopamine) and subsequently with a biocompatible human serum albumin layer and doxorubicin (Dox), an antitumor agent. **B:** Magnetic resonance images showing accumulation of the iron oxide nanoparticles in the tumor area, effectively quenching (darkening) of the MRI signals. Reprinted with permission from Quan et al., *Mol. Pharm.* **8** (2011), 1669–1676, © 2011 American Chemical Society.

patibility, and further as conduit for delivery of *doxorubicin (Dox)*, a known antitumor drug agent. This composite system was shown to target cancerous tissues, enabling their visualization by attenuating the magnetic signal in the MR image (Fig. 4.12B). Furthermore, the experimental data indicated that Dox was gradually released from the NP assemblies, leading to sustained blockage of tumor growth.

While IONPs hold great promise as a biomedical platform and, as mentioned above, are believed to be more physiologically benign compared to other inorganic NP families, there are inherent risks in their expanding uses, specifically with regard to longer-term biological/toxic effects. Studies have indeed pointed to potential hazards associated with IONPs. The formidable oxidizing capacity of iron oxide might lead to the generation of *reactive oxygen species (ROS)* – a major toxic factor implicated in various diseases and pathological conditions. The molecular layers and residues displayed on IONP surfaces are another potential source of adverse biological effects, in particular due to interference with cellular processes and signaling pathways. Several IONP species have been found, for example, to trigger inflammation pathways and induce liver damage in animal studies. It should be noted, however, that the physiological effects of IONPs are strongly dependent on their properties – size, composition, morphology, and biological functionalization. Overall, as in other novel and innovative NP-based therapeutics, research is still needed to systematically characterize the parameters affecting stability, degradation, long-term bio-distribution, and toxicity of IONPs. This knowledge would be essential to assess the future contributions of these fascinating NPs to biomedicine and therapeutics.

4.2 Titanium oxide nanoparticles

Titanium oxide (TiO$_2$ or titania) has long been utilized in industrial applications as a pigment, stabilizer in semi-solid materials such as toothpastes, and as a photocatalyst – for example catalyzing water splitting under ultraviolet (UV) light irradiation. The advent of nanotechnology in recent years has led to the creation of new TiO$_2$ nanostructures exhibiting intriguing properties. At the same time, a wealth of new applications based on TiO$_2$ nanoparticles has been reported. The prominent role of TiO$_2$ NPs in the rapidly growing solar energy field, in particular, has led to considerable research. While nanostructured TiO$_2$ can be found in various devices and technologies, this section focuses on TiO$_2$ NPs, their properties, and uses.

Numerous procedures for TiO$_2$ NP synthesis have been reported. These include chemical vapor deposition (CVD), popular with other types of semiconductor NPs, and direct oxidation of metallic titanium. Most *solution-based* synthesis schemes rely on hydrolysis of titanium ion (Ti^{4+}) precursors, usually of the type Ti(OR)$_4$, in which R corresponds to an alkyl moiety. The precursor molecules are dissolved in aqueous solutions, leading to formation of titanium hydroxide, Ti(OH)$_4$, complexes which are subsequently hydrothermally transformed (usually at high temperatures, >100°C) to titanium oxide. Similar to other NP systems, *surfactant-mediated* synthesis protocols have been introduced, making possible preparation of TiO_2 NPs coated with surfactant layers which prevent aggregation and enable dissolution in different solvents. Attaining NP uniformity is considered a major challenge in TiO_2 NP synthesis, although in some applications the size distribution and NP morphologies do not have a significant impact on the material properties. Indeed, significant progress has been made in recent years towards development of synthetic procedures for the production of monodisperse TiO_2 NPs. A common strategy produces size-tunable TiO_2 NPs by varying the concentration of the *capping agent* (usually the surfactant molecules) as well as the solvent medium in which the reaction takes place.

TiO_2 is an indirect and wide bandgap semiconductor and thus, in principle, the efficiencies of light absorption and transfer of electrons from the valence band to the conduction band are limited. However, similar to semiconductor quantum dots, the energy bandgaps in TiO_2 NPs are shifted compared to the bulk material, enabling optical tuning via modulation of the NP dimensions. Indeed, light absorption can be enhanced as the size of the TiO_2 NPs is reduced, opening up new avenues for electro-optical and photonic applications. Since bare TiO_2, which is a wide bandgap semiconductor, absorbs energy in the ultraviolet (UV) region of the electromagnetic spectrum, the applicability of the pure material is limited in optical and photovoltaic devices. Various approaches have been implemented to overcome this limitation; primary among these is doping of TiO_2 with other molecular species, referred to as sensitizers, which exhibit lower energy bandgaps. The sensitizers absorb light in the visible (or infrared) regions and the excited electrons can then be further transferred into the conduction band of TiO_2 (Fig. 4.13). Maximizing the efficiency of charge transfer from the sensitizer to the TiO_2 NPs is a crucial factor in such coupled sensitizer/TiO_2 systems. Among the parameters contributing to effective transfer are the overlap between energy levels of the TiO_2 and sensitizer and properties of the interface between the two substances.

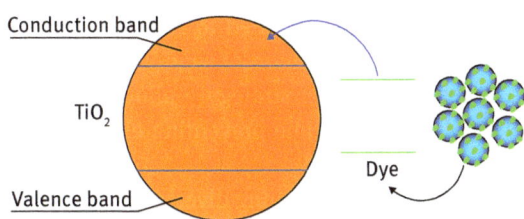

Fig. 4.13: The effect of sensitizers associated with TiO_2 nanoparticles. Excited electrons are transferred from the sensitizer dye exhibiting lower bandgap onto the conduction band of TiO_2.

Organic dyes act as sensitizers in *dye-sensitized solar cells (DSSCs)* – an active research area in which TiO_2 NPs constitute a core electron transport component. The basic DSSC configuration is shown in Figure 4.14. Light is absorbed by organic dyes (usually transition metal complexes) attached to the TiO_2 NPs. The photo-excited electrons are

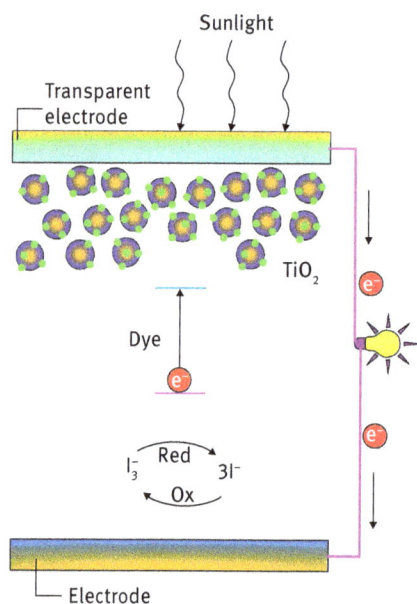

Fig. 4.14: Diagram showing the operation of a dye-sensitized solar cell (DSSC) utilizing TiO_2 nanoparticles. Electrons are excited in the dye molecules (shown as green spots attached to the TiO_2 nanoparticles). From the dye the electrons are transferred to the TiO_2 conduction band and onto the electrode. The electrons are regenerated through the redox reaction of the electrolyte (the I_3^-/I^- pair).

rapidly transferred to the conducting band of the TiO_2 and transported to one of the electrodes. At the same time, reduced/oxidized (redox) pairs (usually I^-/I_3^-, denoted as the *electrolytes*) embedded in the cell are responsible for providing the electrons to reduce the sensitizer to its initial state, simultaneously receiving the electrons from the back electrode and thereby closing the electrical circuit.

While the types of dyes and electrolytes (acting as redox couples) have been shown to play central roles in affecting the efficiency of DSSCs, the TiO_2 NPs also significantly shape solar cell performance. The *interface* between the TiO_2 NPs and the dyes, in particular, is critical, as it determines the extent of electron-transfer from the dye to the conduction band on the one hand, and the occurrence of electron-hole recombination processes which adversely influence cell efficiency on the other. Accordingly, various "interface engineering" strategies designed to reduce or inhibit electron-hole recombination have been reported. Deposition of semi-insulating layers on the TiO_2 NPs, for example, has been a common strategy – allowing charge transfer into the conduction band while at the same time preventing electron-hole recombination.

Numerous DSSC designs have been introduced containing diverse TiO_2 NP layer configurations and morphologies. Most solar cell constructs, in fact, utilize nanostructured or meso-structured TiO_2 films rather than TiO_2 nanoparticles per se. There have been reports, however, of using actual TiO_2 NPs. Figure 4.15 depicts a cell design in which an ordered array of hollow TiO_2 nanotubes was directly synthesized on a transparent electrode and integrated within a conventional DSSC setup. The TiO_2 nanotubes, synthesized by C. A. Grimes and colleagues at Penn State Univer-

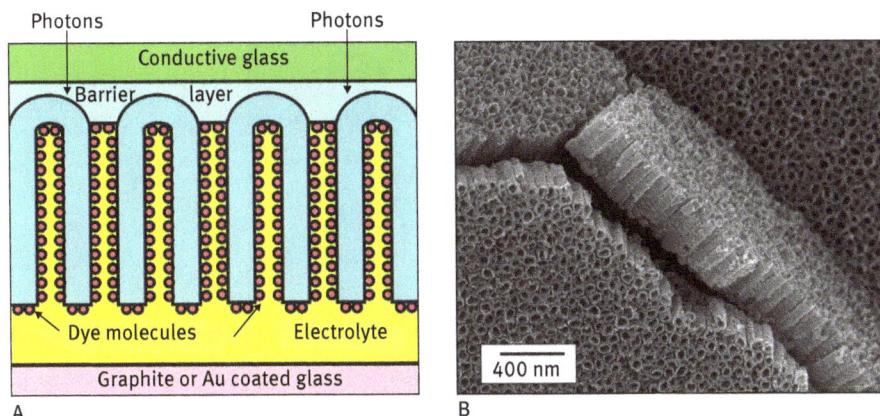

Fig. 4.15: Dye-sensitized solar cell using TiO_2 nanotubes. **A:** Diagram of the cell. The TiO_2 nanotubes (blue) are embedded within the electrolyte layer (yellow) and the organic dye molecules (sensitizer species, brown circles) are adsorbed onto the nanotube surface. **B:** Electron microscopy image of the TiO_2 nanotube array. Reprinted with permission from Mor et al., *Nano Lett.* **6** (2006), 215–218, © 2006 American Chemical Society.

sity through oxidation of a thin titanium film deposited on the electrode surface, were further doped with photon-absorbing metal-organic dyes and exhibited excellent electron transport properties as well as optical transparency – the two critical parameters which affect performance of the solar cell.

As indicated above, the use of bare TiO_2 nanostructures (i.e. those without doping or addition of sensitizer dye molecules) in solar cells is limited because of the large bandgap (in the ultraviolet spectral region), meaning potential absorption of only a small percentage of sunlight. However, the recent discovery of "black" TiO_2 NPs promises to dramatically expand the use of TiO_2 NPs in solar cells (and other applications). Black TiO_2 NPs ("black" refers to the sample appearance, which stands in contrast to white unmodified TiO_2 NPs) have been produced by hydrogen treatment of crystalline TiO_2 NPs. Remarkably, it has been found that the single-step crystallization/hydrogenation process yields NPs with much smaller bandgaps, effectively shifted to the visible region. While the exact mechanistic aspects underlying the unique photophysical properties of black TiO_2 have not been fully elucidated, the altered bandgap properties are believed to arise from formation of a "core-shell" structure in which the crystalline TiO_2 core of the nanoparticle is surrounded by partly reduced hydrogenated TiO_2 layer. Figure 4.16 shows black TiO_2 NPs synthesized by F. Huang and colleagues at the Chinese Academy of Sciences by treatment of conventional TiO_2 NPs with hydrogen plasma. Indeed, as shown in the microscopy image in Figure 4.16A, a thin amorphous layer appears to encase the crystalline TiO_2 core. The hydrogenated TiO_2 NPs were indeed found to exhibit dramatically-enhanced

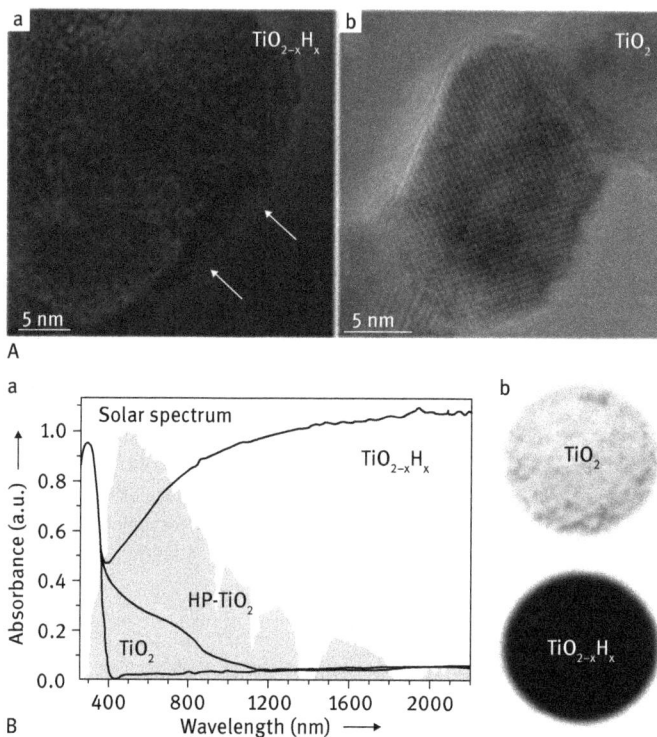

Fig. 4.16: "Black" TiO$_2$ nanoparticles. **A:** Electron microscopy images showing an amorphous layer (indicated by the arrows) presumably corresponding to hydrogenated TiO$_2$ formed around the crystalline TiO$_2$ nanoparticle core. **B:** Physical appearance of the TiO$_2$ nanoparticles (right), and a graph showing the enhanced absorbance of visible and infrared light by hydrogenated (black) TiO$_2$ nanoparticles. The wavelength range of the solar spectrum is shown in grey. Reprinted with permission from Wang et al., *Adv. Funct. Mater.* **23** (2013), 5444–5450, © 2013 John Wiley and Sons.

light absorption in the visible and infrared regions of the electromagnetic spectrum (Fig. 4.16B).

Photocatalysis, mostly aimed at efficient decomposition of organic pollutants, is a broad and commercially viable application in which TiO$_2$ NPs have played a central role due to their low cost, efficiency, and benign environmental impact. The mechanism of photocatalytic reactions induced by TiO$_2$ (and other semiconductors for that matter) is shown in Figure 4.17. Electrons are excited by light from the valence band to the conduction band; the photo-excited electrons (and the holes formed simultaneously in the valence band) can subsequently react with molecular species adsorbed onto the NP surface – electron acceptors can react with the electrons in the conduction band, electron donors with the holes in the valence band – resulting in chemical transformation/decomposition of the adsorbed species. This process can be a powerful means for degrading various organic pollutants adsorbed onto the TiO$_2$ NP sur-

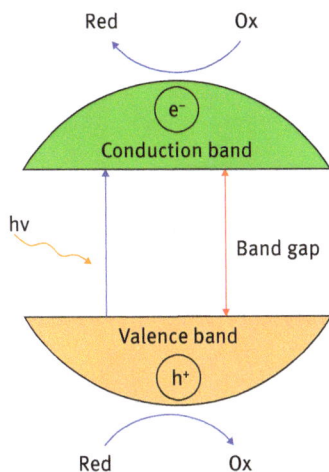

Fig. 4.17: Photocatalysis by TiO_2 nanoparticles. The energy diagram shows the physical/chemical events: photo-excitation of the electrons and subsequent redox reactions with adsorbed species.

face. The main advantage of using TiO_2 nanoparticles as photocatalysts is the significantly large surface area they provide for adsorption and interaction with the desired reactants.

TiO_2 NP photocatalysis has been aided by modifications to the NP synthesis processes. In some cases, doping the TiO_2 NPs with metal ions has been shown to enhance the catalytic activity, presumably by reducing the rate of electron-hole recombination (which minimizes charge transfer to the NP-adsorbed target molecules). Nonmetallic dopants, such as nitrogen or sulfur atoms, were also found to contribute to the photocatalytic properties of TiO_2 NPs. Those dopants are believed to stabilize reaction intermediates at the NP surface, as well as "trap" holes in the semiconductor's valence band and thus inhibit electron-hole recombination, similar to the aforementioned effect of metal dopants.

Water splitting (i.e. decomposing water to O_2 and H_2) is considered one of the most sought after "holy grails" in TiO_2-induced photocatalysis, due to the enormous potential of *hydrogen* as a clean energy source in fuel cells. The generation of O_2 and H_2 by photocatalysis is shown in Figure 4.18A. Essentially, photons with the same or larger energy than the semiconductor bandgap generate electrons and holes; the electrons in the conduction band can reduce the hydrogen in water, yielding H_2, while the holes in the valence band oxidize the oxygen atoms to produce O_2. Various TiO_2 NP structures have been investigated as water-splitting catalysts. Similar to the dye-sensitized nanoparticles discussed above, experimental schemes have explored doping the TiO_2 NPs in order to expand the energy absorbance range. Figure 4.18B, for example, shows TiO_2 nanowires decorated with *gold* nanoparticles, synthesized by Y. Li and colleagues at the University of California, Santa Cruz, which exhibited enhanced water-splitting photocatalysis. The exact mechanism for the Au-induced amplified catalytic properties has not been elucidated, the effect is believed to occur

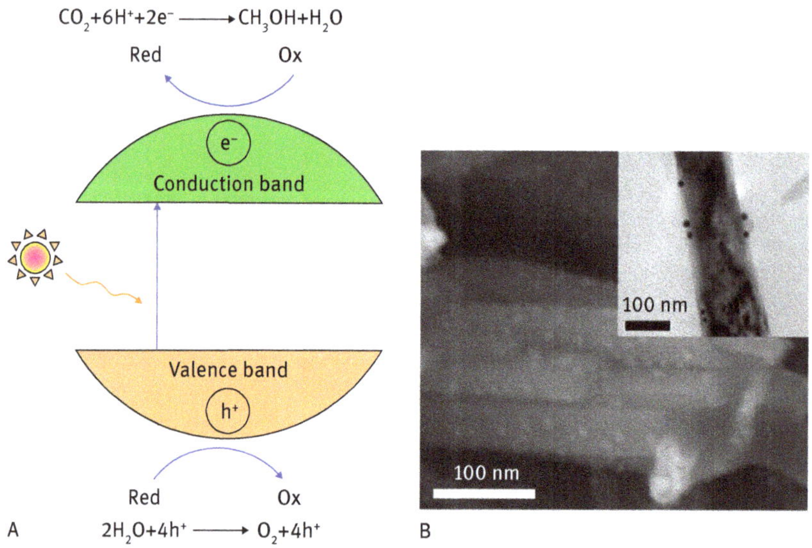

$$CO_2 + 6H^+ + 2e^- \longrightarrow CH_3OH + H_2O$$

Red   Ox

Conduction band

e⁻ (inside circle)

Valence band

h⁺ (inside circle)

Red   Ox

$$2H_2O + 4h^+ \longrightarrow O_2 + 4h^+$$

A

100 nm

100 nm

B

Fig. 4.18: Photocatalysis of the water splitting reaction achieved by TiO_2 nanoparticles. **A:** Energy level diagram depicting the redox reactions induced by the photo-excited electron and hole, respectively. **B:** Electron microscopy image showing TiO_2 nanowires decorated with gold nanoparticles (shown as white specks) used as a water splitting catalyst. Reprinted with permission from Pu et al., *Nano Lett.* **13** (2013), 3817–3823, © 2013 American Chemical Society.

via generation of high-energy surface electrons on light irradiation of the Au NPs, which can subsequently be transferred to the TiO_2 and enhance catalysis of the water splitting reactions.

Hydrogen storage research is closely related to water splitting – both aiding the fledgling efforts towards a future "hydrogen economy". Specifically, storing the hydrogen gas generated in water splitting processes (or in other chemical reactions) is considered a major technical challenge. TiO_2 NPs are particularly amenable for hydrogen storage as they exhibit a very large surface area, are known to efficiently adsorb gas molecules, and are generally stable in different environmental conditions. Figure 4.19 presents experimental results reported by J. Lin and colleagues at the National University of Singapore, demonstrating that elongated TiO_2 nanotubes exhibited significantly higher nitrogen gas adsorption compared to the bulk material, confirming the significance of the expanded surface area of TiO_2 NPs in potential hydrogen storage applications.

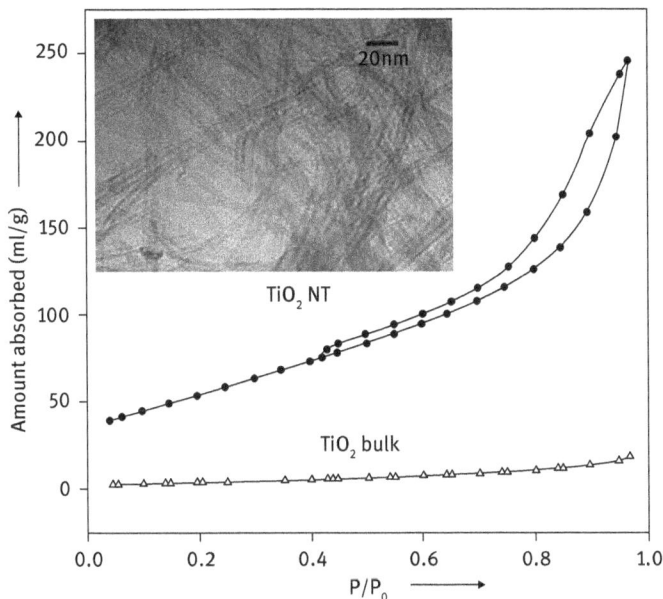

Fig. 4.19: Enhanced gas adsorption on TiO_2 nanotubes. Higher gas adsorption recorded in the TiO_2 nanotube samples (solid circles) compared to bulk TiO_2 (empty circles). The inset shows an electron microscopy image of the nanotubes. Scale bar corresponds to 20 nm. Reprinted with permission from Lim et al., *Inorg. Chem.* **44** (2005), 4124–4126, © 2005 American Chemical Society.

4.3 Zinc oxide nanoparticles

Zinc oxide (ZnO) is used in many industrial applications, primarily as a stabilizer and additive in rubbery materials, ceramics, ointments, and foods. ZnO is a wide bandgap semiconductor, and similar to TiO_2 has been employed in photocatalysis and related applications. ZnO nanoparticles frequently exhibit properties superior to other semiconducting NPs, specifically high chemical and physical stability, low cost, nontoxicity, efficient energy absorbance and light emission, and good electron transport. Representative applications of ZnO NPs are summarized below.

Synthesis of ZnO NPs is usually carried out in nonaqueous (organic) solutions by reacting zinc ions with water (i.e. hydrolysis) at high temperatures. Like other NP systems, amphiphilic protective agents are added to the reaction mixtures to stabilize the resultant NPs and prevent their aggregation. Various reaction parameters have been found to affect ZnO NP sizes and morphologies. Figure 4.20 shows monodisperse ZnO NPs of different morphologies synthesized by R. Brayner and colleagues at Universite Paris Diderot, France, by variation of the ratio between the water molecules and the Zn^{2+} ions in the reaction mixtures (i.e. the "hydrolysis ratio").

Research focusing on ZnO nanorods (NRs) is a particularly active field, since NRs have been found in many instances to exhibit better physical properties than their

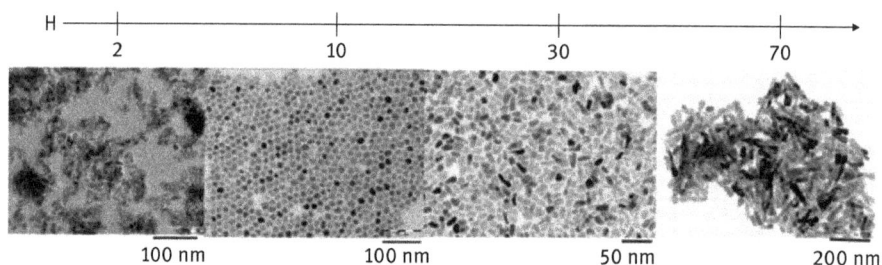

Fig. 4.20: Uniform ZnO nanoparticles of distinct shapes and sizes. Control of nanoparticle structure was achieved by modulation of the hydrolysis ratio (H $= n_{H2O}/n_{Zn2+}$). Reprinted with permission from Brayner et al., *Langmuir* **26** (2010), 6522–6528, © 2010 American Chemical Society.

spherical ZnO NP counterparts. ZnO NRs (and by extension nanowires) are mainly synthesized via directed growth initiated by surface-attached *nucleating sites* (ZnO NPs or gold NPs). Various substrates have successfully been employed as surfaces for ZnO NR growth. Figure 4.21 portrays ZnO nanorods and "nanoneedles" grown on the surface of *cotton fibers*. Specifically, R. R. Ozer and colleagues at the University of Tulsa, Oklahoma, have shown that NR morphologies and dimensions could be modulated by tuning the ratio between the concentrations of the nanoparticle "seeds" and the ZnO precursors in the growth solution.

Fig. 4.21: ZnO nanorods and "nanoneedles". ZnO nanostructures grown on the surface of cotton fibers. Nanorod morphology and dimensionalities are linked to the ratio between the concentrations of the nucleation seeds and nanorod growth-reagents: **A:** seed-to-growth reagents with a ratio of 2; **B:** seed-to-growth reagents with a ratio of 4. Reprinted with permission from Athauda et al., *ACS Appl. Mater. Inter.* **5** (2013), 6237–6246, © 2013 American Chemical Society.

Laser emission has been one of the unique physical phenomena recorded for ZnO NRs (for more in-depth discussion of the lasing properties of semiconductor nanorod lasers see Chapter 2). Practical utilization of ZnO "nanolasers", however, is limited if one aims to utilize single nanorods, and would rather be aided by macro-scale ordering and alignment of NR populations. Figure 4.22 depicts an ordered ZnO NR array displaying dramatic lasing properties. The vertically aligned and uniform NRs were produced by H. Zhou and colleagues at Universitaet Karlsruhe, Germany, by a chemical vapor deposition (CVD) technique employing patterned polymer (polystyrene) nanospheres as nucleating seeds. The ZnO NR assembly shown in Figure 4.22 generated high intensity and sharp lasing emission peaks upon photo-excitation, attesting to the potential practical applications of the material.

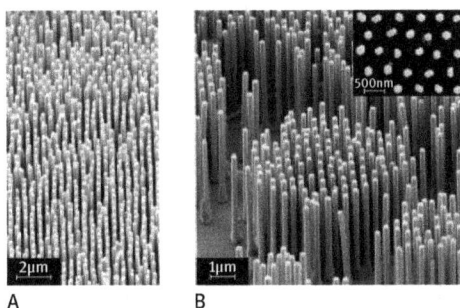

Fig. 4.22: ZnO nanorod array exhibiting lasing properties. Side-views (**A–B**) and top view (*inset*) of the ZnO nanorods grown vertically upon the substrate. The nanorod array produced intense lasing upon excitation. Reprinted with permission from Zhou et al., *Appl. Phys. Lett.* **91** (2007), 181112, © 2007 American Institute of Physics.

Like other semiconductor NPs, ZnO NPs have been widely used in photocatalysis. As discussed above in the case of TiO_2 NPs, photocatalytic reactions involving substances adsorbed onto the NP surface occur through light excitation in wavelengths corresponding to (or exceeding) the semiconductor bandgap; the electrons photoexcited from the valence band to the conduction band (and/or the holes simultaneously formed in the valence band) could react and induce decomposition of the NP-adsorbed species. ZnO, in fact, exhibits excellent *quantum efficiency* (i.e. efficiency at converting absorbed photons into electrons), and as such has often displayed better photocatalytic properties than other commonly used catalysts, such as TiO_2 NPs.

A distinct disadvantage of ZnO in the context of photocatalysis is its bandgap, which is not wide enough to cover parts of the ultraviolet spectral range in sunlight, thus reducing the overall catalytic efficiency of the material. Doping ZnO NPs with other atoms, particularly metal ions, has been successfully employed to overcome this limitation. Mg^{2+} doping, for example, was shown to enhance light absorption and photocatalytic efficiency by increasing the bandgap of ZnO NPs. This phenomenon is thought to occur because some of the zinc ions in the ZnO crystal lattice are substituted with magnesium (noting that the energy gap of MgO is significantly higher than that of ZnO). Studies have demonstrated that ZnO NPs doped with other ions also exhibited higher photocatalytic performance. Metal doping of ZnO NPs gave rise to other interesting physical phenomena. *Aluminum* doping of ZnO NPs, for example, was found to induce plasmon resonance absorbance of infrared (IR) light. The proposed mechanism for this intriguing observation was that the substitution of zinc ions in the ZnO NP lattice with aluminum generated free electrons, which consequently gave rise to light absorption through surface plasmon resonance arising from oscillation of the partly-anchored electrons on the doped ZnO NP surface.

ZnO NPs have been recently examined as ingredients in *sunscreen* lotions. The sun blocking action of ZnO (as well as TiO_2) is related to the relatively significant absorbance of ultraviolet light (due to the wide bandgap of this semiconductor). While a practical limitation for the use of many semiconductor materials in sunscreens is their *opaqueness*, the use of nanoparticles might mitigate this problem, significantly increasing the transparency of the manufactured lotions. The increased use of ZnO NPs in sunscreens (and in other cosmetics and consumer products) has led, however, to heightened awareness and discussion of the biological risks of the material, as NPs can penetrate the skin easily and thus might interfere with various physiological processes (see Chapter 7 for an in-depth discussion of the biological and cellular implications of nanoparticles). In the case of ZnO NPs there are indeed reasons for concern, primarily due to the relatively low stability of ZnO in aqueous solutions, resulting in its dissolution to zinc and oxide ions. While Zn^{2+} is normally present in the body and is involved in numerous biochemical processes, elevated levels of this ion (which might be locally incurred through dissolution of ZnO NPs) might be toxic. Indeed, several studies have pointed to toxic effects associated with high Zn^{2+} concentrations induced through cell uptake of ZnO NPs and their intracellular dissolution.

Another potential toxic factor associated with ZnO NPs in close proximity with cells and tissues is the generation of *radicals* – atomic and molecular species containing *unpaired* electrons. Radicals are widely believed to contribute to toxic processes in cells and tissues, primarily involving oxygen atoms (usually referred to as reactive oxygen species, ROS). While radicals also participate in natural metabolic and signaling processes, ROS have been implicated as toxic agents in various diseases and pathologies – in which toxicity was attributed to the high reactivity of the radicals. In this context, electrons released by photo-excited ZnO NPs might be transferred to biological molecules, generating harmful ROS. Overall, the biological risks associated with ZnO NPs require further research.

4.4 Silicon oxide nanoparticles

Silicon oxide (SiO_2) has been utilized by mankind for thousands of years. This substance is abundant in nature, most notably as quartz and other minerals in bones and microorganism shells. SiO_2 is the basic building block of widely-used materials including glass, ceramics, optical fibers, and others. Silicon has been also part of the "nanorevolution", largely through introduction of "nanoporous" silicon-based mate-

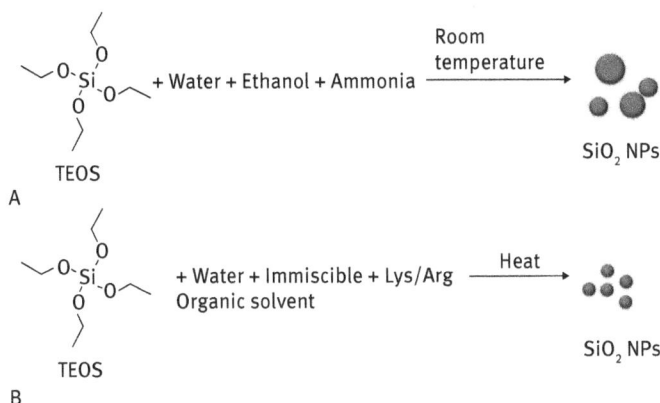

Fig. 4.23: Reaction schemes for synthesis of SiO_2 nanoparticles. **A:** The "Stöber process", generating relatively large nanoparticles. **B:** Reaction pathway using basic amino acids, for fabrication of smaller SiO_2 nanoparticles.

rials used as catalysts, molecular sieves, and in drug delivery. This section focuses on SiO_2 nanoparticles, emphasizing their useful physico-chemical properties, biocompatibility, and applications.

Different solution-phase synthesis routes for the fabrication of SiO_2 NPs have been developed. The "Stöber process", shown in Figure 4.23A, is one of the most widely implemented. In this process, the SiO_2 precursor tetraethyl orthosilicate (TEOS) is treated with ammonia (acting as a catalyst) in a solvent mixture of water and ethanol. The reaction proceeds in two steps. First, TEOS undergoes hydrolysis, in which the organic ligands bound to the silicon are removed by reaction with water molecules, followed by condensation to form the -Si-O-Si network comprising the skeleton of the SiO_2 NPs. The particles formed via this pathway are usually relatively large (>100 nm); smaller SiO_2 NPs can be synthesized by using basic amino acids such as arginine or lysine as stabilizers within an immiscible organic solvent. Subsequent hydrolysis and condensation of TEOS produces SiO_2 NPs with diameters of as low as 10 nm (Fig. 4.23B).

Significant research into SiO_2 NP properties and applications has been carried out at the interface between nanotechnology and biology, primarily due to the biocompatibility and perceived lack of toxicity of these NPs. Further aiding their biological applicability is the observation that the relatively porous matrix of SiO_2 NPs facilitates the loading of molecular guests. In addition, the SiO_2 surface can also be covalently functionalized – enabling display of recognition and targeting residues. These properties have been widely exploited in biological imaging applications. Figure 4.24 portrays applications of *fluorescent* SiO_2 NPs for cell and tissue imaging. The imaging system, designed by K. Wang and colleagues at Hunan University, China, comprised of SiO_2 NPs encapsulating fluorescent dye molecules. The researchers demonstrated cell uptake of the SiO_2 NPs, making visualization of the cells using fluorescence mi-

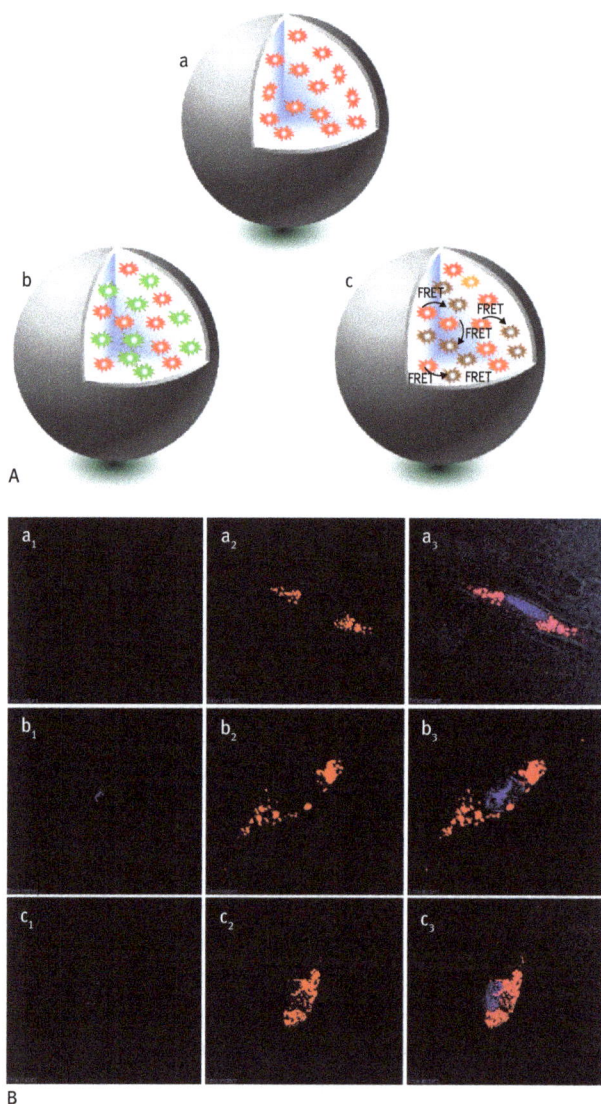

Fig. 4.24: SiO$_2$ nanoparticles encapsulating fluorescent dyes and their use in cell imaging. **A:** Schematic depiction of the nanoparticles containing: a. single fluorescent dye; b. a mixture of two dyes; and c. two dyes between which energy can be transferred by a Forster resonance energy transfer (FRET) process. Reprinted with permission from Wang et al., *Acc. Chem. Res.* **2013** *46*, 1367–1376, copyright (2013) American Chemical Society. **B:** Cell imaging using fluorescently-labeled SiO$_2$ nanoparticles. The panels in **a** correspond to images recorded after 24 hours of incubation; **b** – 48 hours; **c** – 72 hours. Left column: labeling with a nucleus-specific fluorescent dye; middle column: fluorescence emission of the dye-hosted SiO$_2$ nanoparticles; right column: combined images. The fluorescence microscopy data confirm internalization of the SiO$_2$ nanoparticles within the cells. Reprinted with permission from Liu et al., *Bioconjugate Chem.* **21** (2010), 1673–1684, © 2010 American Chemical Society.

croscopy possible. Notably, SiO$_2$ NPs can host more than a single fluorescent dye and also dye mixtures, permitting *multicolor* imaging capabilities. Moreover, pairs of dyes exhibiting energy transfer between them can be encapsulated together within a single NP – an important feature enabling greater flexibility in the choice of excitation wavelengths employed (often a critical issue in physiological and tissue environments) to achieve optimal imaging capabilities.

Mesoporous SiO$_2$ nanoparticles (MSNs) have attracted significant interest as a promising vehicle for biological applications. MSNs belong somewhat to the "mesoscale" world, as their sizes in most cases are in the range of hundreds of nanometers, however they still merit discussion in the context of this section in light of their chemical and biological properties which are dependent on their nanoscale characteristics. The defining feature of MSNs is their *porosity* – providing substantial internal surface area and volume, and thus making the particles available to load large quantities of molecular cargoes. MSNs have additionally been promoted as an effective platform for delivery of biological and therapeutic substances to cells and tissues because of the diverse chemical functionalization routes of the SiO$_2$ framework and its biocompatibility. Furthermore, the porous morphology could provide a "shielding mechanism", protecting the embedded molecular guests from physiological degradation as the NPs are transported to the site of action inside the body.

MSNs in have been examined as conduits for *gene delivery*, a highly sought, albeit challenging therapeutic goal. Figure 4.25 illustrates two distinct gene-transport MSN systems prepared by D-H Min and colleagues at the Korea Institute of Science and Technology. The researchers created two types of MSNs, comprising of small and large pores, respectively, and coated them with a layer of positive residues, designed to bind DNA molecules through electrostatic attraction. Interestingly, the MSNs exhibiting small, nanometer-scale pores were less effective gene carriers, since the DNA molecules could not enter the pores, adsorbed instead upon the MSN surface and degraded by DNA-digesting enzymes before reaching their cellular targets. However, when the MSNs had larger pores, gene delivery was more successful due to protection of the DNA molecules inserted deep within the silica pores (Fig. 4.25B).

Figure 4.26 depicts other MSN morphologies. These NPs were prepared by C. J. Brinker and colleagues at the University of New Mexico via a synthetic method termed "evaporation-induced self-assembly" (EISA). This technique involves mixing an SiO$_2$ precursor with amphiphilic surfactants; the surfactant/SiO$_2$ constituents self-assemble into small spherical structures (i.e. "micelles") upon slow evaporation of the solvent, ultimately producing organized mesoporous NP structures. Inclusion in the reaction mixtures of different surfactants in different mole ratios yielded diverse morphologies, such as "onion-shaped" lamellar structures and domain formation within the MSNs.

A central issue in MSN research concerns devising mechanisms for controlled release of the molecular cargo transported within the particles when they reach their physiological targets. Numerous strategies have been developed to achieve this goal

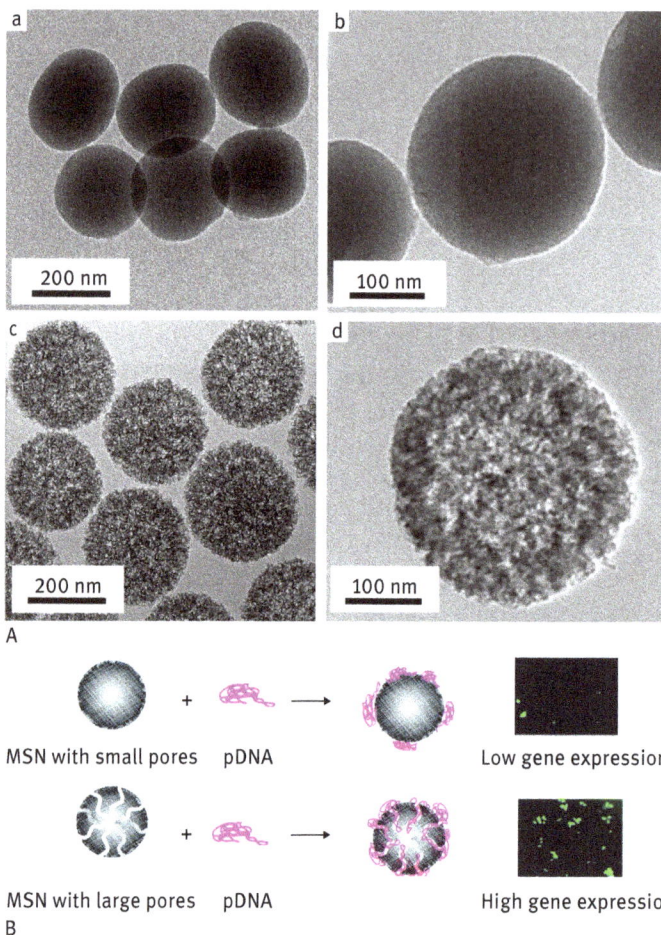

Fig. 4.25: Mesoporous silica nanoparticles (MSNs) for gene delivery. **A:** Electron microscopy images showing MSNs exhibiting small pores (and corresponding smooth surface area, **a–b**), and "swelled" MSNs containing larger pores (**c–d**). **B:** Illustration showing usage of the two MSN types for drug delivery. The small-pore MSNs do not allow insertion of DNA molecules, which instead are attached to the particle surface whereby they are enzymatically degraded. The large-pore MSNs allow encapsulation of the DNA deeper within the particle. The DNA molecules are thus more protected, giving rise to efficient delivery. Reprinted with permission from Kim et al., *ACS Nano* **5** (2011), 3568–3576, © 2011 American Chemical Society.

and only a few examples can be discussed here; in general, chemical modification of MSNs has been the preferred approach for endowing controlled release capabilities to the particles. Key aspects of controlled release mechanisms include the choice of external stimuli triggering release, feasibility and efficiency of the release processes in actual physiological scenarios, and the extent of target selectivity.

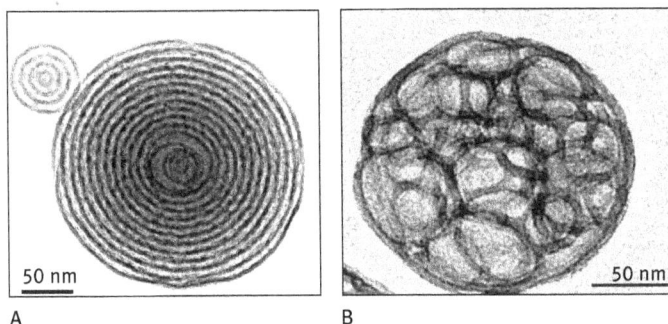

Fig. 4.26: Mesoporous silica nanoparticles. Electron microscopy images showing distinct morphologies produced via evaporation-induced self-assembly (**A**), and solution-based self-assembly (**B**). The strikingly different nanoparticle organizations were ascribed to the presence of different surfactants and reactant concentrations. Reprinted with permission from Tarn et al., *Acc. Chem. Res.* **46** (2013), 792–801, © 2013 American Chemical Society.

Figure 4.27 depicts an elegant MSN system in which temperature increase was the triggering event for cargo release. The "tailor-made" particles, produced by R. Martinez-Manez and colleagues at the Universitat de Valencia, Spain, were functionalized with an amphiphilic molecule (alkylsilane), which in turn enabled further coating of the particles with a hydrophobic paraffin layer (Fig. 4.27A). When the temperature was increased to above the melting point of paraffin, the hydrophobic paraffin capping layer disintegrated, releasing the entrapped guest molecules. To demonstrate the therapeutic potential of the triggering mechanism, the researchers released a MSN-embedded fluorescent dye within living cells through raising the temperature from 37 °C to 42 °C (Fig. 4.27B).

Acidity (e.g. *pH change*) has been another external stimulus for controlled release of therapeutic cargo from MSNs. pH is, in fact, a pertinent parameter for targeted delivery applications because acidity varies both among different tissues as well as in disease conditions. In particular, it is known that cancer cells induce higher acidity in their environment – thus opening the way for possible therapeutics through pH-triggered controlled release of drugs in tumor areas. Moreover, another fundamental physiological parameter underlying pH-based release mechanisms has been the realization that MSNs (like many other foreign substances) enter the cells via transport to the *lysosomes* (i.e. *endocytic pathways*), which exhibit high internal acidity. Accordingly, pH-sensitive release would be an efficient cell-targeting approach, since the pH outside the cell (e.g. in the bloodstream) is neutral.

Figure 4.28 illustrates such a pH-based release mechanism in which proteins adsorbed within MSN pores were delivered across cellular membranes and then released inside the cell. In the experimental scheme, demonstrated by V. S. Y. Lin and colleagues at the Iowa State University, a representative enzyme (cytochrome-C) was encapsulated in the MSN pores by electrostatic interactions between the enzyme and

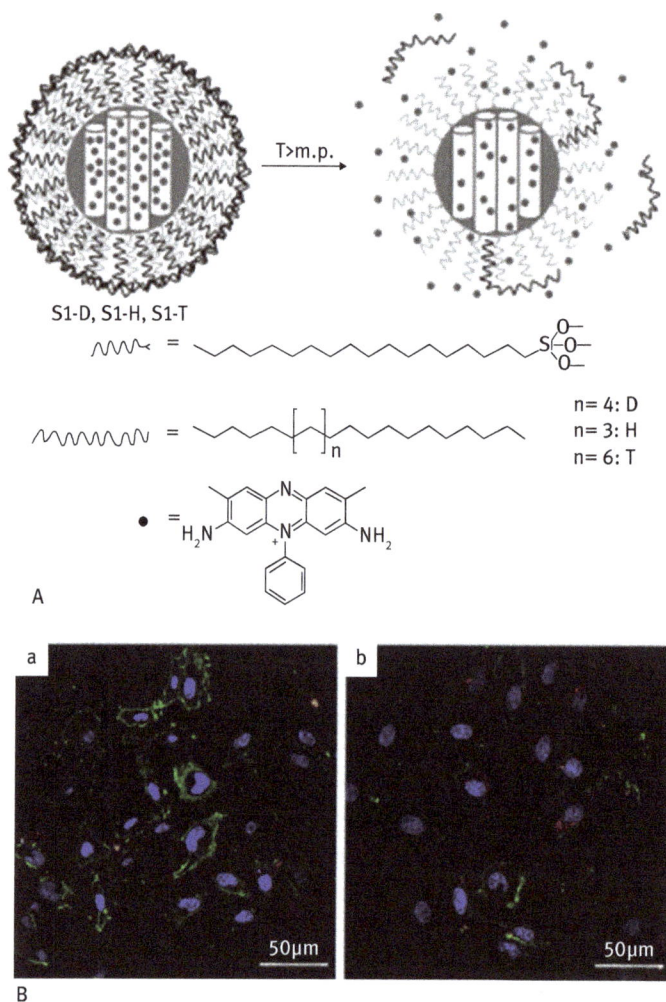

Fig. 4.27: Temperature-triggered release mechanism of molecular cargo within mesoporous silica nanoparticles. **A:** The experimental scheme: mesoporous silica nanoparticles encapsulating a fluorescent dye (depicted as black circles and representing a molecular guest) are coated with a hydrophobic layer (alkylsilane) and paraffin. On increasing the temperature the paraffin cap disintegrates, resulting in release of the fluorescent dye. **B:** Confocal fluorescence microscopy images showing temperature-triggered intracellular release of a fluorescent dye modeling a guest cargo. **(a)** 37 °C – no release of the red fluorescent dye; **(b)** 42 °C – the dye is released inside the cells, resulting in red staining of the intracellular space. The cell nuclei are stained by a blue dye to help identify cell positions, and a green dye marks the plasma membrane of the cells. Reprinted with permission from Aznar et al., *Angew Chem.* **123** (2011), 11368–11371, © 2011 John Wiley & Sons.

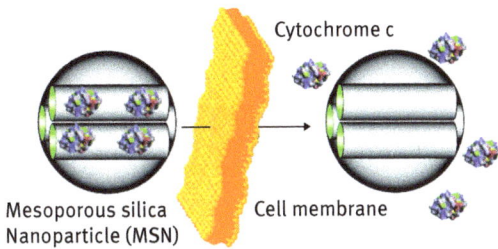

Fig. 4.28: pH-triggered release from mesoporous silica nanoparticles. Cytochrome-C molecules are embedded within the nanoparticle pores by electrostatic attraction. Acidic pH in the intracellular compartments minimizes the electrostatic interactions, resulting in release of the encapsulated cytochrome-C guest molecules. Reprinted with permission from Slowing et al., *JACS* **129** (2007), 8845–8849, © 2007 American Chemical Society.

pore walls. Importantly, the difference between the (neutral) pH in the extracellular space and the acidic pH within cell compartments associated with cell transport (i.e. endosomes, lysosomes) led to a change in the electrostatic charge on the MSN surface. As a consequence, the attraction between the enzyme molecules and the MSN pores became weaker, thereby triggering their release. The study highlighted in Figure 4.28 is additionally noteworthy because it demonstrates that MSNs could be an effective platform for delivering biological molecules *across physiological barriers* (i.e. the cell membrane).

A somewhat more complex system demonstrating a pH-triggered release is depicted in Figure 4.29. The design concept, developed by J. I. Zink and colleagues at the University of California, Los Angeles, is based on the display of pH-sensitive "molecular rods" on the MSN surface. The rods comprised of an elongated carbohydrate chain displaying positively-charged amine residues at two locations close to the rod "base", and a nitrogen atom close to the rod terminus which can be *protonated*, depending on the pH of the solution (Fig. 4.29A). As shown in Figure 4.29, the rods attached to the pore surface were further partly coated with pumpkin-shaped "molecular beads" which, importantly, were negatively-charged. Accordingly, in neutral pH the beads were located at the lower ends of the rods – electrostatically attached to the positive amine residues. In this configuration the bulky beads inhibited release of the entrapped substance, as they were positioned right at the MSN surface blocking the pore "exits" to the solution. However, in (slightly) acidic environments, the nitrogen at the top of the rod became protonated and thus positively-charged; the negative beads were consequently repositioned and moved towards the charged nitrogen, "opening" the pores and allowing cargo release.

Enzyme-triggering is a popular controlled-release mechanism in MSNs, primarily because of its intrinsic specificity (through enzyme-substrate recognition) and biological relevance. Figure 4.30 illustrates such an enzyme-triggered release system, in which the MSNs were functionalized with chemical moieties which could be

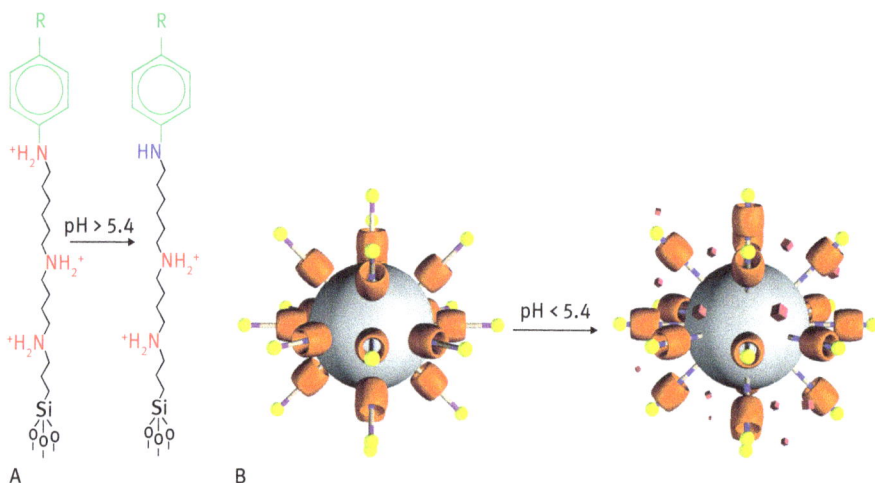

Fig. 4.29: pH-triggered release of cargo from mesoporous silica nanoparticles by a "molecular motion" mechanism. **A:** Structure of the molecular rod; note in particular protonation of the nitrogen close to the aromatic moiety in low pH. **B:** Schematic depiction of the pH-release mechanism. In neutral pH (*left*) the negatively-charged "beads" (shown in orange) placed around the molecular rods (displaying the yellow tips) are located close to the nanoparticle surface, thereby blocking cargo release from the nanoparticle pores. In acidic pH (*right*) the top nitrogen within the molecular rods is protonated, pulling the beads "upwards", thereby unblocking the pores and releasing the cargo. Reprinted with permission from Angelos et al., *JACS* **131** (2009), 12912–12914, © 2009 American Chemical Society.

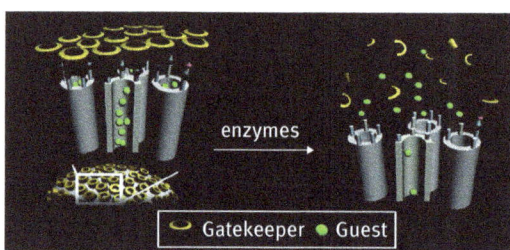

Fig. 4.30: Enzyme-triggered release of molecular guests within mesoporous silica nanoparticles. Cyclodextrin (CD) residues (shown as yellow rings) are attached to the nanoparticle surface blocking the pores and preventing release of the encapsulated cargo (left). Enzyme catalysis degrades the CD "gatekeepers", resulting in pore opening and cargo release. Reprinted with permission from Park et al., *JACS* **131** (2009), 16614–16615, © 2009 American Chemical Society.

recognized (and digested) by enzymes. Specifically, C. Kim and colleagues at Inha University, Korea, functionalized the porous surface of the MSNs with cyclodextrin (CD) moieties. CD is a bulky, toroid-shaped sugar residue acting in this system as a "gate-keeper", preventing release of the MSN-encapsulated guest molecules. The researchers, however, showed that two common enzymes could react with the CD-functionalized MSNs, degrading the CD "pore-blockers" and inducing release of the embedded molecules: *amylase* which specifically binds and hydrolyzes cyclodextrin, and *lipase* which recognizes and cleaves ester bonds (present here within the "arm" connecting the CD to the MSN surface).

Light activation is another external triggering strategy for controlled release of MSN-entrapped molecules. Figure 4.31 portrays such a mechanism. This MSN assembly, designed by J. I. Zink and colleagues, was functionalized with light-sensitive *azobenzene* molecules. The "azo" unit (benzene ring attached to the N=N bond; see Fig. 4.31) undergoes back-and-forth transformations between the "cis" and "trans" positions when illuminated (i.e. *photo-isomerization*). Initially, the azo-containing molecules covalently attached to the MSN surface (and close to the pore opening) do not allow passage of the encapsulated molecules. However, while the bulky benzene residues constitute effective blockers of the MSN pores, their light-induced flipping (or "wagging") allowed slow release of guest molecules.

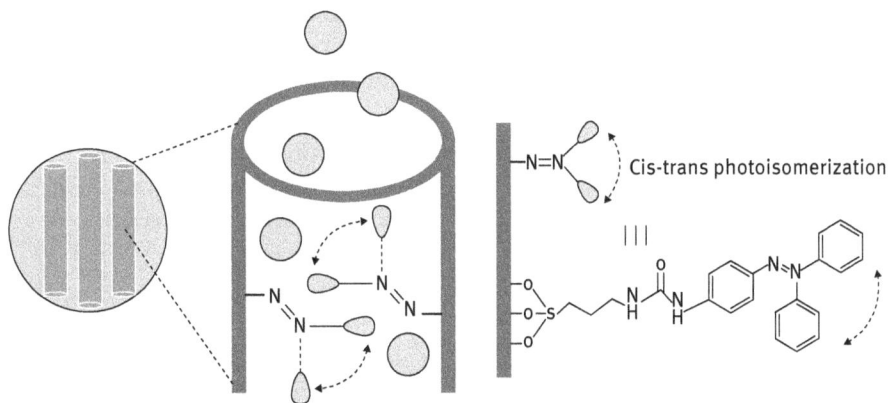

Fig. 4.31: Light-activated release of encapsulated guests in mesoporous silica nanoparticles. The azo-containing residues initially block the pore entrance, preventing cargo release. Light-induced flipping of the azo unit (the benzene ring) enables passage of encapsulated guests and their release from the pores. Reprinted with permission from Lu et al., *Small* **4** (2008), 421–426, © 2008 John Wiley & Sons.

Despite the perceived nontoxicity of SiO_2 NPs, considerable research has been carried out with the goal of assessing their cellular impact and biological safety. Indeed, concerns have been raised regarding the long-term physiological effects of SiO_2 NPs. *Silicosis*, in particular, is a debilitating lung disease associated with exposure to SiO_2 dust. There have been, however, no studies demonstrating a definitive causative link between silicosis and SiO_2 NPs. Other risk factors need to be considered as well. Based on in vitro studies, it is believed that SiO_2 NPs might induce cell membrane damage through electrostatic affinity between negatively-charged domains within the SiO_2

framework (such as SiO^-) and positive moieties in membranes (for example the abundant choline unit in many phospholipids comprising membrane frameworks).

While the majority of reported applications of SiO_2 NPs have been biological, other potential uses have been proposed, taking advantage of the hosting capabilities of the particles. An intriguing application of SiO_2 NPs as a conduit for *white light generation* is shown in Figure 4.32. Fabrication of white light emitters (also referred to as "white fluorescence" emitters) is, in fact, quite a challenging task for chemists. Essentially, white light emitters include (at least) three different dyes producing blue, green, and orange (or yellow) light, respectively; mixing the three dyes yields white light. The dye concentrations and physical distributions of the emitters need to be carefully optimized to achieve a balanced white light and prevent photophysical processes such as energy transfer among the dye molecules. SiO_2 NPs constitute, in principle, an excellent platform for white light generation due to their intrinsic transparency and the possibility of encapsulating different guest molecules within the same particle.

The white light emitting system designed by P. Audebert and colleagues at the Ecole Normale Superieur de Cachan, France, consisted of SiO_2 NPs doped with two fluorescent dyes of distinct colors: *naphthalimide* (blue) and *tetrazine* (yellow-orange). Importantly, the researchers showed that changing the mole ratio between the two dyes encapsulated within the SiO_2 NPs and tuning the preparation protocols generated three colors: blue (free naphthalimide), green (naphthalimide forming excited dimers, or "excimers" within the NPs), and yellow/orange (tetrazine bound to the NP surface). As demonstrated in Figure 4.32, different emission colors including white light could be produced from SiO_2 NPs of different dye combinations.

Encapsulation of fluorescent dyes within SiO_2 NPs has been carried out via other means, producing composite SiO_2 NPs exhibiting interesting photophysical properties. Figure 4.33, for example, outlines an elegant method for synthesis of dye/SiO2 core-shell NPs exhibiting extremely bright and tunable fluorescence. The two-step synthesis procedure, developed by U. Wiesner and colleagues at Cornell University, was based on forming the NP "core" using SiO_2 precursors which were covalently attached to the desired fluorescent dye; the "shell" was subsequently assembled by conventional NP growth processes, such as the Stöber route (Fig. 4.23, above). The NPs produced using this approach exhibited excellent stability and brightness ascribed to immobilization of the fluorophore within the SiO_2 core and the shielding effect of the shell. Furthermore, as demonstrated in Figure 4.33, the technique allowed encapsulation of diverse dyes within the NPs, forming NPs exhibiting a broad range of colors.

4.5 Rare earth oxide nanoparticles

Rare earth elements have electrons in the *f orbitals* which are relatively far from the nucleus, endowing the atoms with distinct physico-chemical properties. Rare earth *oxides* (oxygen-containing molecules of *lanthanides, scandium,* and *yttrium*) have

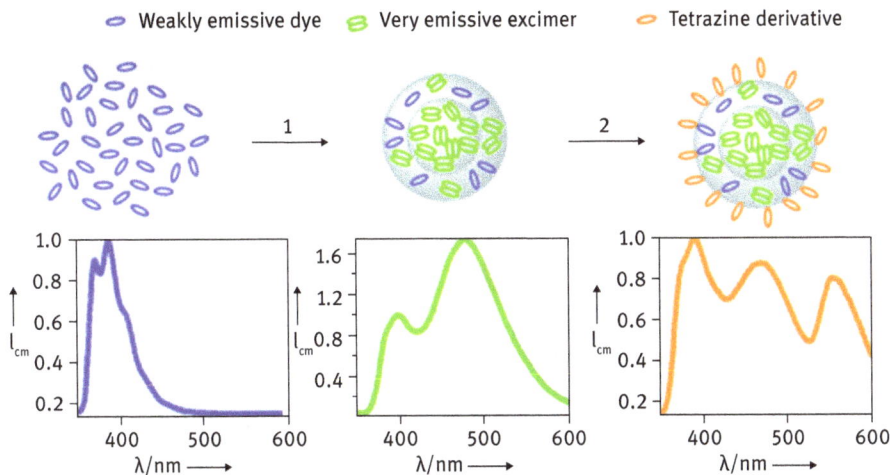

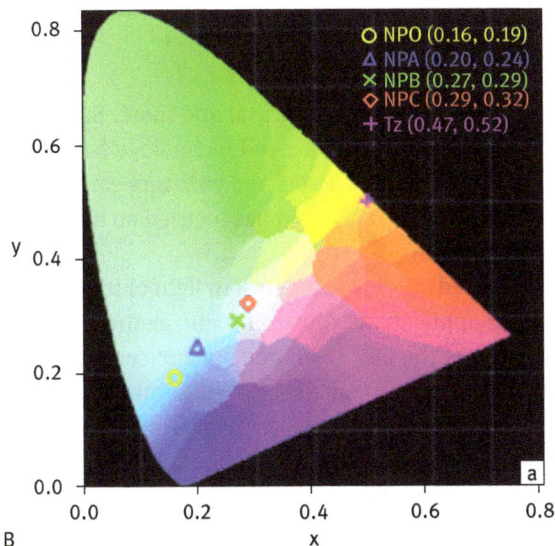

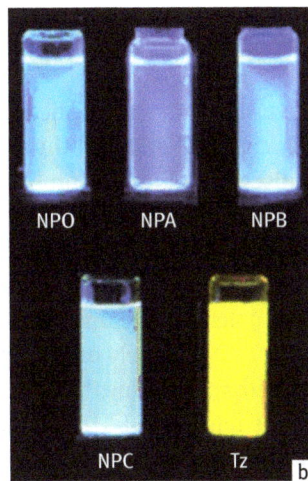

Fig. 4.32: Silicon oxide nanoparticles form a white light emitter. Different colors generated by association of two fluorescent dyes within SiO$_2$ nanoparticles. **A:** Schemes of the dye/nanoparticle assemblies and corresponding emission spectra (i.e. colors): nanoparticle-free naphthalimide (blue); naphthalimide embedded within the SiO$_2$ nanoparticles forming excimers (green); tetrazine bound on the SiO$_2$ nanoparticle surface (yellow/orange). **B:** Chromaticity diagram (**a**) and specific colors (**b**) generated by SiO$_2$ nanoparticles containing different ratios between naphthalimide and tetrazine. Note the SiO$_2$ nanoparticles denoted NPC, which contained approximately 2:1 ratio between naphtalimide and tetrazine, generating almost pure white color. Reprinted with permission from Malinge et al., *Angew. Chem. Int. Ed. Eng.* **51** (2012), 8534–8537, © 2012 John Wiley & Sons.

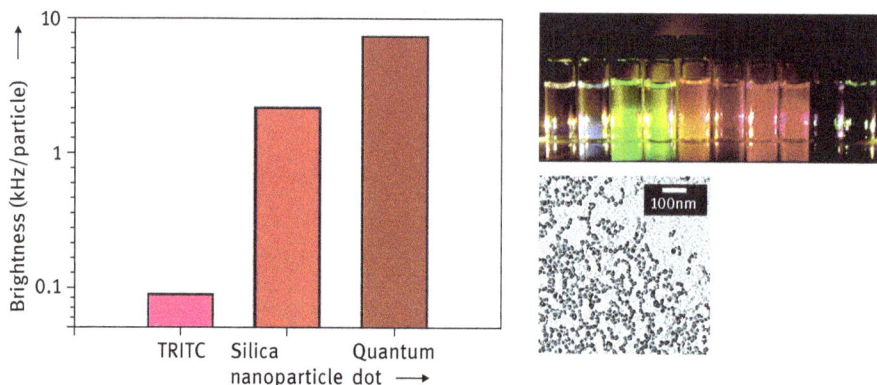

Fig. 4.33: Bright colors of core-shell silicon oxide nanoparticles containing different fluorescent dyes. **Left:** comparison of the brightness of dye-encapsulated silicon oxide nanoparticles (middle column) to the bare dye (left column) or conventional quantum dots (right column). **Right:** Encapsulation of different dyes in the core shell silicon oxide nanoparticles produces different colors; an electron microscopy image of the nanoparticles (bottom). Reprinted with permission from Ow et al., *Nano Lett.* **5** (2005), 113–117, © 2005 American Chemical Society.

been used in diverse applications as catalysts, sensors, filters, and more. Similar to other chemical systems, the advent of nanoscale science and technology has led to the discovery of unique properties and applications associated with rare earth oxide nanoparticles. In particular, significant activity in this field has focused on biological and biomedical applications, discussed below.

Cerium oxide (CeO$_2$) NPs have garnered significant interest in light of their therapeutic potential and diverse biological applications. Most synthetic methods of producing CeO$_2$ NPs rely on the prevalence of the *Ce^{4+}* oxidation state of cerium, using precursors containing this ion. An elegant micro-emulsion CeO$_2$ NP synthesis method demonstrated by J. J. Hickman and colleagues at the University of Central Florida is illustrated in Figure 4.34. The Ce^{4+} precursors were injected into the solution of a nonpolar solvent and surfactants, yielding micelles stabilized by the surfactant molecules and enclosing aqueous solutions containing the precursors. The micelles then served as "nanoreactors" for slow assembly of the crystalline CeO$_2$ NPs.

CeO$_2$ NPs display unique biocatalytic properties which correlate with the ready conversion between the Ce^{4+} and Ce^{3+} oxidation states. Cerium predominantly exists as Ce^{4+}, and accordingly most crystal lattice sites within CeO$_2$ NPs are occupied by this ion. However, a certain portion of cerium ions in the crystal adopt the *Ce^{3+}* oxidation state. As a consequence, "lattice defects" (also referred to as oxygen "vacancies") occur within crystalline CeO$_2$ NPs, since fewer oxygen molecules are required in the crystal lattice for electrostatic balance (as oxygen exhibits the negative charge compensating for the positive cerium ions; see Fig. 4.35).

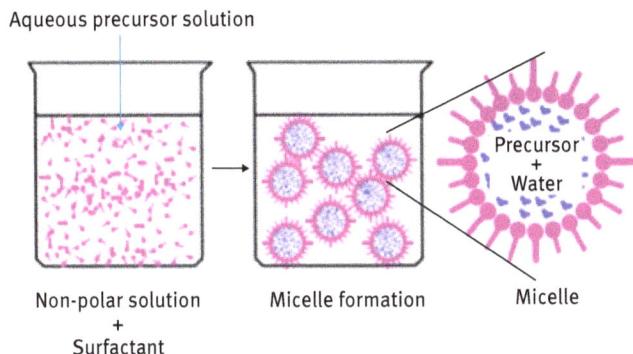

Fig. 4.34: Cerium oxide nanoparticles produced via a micro-emulsion methodology. Micelles containing aqueous solution and the CeO_2 nanoparticle precursors act as "nanoreactors" for nanoparticle formation. Reprinted from Das M. et al., Auto-catalytic ceria nanoparticles offer neuroprotection to adult rat spinal cord neurons, *Biopolymers* **28** (2007), 1918–1925, with permission from Elsevier.

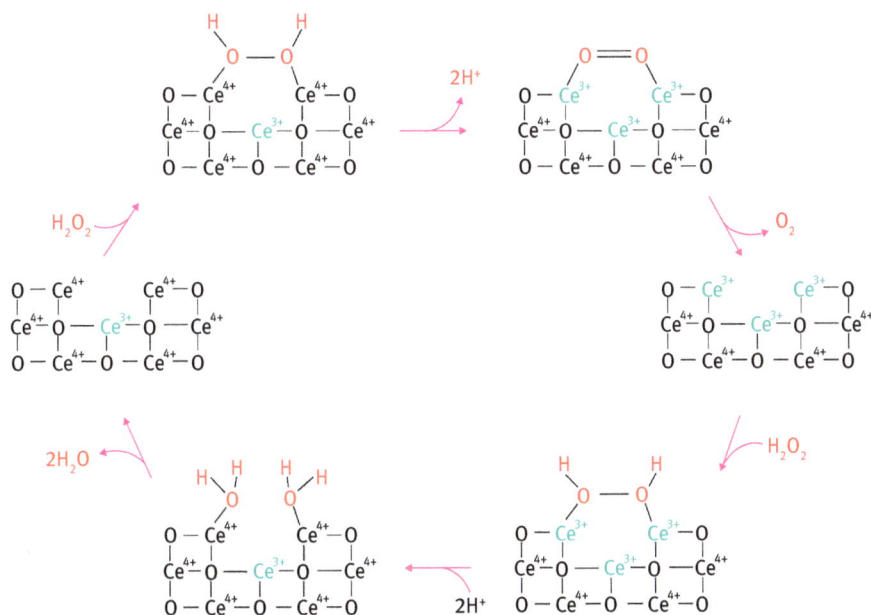

Fig. 4.35: Catalysis of hydrogen peroxide degradation by cerium oxide nanoparticles. A proposed cycle of oxidative disintegration of H_2O_2 through docking of the molecules at oxygen "vacancies" and subsequent electron transfer to Ce^{4+} at the CeO_2 nanoparticle framework. Ce^{4+} ions in the crystal lattice are indicated in black and Ce^{3+} ions are shown in green. Figure inspired by Celardo et al., *Nanoscale* **3** (2011), 1411–1420.

The oxygen vacancies and Ce^{3+}/Ce^{4+} redox transformations constitute the key factor underlying the catalytic properties of CeO_2 NPs. In particular, these vacancies, prevalent at the NP surface, promote adsorption (or "scavenging") of cell-damaging oxygen radicals (generally referred to as reactive oxygen species, ROS) and oxidizing species (e.g. molecules which tend to inflict physiological damage through electron extraction), which are then catalytically decomposed. Figure 4.35 depicts a proposed catalytic mechanism for removal of H_2O_2 by CeO_2 NPs. H_2O_2 (hydrogen peroxide) is a potent oxidant believed to contribute to oxidative stress pathways (for example neuronal cell destruction occurring in ischemic stroke). In essence, the oxygen vacancies within the CeO_2 NP framework serve as binding sites for the hydrogen peroxide molecules (or other oxygen-containing radical species). As shown in Figure 4.35, subsequent oxidation/reduction steps eliminate the bound H_2O_2 via generation of protons and water molecules released to the solution. It should be emphasized that the cycle depicted in Figure 4.35 is *autocatalytic* – the CeO_2 NPs are continuously regenerated by water and oxygen release. Furthermore, the redox process catalyzed by the CeO_2 NPs is chemically similar to enzymatic reactions responsible for H_2O_2 degradation, underscoring the "biomimetic" features and therapeutic potential of CeO_2 NPs.

While the anti-oxidant and radical scavenging activities of CeO_2 NPs as depicted in Figure 4.35 have been intensively studied in various model systems, the "redox cycling" mechanism (and other putative anti-oxidative processes involving the CeO_2 NPs) has not been unequivocally proven in vivo (e.g. in actual situations occurring in the human body). Furthermore, the toxicity of CeO_2 NPs themselves needs to be assessed, as the material could induce electron transfer and harmful side reactions with other bio-molecules. Moreover, the anti-oxidant properties of CeO_2 NPs might actually aid survival of cell species proliferating in oxidative stress conditions, for example cancer cells. Overall, however, CeO_2 NPs are considered promising candidates for new therapeutic treatments of diverse diseases and pathological conditions such as ischemic stroke, and their introduction to mainstream biomedicine is steadily progressing.

The "scavenging" mechanisms associated with CeO_2 NPs have been exploited not only as potential therapeutic platforms, but also for applications such as biosensing. Figure 4.36, for example, illustrates a colorimetric paper-based glucose sensor based on the color changes induced on H_2O_2 adsorption onto CeO_2 NPs. The sensing concept, introduced by S. Anreescu and colleagues at Clarkston University, US, utilized CeO_2 NPs coupled to the enzyme *glucose oxidase* immobilized on a paper-supported matrix (Fig. 4.36). Oxidation of glucose (the target analyte) by the embedded glucose-oxidase yielded hydrogen peroxide (H_2O_2), which in turn was adsorbed onto the CeO_2 NPs. Importantly, CeO_2 NPs containing surface-adsorbed H_2O_2 exhibit a darker orange-brown color (in comparison with the yellowish appearance of the as-synthesized CeO_2 NPs), making visual detection of glucose by the CeO_2 NPs/glucose-oxidase/paper assay possible.

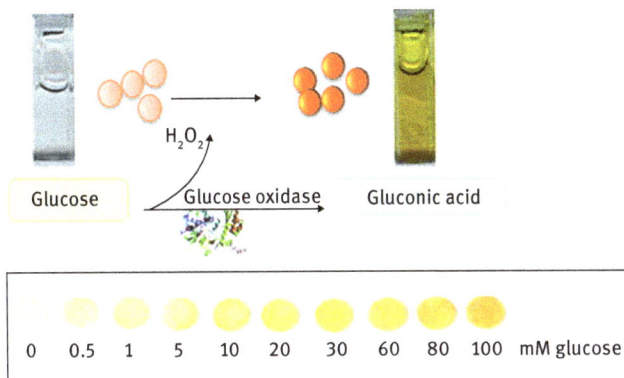

Fig. 4.36: Glucose sensing by cerium oxide nanoparticles. The sensor assembly comprises of CeO_2 nanoparticles and the enzyme glucose oxidase, both immobilized within a paper matrix. The experimental scheme (top) is based on oxidation of glucose by the glucose oxidase enzyme, producing H_2O_2 which gets adsorbed onto the CeO_2 nanoparticles. Following adsorption of H_2O_2, the CeO_2 nanoparticles undergo a visible yellow-brown color change. The scanned image shows the visible color transitions of the CeO_2 nanoparticle paper sensor in the presence of different glucose concentrations. Reprinted with permission from Ornatska et al., *Anal. Chem.* **83** (2011), 4273–4280, © 2011 The American Chemical Society.

Gadolinium oxide (Gd_2O_3) NPs have been the most effective *contrast agents* in magnetic resonance imaging (MRI), and as such play an extremely important role in biomedicine. Gadolinium is the element with the largest number of unpaired electrons in the outer shell (seven electrons), allowing efficient magnetic coupling to protons in close proximity and consequent faster relaxation of the proton magnetization. Accordingly, contrast agents containing Gd^{3+} (generally referred to as "positive-T_1" contrast agents) enhance MRI images by increasing the rate of magnetic relaxation of adjacent water protons in tissues (and thus enabling application of faster magnetic pulses and consequently higher signals). While Gd^{3+} ions by themselves are toxic, chelated (i.e. complex-forming) gadolinium is considered safe, and is the most routinely and widely used MRI contrast agent. Gd_2O_3 NPs, in this context, also offer distinct advantages as effective and medically-safe contrast agents.

"Ultra-small" Gd_2O_3 NPs (US-Gd_2O_3; in the range of 2–3 nm) have attracted particular interest as MRI contrast agents. These NPs contain relatively large numbers of gadolinium ions within the very small volume of the particle, thus exhibiting, in principle, good contrasting capacity. Figure 4.37 highlights the fact that Gd^{3+} ions are abundant on the surface of US-Gd_2O_3 NPs, thus affecting the magnetic relaxation of protons in neighboring water molecules to a greater extent than conventional water-soluble Gd^{3+} complexes. Furthermore, the size of US-Gd_2O_3 NPs (significantly larger than chelated Gd^{3+} complexes) and their stability in physiological environments leads to slower removal from the targeted tissues and longer circulation times within the bloodstream, thereby prolonging imaging times and efficiencies.

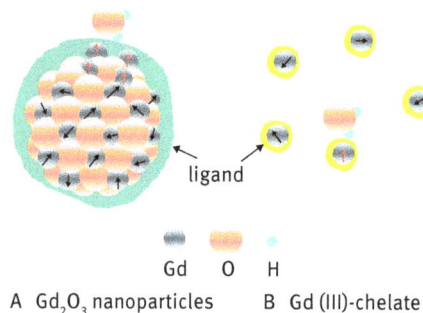

Fig. 4.37: Gadolinium oxide nanoparticles as contrast agents in magnetic resonance imaging (MRI). The large number of Gd^{3+} ions at the surface of Gd_2O_3 nanoparticles induce more effective relaxation of protons within adjacent water molecules as compared to conventional Gd^{3+}-containing complexes, which exhibit smaller surface area and lower interface with water molecules. Reprinted with permission from Park et al., *ACS Nano* **3** (2009), 3663–3669, © 2009 The American Chemical Society.

A Gd_2O_3 nanoparticles   B Gd (III)-chelate

As with other nanoparticle species, optimizing biocompatibility and endowing targeting capabilities are considered critical issues to make Gd_2O_3 NPs an effective imaging tool. Research in this field generally focuses on creating "hybrid" Gd_2O_3 NPs which comprise of additional (usually bio-active) components. A common approach has been to coat the Gd_2O_3 NP core with layers containing biological recognition elements, and/or additional reporting molecules such as fluorescence dyes. Figure 4.38, for example, shows the imaging capabilities of Gd_2O_3 NPs coated with a polysiloxane shell (a silicon-oxide matrix which further hosted fluorescent dyes). The nanoparticles, synthesized by S. Roux and colleagues at Universite Claude Bernard, Lyon, France, provided dual imaging platforms – both fluorescence and MRI enhancement. In particular, the MRI data in Figure 4.38 show that image enhancement was inversely correlated to NP diameter; consistent with the contribution of the NP/water interface to proton magnetic relaxation and signal enhancement (i.e. larger nanoparticles provide less overall surface area). The experiments further indicate that the silicon-oxide shell did not adversely affect the image enhancement capabilities of the NPs.

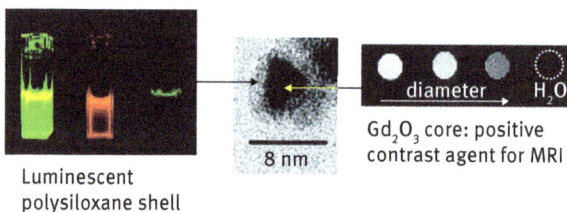

Luminescent
polysiloxane shell

Fig. 4.38: Gadolinium oxide nanoparticles coated with luminescent shells. The core-shell nanoparticles comprise of Gd_2O_3 cores and polysiloxane shells which further embed a fluorescent dye. Enhancement of magnetic resonance imaging (MRI) signal by the core-shell nanoparticles is shown in the right photograph. Greater enhancement is recorded in solutions of smaller-diameters particles, consistent with the significance of surface-to-volume ratio in affecting the magnetic relaxation of water protons and corresponding MRI signal enhancement. Reprinted with permission from Bridot et al., *JACS* **129** (2007), 5076–5084, © 2007 The American Chemical Society.

4.6 Other metal oxide nanoparticles

Other metal oxide NP systems have been developed. *Tin dioxide (SnO_2)* is a wide bandgap semiconductor material used in varied applications, including solar cells, photocatalysis, gas sensing, and others. Similar to the ZnO and TiO_2 NPs, discussed above, SnO_2 nanoparticles may have useful roles in photocatalysis and sensing applications. Indeed, SnO_2 NPs are particularly advantageous for such applications since they are chemically and physically stable over time, and exhibit high surface area – favorable for large-scale adsorption of substrates for catalytic degradation or target analytes in sensing applications (particularly gas sensing).

The operation principle of SnO_2 NP-based gas sensors is modulation of the electrical resistance of the sensor films induced by surface adsorption of gas molecules. In this context, modifying the surface morphology and surface areas of SnO_2 NPs has been found to intimately affect the performance of such sensors. Figure 4.39 shows electron microscopy images of SnO_2 NPs with octahedral morphology exhibiting higher gas detection sensitivity, presumably due to greater adsorption of gas molecules onto the flat crystalline facets of the particles. Furthermore, Z. Xie and colleagues at Xiamen University, China, who synthesized the NPs, discovered that the sensor response was lower in cases where the NPs had an *elongated* morphology (Fig. 4.39B); this observation was ascribed to the distinct atomic organization of the exposed facets within each NP configuration. Specifically, the researchers hypothesized that the elongted NP structure has a lower abundance of tin ions with "dangling bonds" – i.e. ions which are not bonded to oxygen – available to anchor gas molecules.

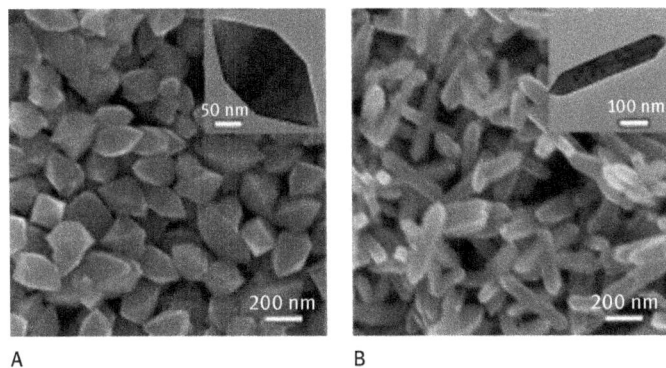

A   B

Fig. 4.39: Octahedral tin oxide nanoparticles. The elongated SnO_2 nanoparticles adsorb smaller numbers of gas molecules, presumably due to lower abundance of nonbonded Sn^{4+} available for binding. Reprinted with permission from Han et al., *Angew. Chem Int. Ed.* **48** (2009), 9180–9183, © 2009 John Wiley & Sons.

Other types of SnO_2 NP-based gas sensors have been reported. Figure 4.40 illustrates a gas sensor based on a porous matrix comprising of SnO_2 NPs coupled to glucose molecules. The glucose residues functioned here as stabilizers of the porous framework, essential for facilitating adsorption of the target analytes. This composite SnO_2 NP/glucose framework, reported by S. V. Manorama and colleagues at the Indian Institute of Chemical Technology, Hyderabad, displayed superior photocatalytic properties – attributed to the high surface area of the NPs available for adsorption, interspersed within the porous and transparent matrix.

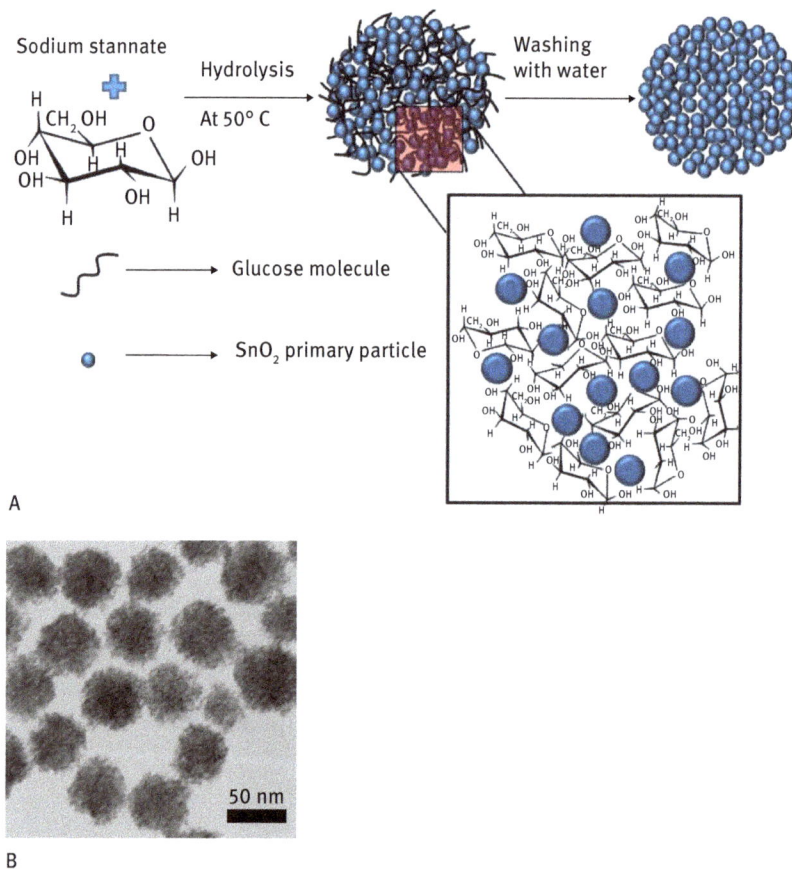

Fig. 4.40: Tin oxide nanoparticles/glucose porous assembly for gas sensing. Coupling of the nanoparticles and glucose molecules yields a porous framework promoting adsorption of gas molecules. The electron microscopy image shows the sponge-like morphology of the nanoparticles. Reprinted with permission from Manjula et al., *ACS Appl. Mater. Interfaces* **4** (2012), 6252–6260, © 2012 American Chemical Society.

SnO_2 nanowires (NWs) have also been employed as platforms for gas sensing. Interestingly, several studies have reported significant enhancement of the sensing capabilities of SnO_2 NWs which were doped with *metal* NPs. Figure 4.41, for example, illustrates a gas sensor setup constructed by A. Kolmakov and colleagues at the University of California, Santa Barbara, in which a SnO_2 NW placed between the source and drain was doped with palladium (Pd) nanocrystals. The doped SnO_2 NWs were found to be much more sensitive as sensors for oxygen gas compared to bare SnO_2 nanowires. This phenomenon was ascribed to the catalytic properties of the Pd nanoparticles (see Chap. 3), resulting in rapid disintegration of surface-adsorbed O_2 molecules and subsequent diffusion of the resultant atom products which alter the resistivity of the NWs.

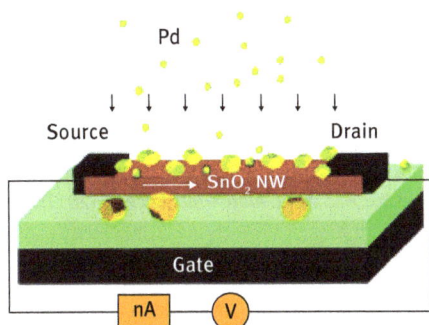

Fig. 4.41: Metal-doped tin oxide nanowire gas sensor. Doping of the SnO_2 nanowire by palladium nanocrystals is carried out following placement of the nanowire in the transistor device. The doped nanowire displayed higher sensitivity as a gas (oxygen) sensor. Reprinted with permission from Kolmakov et al., *Nano Lett.* **5** (2005), 667–673, © 2005 American Chemical Society.

Indium tin oxide (ITO) NPs have garnered significant interest due to the prominence of ITO in numerous electro-optical devices and applications. ITO is at present the main material in transparent conductive electrodes (TCEs) and transparent conductive films, essential components in solar cells, electro-optical devices, and more. ITO usage in such devices is due to the wide bandgap of In_2O_3, a semiconductor, which minimizes light absorbance in the visible spectral region (thus making ITO transparent). In addition, the "crystal defects" – usually oxygen vacancies – within the In_2O_3 lattice and distortions of crystal organization induced by doping with tin ions (Sn^{4+}) promote electron transport and contribute to the enhanced electrical conductivity of ITO.

Research focused on ITO nanoparticles partly stems from practical reasons. ITO NPs can easily be synthesized in solution via a thermally-induced reaction between indium and tin precursors, yielding particles with tunable size distributions. ITO NPs can then be deposited on varied substrates, forming uniform films. Indeed, the availability of simple, inexpensive, and scalable solution-phase NP deposition techniques such as *spin casting* (deposition of a NP-containing solution on a spinning surface, followed by evaporation of the solvent) could provide a powerful alternative for existing ITO film fabrication methods which are technically-demanding, expensive, and often environmentally hazardous. Figure 4.42 illuminates the fabrication of a highly transparent conductive film comprising ITO NPs using a polymer substance as a tem-

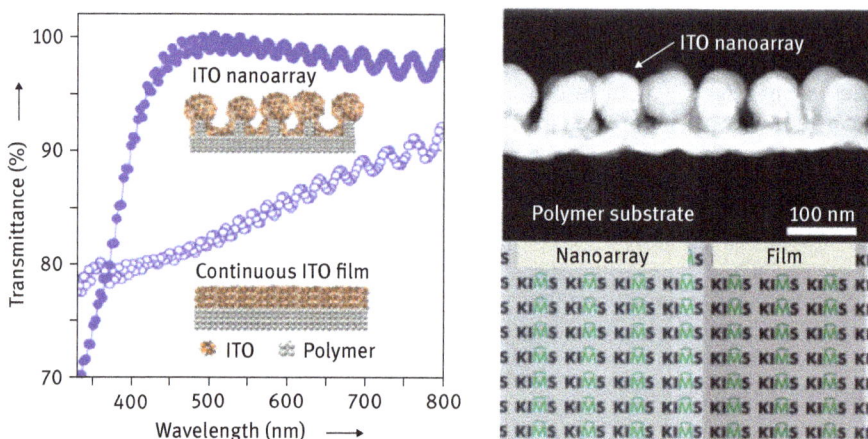

Fig. 4.42: Transparent conductive films from indium thin oxide (ITO) nanoparticles. The ITO nanoparticles were deposited on a patterned polymer substrate from the vapor phase creating an ITO nanoarray (microscopy image on the right). The transmittance graph on the left demonstrates that the nanoparticle film featured significantly higher optical transparency compared to conventional continuous ITO films. Reprinted with permission from Yun et al., *ACS Appl. Mater. Interfaces* **5** (2013), 164–172, © 2013 American Chemical Society.

plate. The films were prepared by J. Yun and colleagues at The Korea Institute of Materials Sciences by vapor deposition (sputtering) of the ITO NPs on a polymer substrate which was sculpted to include protruding "pillars". The patterned ITO NP films exhibited excellent conductivity and transparency ascribed to the "openings" between the deposited NPs.

Tunable surface plasmon resonance (SPR) has been another interesting phenomenon observed in ITO NPs. Described in more detail in Chapter 3, SPR results from interaction between light and partly-free electrons present at NP surfaces, producing in many instances visible colors of NP suspensions (depending on the wavelength of the plasmon resonance). While SPR has mostly been recorded in metallic NPs (primarily Au NPs), it is also generated in metal oxide NPs such as ITO NPs, attributed in such semiconductor NPs to free electrons excited by light to the conduction band. Interestingly, the percentage of tin in ITO NP compositions was shown to be the likely main parameter for tuning SPR in ITO NPs, essentially determining the color of the NP suspensions. Figure 4.43 depicts suspensions of ITO NPs created by S. Sun and colleagues at Brown University, displaying different colors which were dependent upon the mole percentage of tin in the particles.

Transition metal oxide NPs are among the most diverse NP families in terms of structure, chemical and physical properties, and macro-scale organization. This variety emanates from the diversity of transition metal elements themselves – their oxidation states, electronic structures, and crystal organizations. *Manganese oxide NPs* appear in diverse morphologies, in large part because of the several oxidation

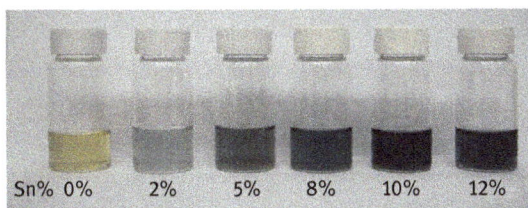

Fig. 4.43: Color of indium tin oxide (ITO) nanoparticle suspensions depends on percentage of tin in the nanoparticles. The solution color reflects the surface plasmon resonance (SPR) associated with the particles. Reprinted with permission from Lee et al., *JACS* **134** (2012), 13410–13414, © 2012 American Chemical Society.

states of manganese available. Figure 4.44 depicts MnO_2 nanowires self-assembled into a dense "free-standing" (i.e. non solid-supported) membrane. These extremely long nanowires, synthesized by L. Yu and colleagues at Guangdong University of Technology, China, were homogeneous, could be dissolved in water, and formed intertwined stable membranes after drying in a vacuum. Notably, the MnO_2 nanowire membrane could be further functionalized. For example, incorporation of *magnetic nanoparticles* within the nanowire matrix (Fig. 4.44B), gave rise to formation of a functional "magnetic membrane".

Fig. 4.44: Manganese oxide nanowires. **A:** Dense free-standing mesh of MnO_2 nanowires; **B:** MnO_2 nanowires doped with magnetic nanoparticles (shown as white specks). The doped nanowire matrix becomes magnetic. Reprinted with permission from Lee et al., *ACS Appl. Mater. Interfaces* **5** (2013), 7458–7464, © 2013 American Chemical Society.

While MnO_2 is the most abundant, naturally-occurring manganese-containing mineral, the Mn^{2+} oxidation state is actually more common in manganese compounds. MnO NPs, in particular, have displayed intriguing properties and structures. Figure 4.45A shows electron microscopy images of multipod MnO NPs, synthesized by D. Zitoun and colleagues at Universite Montpellier II, France. These unconventional NP structures, observed also in binary semiconductor NPs (see Chap. 2), are believed to occur by a two-step crystallization process. First, the multipod core is formed around nucleating seeds in solution. The subsequent anisotropic growth of the "arms" is

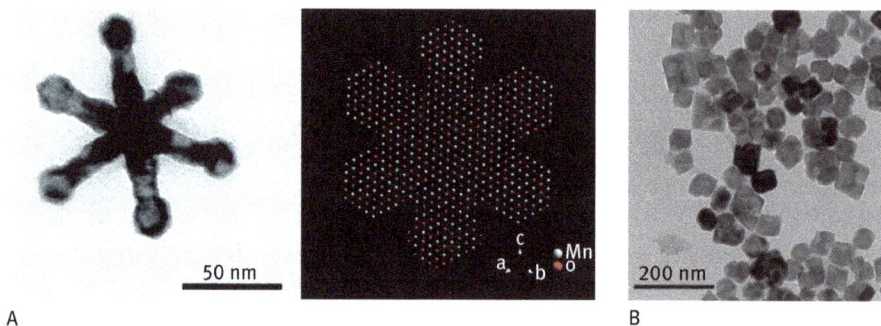

Fig. 4.45: Multipod manganese oxide nanoparticles. **A:** Electron microscopy image of MnO hexapod (left) and a schematic depiction of manganese and oxygen atoms in the crystal lattice (right); **B:** Cubic MnO nanoparticles produced following a short reagent incubation time. This result supports the mechanism in which the multipod synthesis starts with crystallization of the core, followed by growth of the arms. Reprinted with permission from Zitoun et al., *JACS* **27** (2005), 15034–15035, © 2005 American Chemical Society.

governed by differences in reactivity of the crystal facets of the core. Corroborating this hypothesis was the observation that polyhedral "cores" without protruding arms were produced following shorter synthesis duration.

Copper (cupric) oxide (CuO) NPs constitute a class of semiconductor NPs distinctive as low bandgap materials (in contrast to most metal oxides which exhibit high energy bandgaps). The *p-type* semiconductor CuO NPs exhibit useful properties, such as sensing when coupled to *n-type* semiconductor oxides (e.g. ZnO or SnO_2), related to the formation of "p-n junctions" in such constructs. The electric field which develops in p-n junctions – e.g. the interface between the p-type semiconductor contributing holes and the electron-providing n-type semiconductor – usually translates into narrower conduction bands of the semiconductor sensor leading to enhanced sensitivity following adsorption of (gas) molecules. An example of such a sensor device is shown in Figure 4.46. The sensor, developed by J. Chen and colleagues at the University of

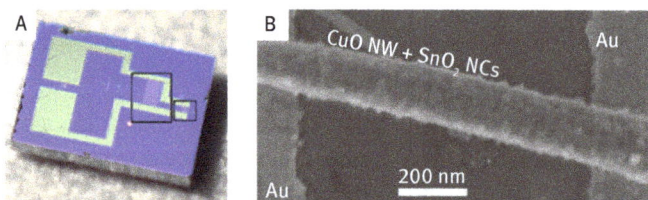

Fig. 4.46: Cupric oxide nanowire doped with tin oxide nanocrystals used in a sensor device. Doping of the p-type CuO nanowire with n-type SnO_2 nanoparticles increases the sensitivity of the sensor due to formation of p-n junctions. Reprinted with permission from Lee et al., *ACS Appl. Mater. Interfaces* **4** (2012), 4192–4199, © 2012 American Chemical Society.

Wisconsin, comprised of a single CuO nanowire doped with SnO_2 NPs (which form p-n junctions with the CuO NW). Adsorption of gas molecules onto the NW surface gave rise to greater modulation of the resistance of the NW device in comparison with un-doped CuO nanowires.

Vanadium oxide NPs have been used in various applications, from catalysts to antibacterial agents. *Vanadium pentoxide (V_2O_5) nanowires*, in particular, have been touted as promising "anti-fouling" agents, on account of their capacity to prevent formation of bacterial biofilms – the resilient impenetrable matrix formed by many bacterial species on surfaces. This inhibitory action has been ascribed to interference with bacterial communication pathways important in biofilm formation. As an example, W. Tremel and colleagues at the Johannes Gutenberg-University, Germany, demonstrated that V_2O_5 nanowires could act as anti-fouling agents (Fig. 4.47).

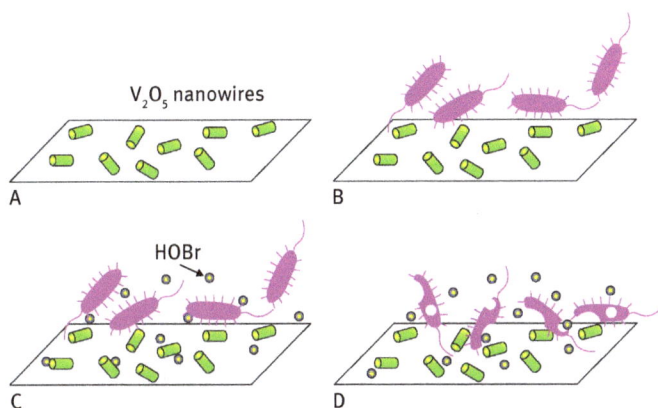

Fig. 4.47: Vanadium oxide nanowires acting as anti-bacterial agents. **A:** V_2O_5 nanowires are deposited on a surface. **B:** Bacteria near the surface aim to establish colonies. **C:** In the presence of Br⁻ and H_2O_2, the V_2O_5 nanowires catalyze production of HOBr (yellow spheres), an antibacterial agent. **D:** The bacterial cells are destroyed. Figure inspired by Natalio et al., *Nature Nanotech.* **7** (2012), 530–535.

Specifically, the V_2O_5 NWs catalyze the oxidation/reduction reaction between hydrogen peroxide (H_2O_2) and Br⁻ ions according to the equation

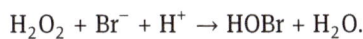

$$H_2O_2 + Br^- + H^+ \rightarrow HOBr + H_2O.$$

HOBr (hypobromous acid) is known as an effective anti-bacterial compound, as it adversely affects intercellular bacterial signaling processes which play crucial roles in generating bacterial biofilms (also known as "quorum sensing" processes). Indeed, researchers demonstrated that surface deposition of V_2O_5 nanowires and co-addition of bromide salt resulted in pronounced inhibition of bacterial adhesion and blocking of biofilm formation (i.e. the nanowires acted as anti-fouling agents).

5 Organic and biological nanoparticles

The organic and biological worlds have contributed a variety of materials to the "nanoparticle universe". Some of these nanostructures occur in natural/physiological environments (for example peptide "nanofibers"), and many are synthetic entities, including polymeric NPs or NPs built from organic compounds. For reasons of space and scope, this chapter focuses on biological and organic molecule assemblies which can be categorized as *"synthetic nanoparticles"*; many other fascinating biological, polymer, and organic molecular systems exhibiting nanostructural features cannot be discussed here and the reader is referred to the pertinent literature.

5.1 Biological nanoparticles

From proteins to deoxyribonucleic acid (DNA) to lipids, biology provides a plethora of molecules which tend to self-assemble, and as such often produce nanoscale aggregates. Various *biological nanostructures* have been identified in nature, some of which in fact are implicated in devastating diseases, such as amyloid nanofibers, which are among the pathological hallmarks of Alzheimer's disease (Fig. 5.1). Furthermore, the diverse synthetic and biochemical methods available for modifying and functionalizing biomolecules provide ample means to control the structural outcomes of biological self-assembled nanostructures, creating varied biomimetic nanostructures. This section focuses on aggregates which are more closely defined as "nanoparticles" – entities which conform to the generic classification of stand-alone structures being small, from a few to tens of nanometers.

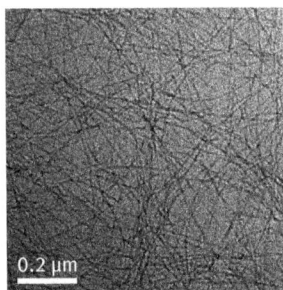

Fig. 5.1: Amyloid nanofibers. Electron microscopy image showing long fibers assembled spontaneously by beta-amyloid, the ubiquitous peptide which forms amyloid plaques in the brains of Alzheimer's disease patients.

Tubular peptide nanoparticles (also referred to as *"peptide nanotubes"*) have attracted significant interest from several angles. These fascinating morphologies depend to a large degree on the amino acid sequences of the constituent peptides (and thus can be tuned by modifying the peptide sequence). Furthermore, as these anisotropic NP structures mimic physiologically-encountered tubular and fibrillar peptide aggregates (e.g. the amyloid fibers shown in Fig. 5.1), they could help to better understand the pathologies of diseases such as Alzheimer's. Also important, peptide nanotubes and nanofibers can serve as templates for functional materials – for example as scaffoldings for fabrication of metal nanowires and nanotubes.

Figure 5.2 illustrates the use of a synthetic peptide nanotube as a template for metal nanowire formation. A short dipeptide comprising of two phenylalanine residues (i.e. an "FF" peptide sequence, using the single letter notation for phenylalanine) was shown by E. Gazit and colleagues at Tel Aviv University, Israel, to adopt elongated nanotubular morphology (Fig. 5.2A). Self-assembly of such nanotubes is presumed to occur through interactions between adjacent *aromatic* residues of amino acids such as phenylalanine and tyrosine (i.e. "π stacking"; Fig. 5.2B). The peptide nanotubes were subsequently used as templates to cast metal nanowires through embedding metal (silver) ions within the nanotubes, which generated silver nanowires after reduction (Fig. 5.2C). The peptide nanotube scaffold could be dissolved by simple enzymatic degradation, not requiring harsh conditions or environmentally harmful reagents, underscoring a potential advantage of this "biomimetic" strategy for fabrication of metallic nanowires.

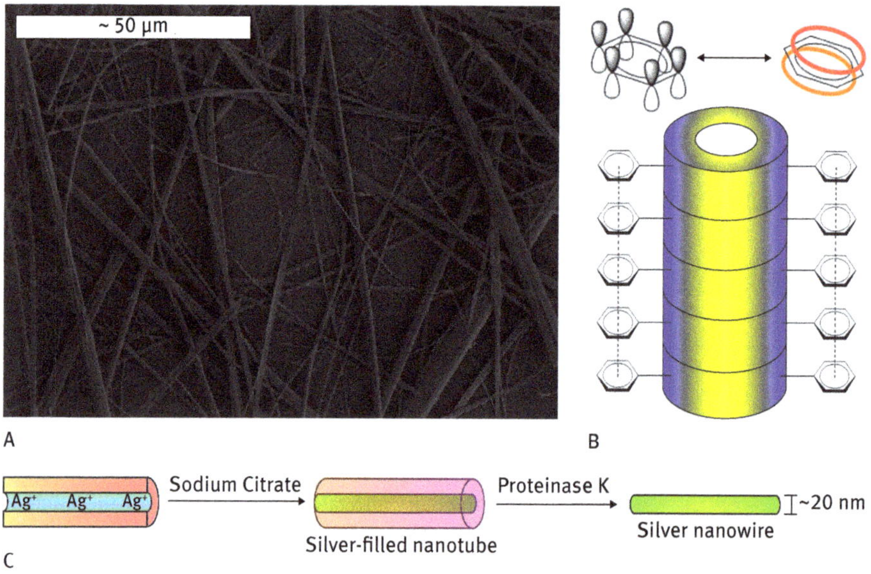

Fig. 5.2: Dipeptide nanofibers as templates for metal nanowires. **A:** Electron microscopy image of diphenylalanine (Phe-Phe) nanofibers. Image courtesy of E. Gazit, Tel Aviv University, Israel. **B:** Model of the π-stacking mechanism hypothesized to produce the nanofibers via interactions between the aromatic rings of phenylalanine. **C:** Use of the peptide nanotube as a template for construction of silver nanowires through reduction of Ag$^+$ ions embedded inside the tubes.

Short peptide motifs were shown to produce other morphologies. Figure 5.3 depicts intriguing spherical "caged" NPs comprising *diphenylglycine*. While diphenylglycine is structurally close to diphenyl*alanine* (which produced elongated nanotubes; Fig. 5.2), it exhibits a more rigid conformation due to the absence of a C-CH$_2$-C bond between the backbone and aromatic residues (giving rise to less rotational degrees of freedom; Fig. 5.3B). This structural feature is likely responsible for the formation of spherical NPs. Remarkably, E. Gazit and colleagues at Tel Aviv University, Israel, have shown that the spherical peptide NPs were not degraded even in extreme conditions such as elevated temperatures or highly acidic and basic solutions, pointing to the putative role of π-stacking interactions in nanoparticle stability.

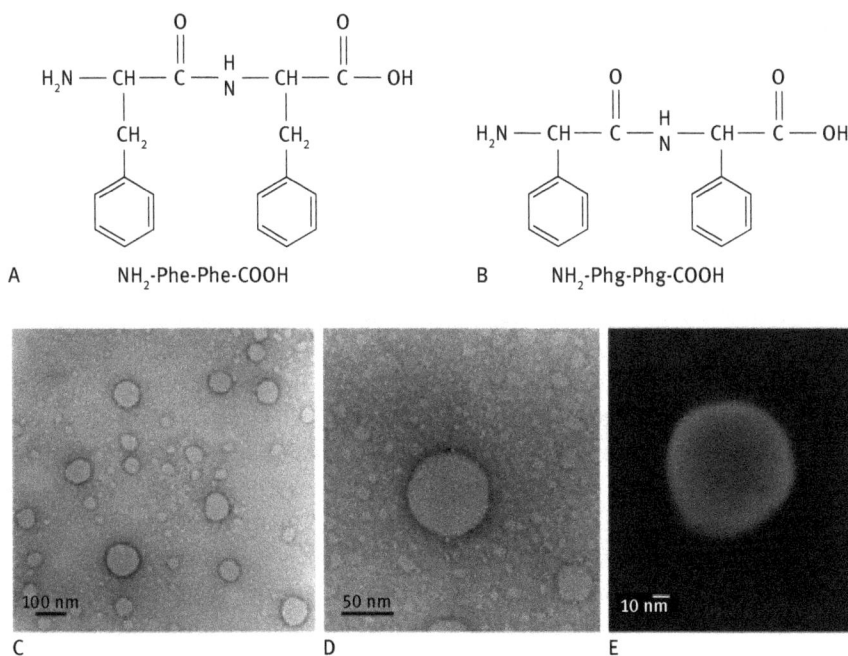

Fig. 5.3: Spherical "caged" nanoparticles comprising of diphenylglycine. **A:** Diphenylalanine and. **B:** diphenylglycine chemical structures. **C–E:** Electron microscopy images showing spherical diphenylglycine nanoparticles. Reprinted with permission from Reches and Gazit, *Nano Lett.* **4** (2004), 581–585, © 2004 The American Chemical Society.

The dramatic differences in nanoparticle morphologies between the two closely-related dipeptides – nanotubes vs nanospheres (Figures 5.2 and 5.3) – triggered efforts to further exploit the structural variability in order to develop new dipeptide-based NP assembly processes. Figure 5.4 highlights experimental data showing a reversible transformation between diphenylalanine spherical NPs and nanotubes achieved by changing the *solution properties*. Specifically, G. Rosenman and colleagues at Tel

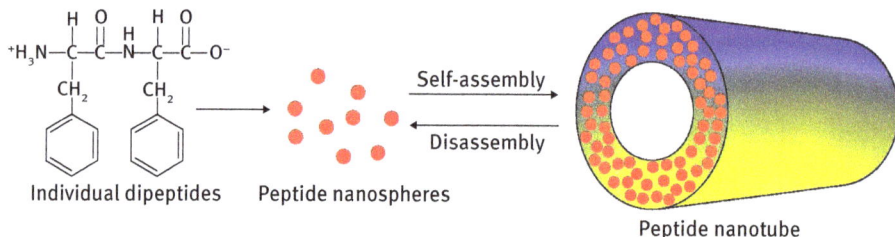

Fig. 5.4: Reversible transition between dipeptide nanospheres and a nanotube. Diphenylalanines spontaneously assemble into spherical nanoparticles in methanol, while changing the solvent to water induces formation of nanotubes.

Aviv University, Israel, discovered that concentrated solutions of diphenylalanine in methanol produced nanosize spherical NPs (reminiscent of "quantum dots"), while changing the solvent to water initiated self-assembly of the nanospheres into *nanotubes*. This nanosphere-to-nanotube transformation was fully reversible, induced simply by changing the solvent. The researchers hypothesized that the polar organic solvent aided stabilization of the peptide "dots", while in the aqueous environment the individual NPs made of hydrophobic residues were drawn together, forming nanotubes.

Modulation of the amino acid *sequences* has been the predominant strategy to control peptide self-assembly properties, making formation of nanofibers, nanotubes, and nanospheres with tunable dimensionalities and morphologies possible. Other strategies have added *nonpeptide* components to the biological nanoparticle construction toolbox. Figure 5.5 portrays an anti-tumor peptide covalently bonded to a hydrophobic hydrocarbon moiety. Even though the peptide itself is hydrophilic and dissolves in water, attachment of the lipid-like residue promoted self-assembly of the "peptide amphiphiles" into nanotubular aggregates (which can also be characterized as "micellar assemblies"). Such nanotubes offer distinct advantages as therapeutic vehicles. The tubular structure enables high loading of the biologically-active peptide cargo through condensation, and the amphiphilic layer protects the peptide from enzymatic degradation in physiological environments (e.g. the bloodstream) until it reaches its destination. Furthermore, the hydrophobic coating allows docking of the nanotube onto the cell membrane and subsequent injection of the peptide molecules into the cell. The peptide nanotube could also undergo gradual disintegration enabling slow release of the therapeutic molecules.

Integrating peptides and lipid molecules within composite nanoparticles has been pursued as a promising therapeutic strategy. In addition to covalent binding between peptides and lipids (as in the peptide amphiphile system discussed above), studies have shown that *noncovalent association* can also lead to new NP species exhibiting useful functionalities. Figure 5.6, for example, depicts "core-shell" spherical NPs produced by mixing a therapeutic drug, phospholipids, and a helical peptide.

A

Self-assembly

B

C

D

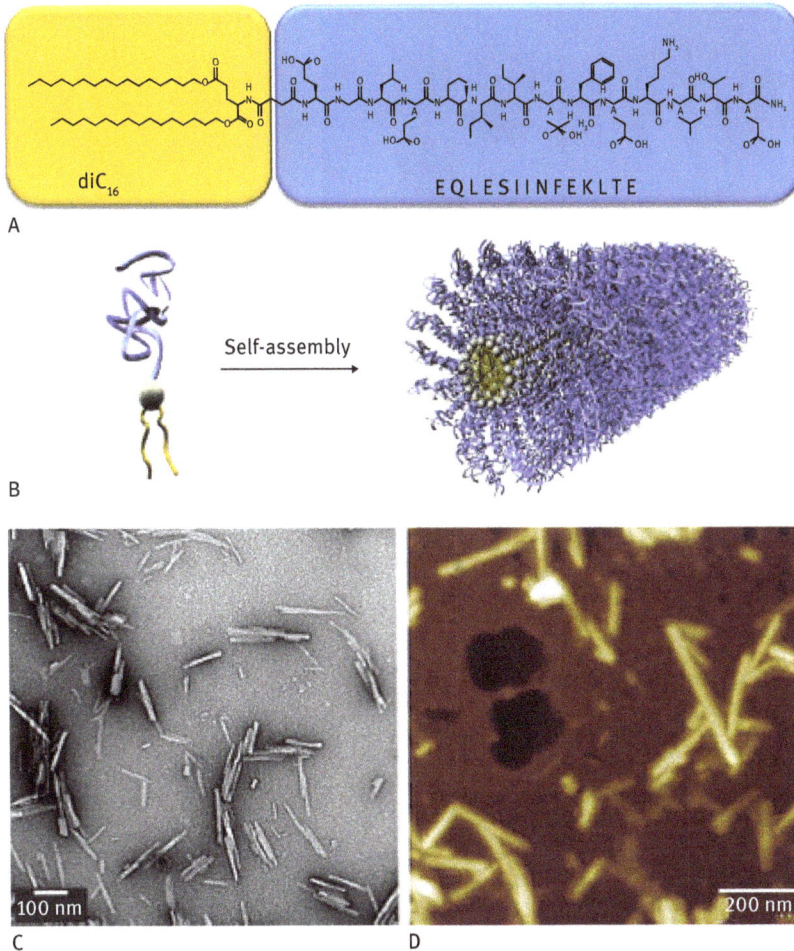

Fig. 5.5: Nanotubes assembled from peptide-amphiphiles. **A:** Chemical structure of the peptide amphiphile, comprising of a hydrophobic hydrocarbon chain (yellow background) covalently linked
to a short hydrophilic peptide (blue background). **B:** Schematic depiction of nanotube formation
through self-assembly of the peptide amphiphiles. **C:** Electron microscopy. **D:** Atomic force microscopy images showing the nanotubes. Reprinted with permission from Black et al., *Adv. Mater.*
24 (2012), 3845–3849, © 2012 John Wiley & Sons.

The objective underlying the choice of molecular composition and overall structure
of these NPs was to achieve a delivery system that would both protect the bio-active
cargo from degradation prior to reaching its target and also provide the means for delivery into the cell interior. In general, particles made of only phospholipids (i.e. lipid
vesicles or *liposomes*) are not considered effective drug delivery vehicles since they are
not stable for extended periods in the bloodstream as they spontaneously disintegrate
and/or are degraded by lipophilic enzymes. To overcome this barrier, G. Zheng and

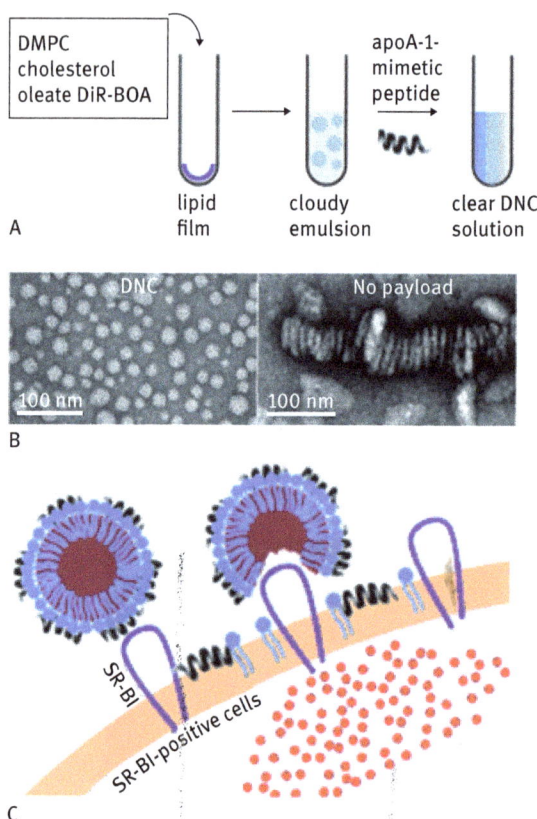

Fig. 5.6: Drug nanocarrier (DNC) constructed from a mixture of lipid/peptide/drug. **A:** DNC preparation procedure; noncovalent incorporation of the peptide induces formation of water-soluble nanoparticles, recognized by cell-displayed receptors. **B:** Electron microscopy images showing the compacted DNC nanoparticles (left), and the discoidal structures formed when no drug molecules were present in the mixture (right). The images highlight the crucial role of the drug compounds for nanoparticle assembly. **C.** Mechanism of drug cargo cell uptake. The DNCs latch onto the cell surface through interaction with specific receptors (the SR-BI protein receptor). Interactions between the surface peptides of the DNCs enable delivery of the encapsulated drug cargo (red spheres) into the cell interior. Reprinted with permission from Zhang et al., *Angew. Chem. Int. Ed.* **48** (2009), 9171–9175, © 2009 John Wiley & Sons.

colleagues at the University of Toronto encapsulated a drug cargo within a phospholipid layer and added an amphipathic helical peptide to the vesicles which compacted and rigidified the resultant NPs. The NP surface-displayed helical peptide also contained a recognition sequence designed to dock onto specific cell-surface receptors and thus initiate cell uptake. The NP system depicted in Figure 5.6, denoted "drug nanocarrier" (DNC), represents a hybrid peptide/lipid/drug assembly in which the "whole is greater than the sum of its parts", providing a potentially versatile biological delivery platform.

Lipoproteins are a class of biomolecules which form nanoparticle aggregates in their natural environment – within the bloodstream. In fact, the critical function of capturing and transporting cholesterol and other lipids in blood vessels is accomplished via adsorption of the target lipids within naturally-assembled *lipoprotein NPs*. For example, the notorious *low-density lipoprotein (LDL,* popularly known as the "bad cholesterol") constitutes, as its name implies, NPs which are less compact compared to *high-density lipoprotein (HDL)*. Attempts have been made to "engineer" lipoprotein NPs with the aim of constructing biomimetic functional materials.

Figure 5.7 depicts a process in which natural LDL NPs were coupled with *gadolinium ions* (common contrast agents in magnetic resonance imaging) and used for imaging of *atherosclerotic plaques*, the pathological hallmarks of coronary heart disease.

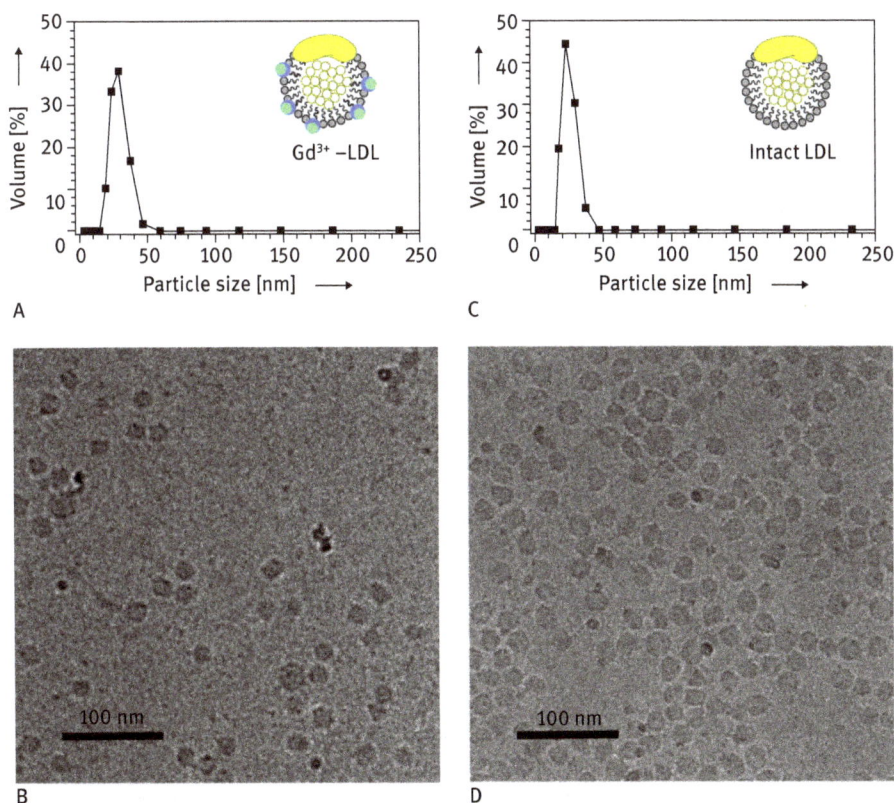

Fig. 5.7: Gadolinium-labeled low-density lipoprotein (LDL) nanoparticles. **A, B:** Gd^{3+}-labeling does not alter the size distribution of the LDL nanoparticles. **C, D:** Electron microscopy images further confirm that both nanoparticle species exhibit similar morphologies. Reprinted with permission from Lowell et al., *Bioconjugate Chem.* **23** (2012), 2313–2319, © 2012 American Chemical Society.

Conjugation of the Gd^{3+} ions to the LDL NPs was carried out by Y. Yamakoshi and colleagues at the University of Pennsylvania by incubating the nanoparticles with a lipid-like Gd^{3+} "chelating residue". LDL NPs are known to deposit atherosclerotic plaques in blood vessels and contribute to their growth. As a consequence of the accumulation of the Gd^{3+}-labeled NPs, significant enhancement of MRI signals from atherosclerotic plaques was achieved in a mouse model by the researchers. The composite NPs in Figure 5.7 provide a vivid example of the multifunctionality and versatility inherent to lipoprotein-based NPs.

Peptide-based NPs might serve as vaccines. Generally, a vaccine comprises of the *immunogen* – substance recognized by the host which elicits an immune response – and *adjuvant*, a material enhancing immune action, for example through surface display and stabilization of the biologically-active conformation of the immunogen. NPs could fulfill these prerequisites since they constitute an effective platform for delivery and presentation of biological molecules on their surface. Figure 5.8 depicts an NP vaccine design in which two distinct proteins were linked together – *hemagglutinin (HA)*, functioning as the immunogen, and *ferritin*, a ubiquitous protein which assembles into compact NPs, serving as the delivery scaffold. HA is a prominent protein displayed on the surface of the *influenza* virus and is known to generate an immune response against the virus. In the system depicted in Figure 5.8, HA was fused to ferritin via genetic engineering, with the ferritin NPs stabilizing the bio-active conformation of HA extending from the NP surface. The composite ferritin-HA protein NPs were found by G. J. Nabel and colleagues at the National Institutes of Health, US, to elicit a broad immune response against the influenza virus.

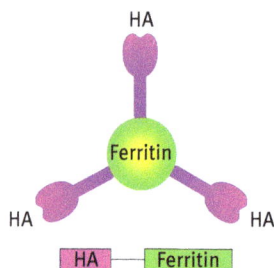

Fig. 5.8: Biological nanoparticle vaccine. Hemagglutinin (HA), an immune-triggering protein displayed on the surface of the influenza virus, was linked to ferritin, a protein forming compact nanoparticles. The HA units elicited an immune response when injected, indicating that HA adopted biologically-active conformation when attached to the ferritin nanoparticle.

DNA is the fundamental information-bearing molecule in the cellular world. In the context of nanotechnology, the modular organization of DNA, i.e. base-pair complementarity underlying the double helix configuration, makes DNA a useful building block to construct a variety of structures, both in micro- and nanoscales. In particular, the elongated DNA strands are amenable to nanowire formation; *DNA-based nanowires* have been prepared via various routes. In aqueous solutions, however, DNA tends to adopt "coiled" conformations; accordingly efforts in this field have been directed to a large extent at "stretching" DNA NWs, usually by deposition on surfaces.

Figure 5.9 illustrates an elegant method for aligning DNA NWs on polymer surfaces. The phenomenon exploited by Z. Lin and colleagues at the Georgia Institute of Technology to create the oriented DNA assemblies is the accumulation and subsequent precipitation of DNA molecules at the edge of slowly evaporating water meniscus (conceptually similar to grains of sands deposited at the edges of receding waves on the beach). Accordingly, the researchers pressed a cylindrical lens upon a DNA solution placed on a flat polymer substrate. The interface between the curved cylinder and the flat surface created a confined geometry which reduced random solution flow and minimized instabilities, resulting in gradual deposition of oriented DNA NWs at the receding fronts of the evaporating aqueous solution. Intriguingly, the evaporative-alignment approach also enabled formation of *cross-shaped* nanowire structures via 90-degree rotation of the cylindrical lens (Fig. 5.9B).

DNA NWs have often been used as templates in material design. This line of research is based on the versatile routes for chemical modification of oligonucleotides, allowing display of different functional units on DNA strands, such as thiols (for coupling with gold substances, for example in conducting electrodes), silanes (allowing conjugation with glass surfaces), amines, and more. Chemically speaking, one can perceive DNA molecules as programmable polymers whose structures can be modulated in the nanoscale by the intrinsic information-bearing features of oligonucleotide complementarity. Moreover, studies such as the one depicted above demonstrating DNA *ordering* (i.e. Fig. 5.9) indicate that DNA might serve as a useful template for metallic structures and devices.

Figure 5.10 depicts an experiment in which a DNA network was used as a scaffold for fabrication of metal nanowires. The experiment, designed by Q. Gu at Pacific Nanotechnology Inc., exploited the electrostatic attraction between the negatively-charged DNA and positive Pd^{2+} ions. The palladium ions bound to the DNA NW template were subsequently reduced to Pd^0 particles, which in turn constituted nucleation sites for cobalt deposition, ultimately yielding continuous Co NWs. This generic approach was employed to create various metal nanowire arrays, pointing to potential use of DNA NWs in *nanoelectronics* applications.

DNA templates which are not "wire-like" (i.e. linear) have also been reported. Figure 5.11 highlights a simple method for assembling *circular* DNA rings and their use as templates for deposition of metallic silver. The experimental system designed by A. A. Zinchenko and colleagues at Kyoto University, Japan, exploited electrostatic attraction between negatively-charged DNA and a highly positively-charged *polycation*, resulting in the DNA folding into a compact "nanoring" structure. Subsequently, similar to the metal NW scheme discussed above (Fig. 5.10), positive Ag^+ ions were embedded within the DNA template via binding to the negative DNA units. The Ag^+ ions were then reduced, yielding "doughnut-shaped" metallic silver NPs. Notably, the silver nanorings gave rise to distinctive plasmon resonance light absorbance, pointing to possible use of the NP templating technique in optical modulation applications.

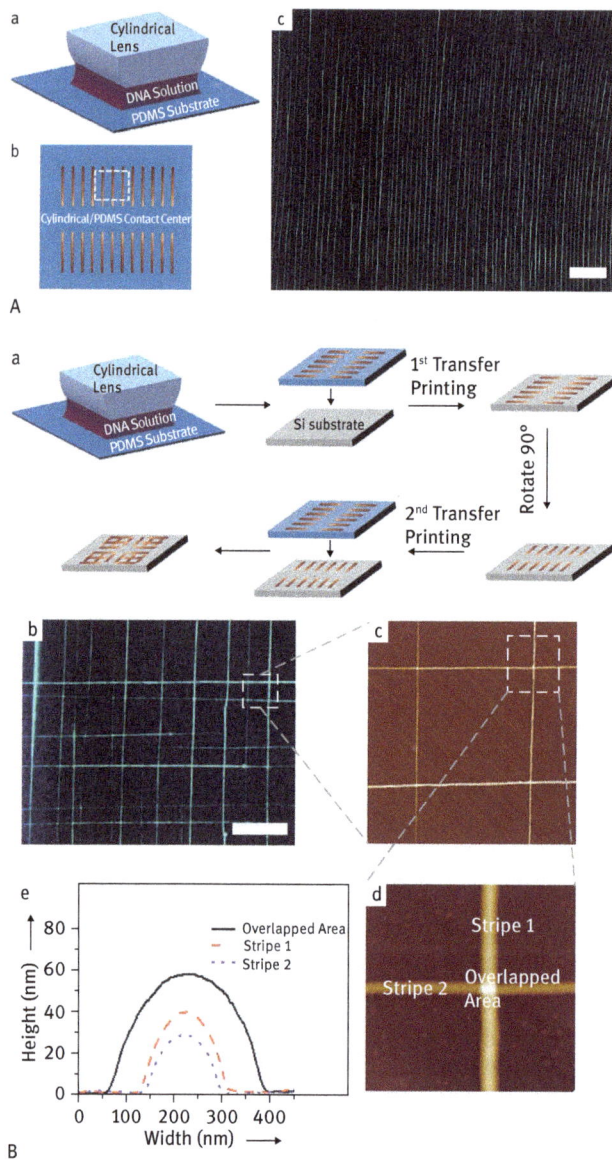

Fig. 5.9: Alignment of DNA nanowires. **A:** Alignment of DNA nanowires on a flat polymer surface achieved via slow evaporation of the DNA solution underneath a cylindrical "stamp". (**a**) Schematic diagram of the experiment and (**b**) the DNA nanowire array generated. The microscopy image in (**c**) shows a representative DNA nanowire array deposited on the surface, indicated by a broken line in (**b**). **B:** Production of a "crossed nanowire" array: (**a**) The experimental scheme, involving two consecutive alignment steps in which the cylindrical stamp is rotated by 90° between the two evaporation stages; (**b–d**) microscopy images depicting the perpendicular DNA nanowires; (**e**) a cross-section of the DNA nanowire intersection showing the overlapped area. Reprinted with permission from Li et al., *ACS Nano* **7** (2013), 4326–4333, © 2013 American Chemical Society.

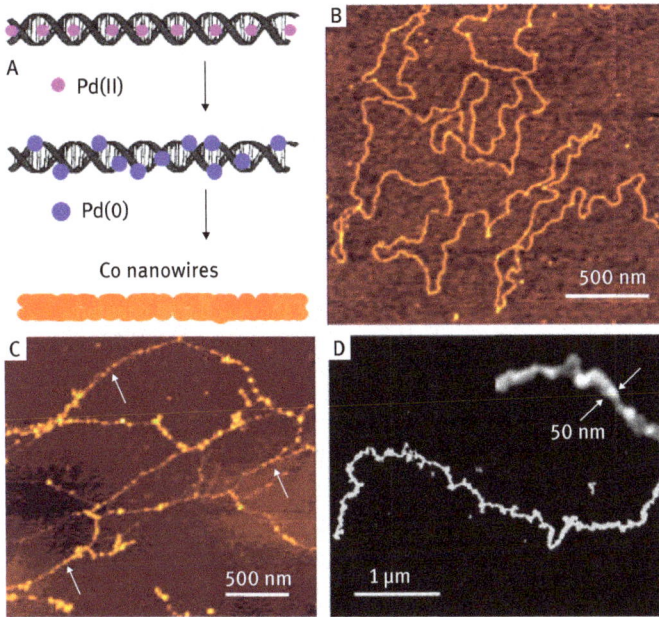

Fig. 5.10: DNA as a template for metal nanowires. **A:** Schematic diagram of the nanowire deposition method. Palladium ions are electrostatically attracted to the DNA strand and subsequently reduced. The Pd^0 metal seeds constitute nucleation sites for deposition of cobalt, producing metallic cobalt nanowire. **B:** Atomic force microscopy (AFM) image showing the DNA wire. **C:** AFM image of the DNA wire after reduction of the embedded palladium ions with apparent individual Pd nanoparticles. **D:** Scanning electron microscopy (SEM) image of the continuous cobalt nanowire; the inset shows a magnified view. Reprinted from *Mater. Lett.* Vol 62, Q. Gu and D. T. Haynie, *Palladium nanoparticle-controlled growth of magnetic cobalt nanowires on DNA templates*, pp. 3047–3050, © 2008, with permission from Elsevier.

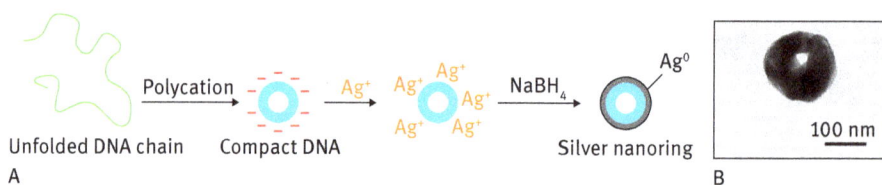

Fig. 5.11: Circular DNA as a template for silver deposition. **A:** Schematic diagram of the experiment: unfolded DNA chain is compacted into a "nanoring" structure via addition of polycations; Ag^+ ions are attracted to the DNA template by electrostatic attraction; subsequent reduction yields "doughnut-shaped" silver nanorings. **B:** Electron microscopy image showing the Ag nanoring. Reprinted with permission from Zinchenko et al., *Adv. Mater.* **17** (2005), 2820–2825, © 2005 John Wiley & Sons.

Similar to DNA, *ribonucleic acid (RNA)* molecules exhibit base-pair complementarity (the nucleotide *thymine* in DNA is substituted in RNA sequences with *uracil*), enabling design of intriguing programmable nanoparticles. RNA, however, exhibits important differences to DNA which could render it particularly attractive for nanoparticle design applications. Since RNA usually occurs in single strands, rather than the double strand configurations abundant in DNA, it can adopt a variety of tertiary structures, including helices, loops, bulges, and stems, which could be employed as building blocks for diverse nanoparticle architectures. Examples of RNA-based nanoparticles are presented in Figure 5.12. These unique nanostructures were designed by L. Jaeger and colleagues at the University of California, Santa Barbara, using L-shaped RNA modules containing "reactive loops" and "stems" which assembled into distinct square-shaped nanoscale configurations. Uniform nanosquares exhibiting distinct dimensions were produced using different RNA motifs. Links between the RNA units forming the squares depended on the properties of the RNA building blocks, i.e. positioning of the reactive loops, size of the RNA "stems", etc.

The design strategy for DNA and RNA nanoparticles is intimately linked to the intrinsic information-bearing properties of these two molecules. First, the structure of the desired nanoparticle and its functional objectives are defined. A computational scheme is then applied to determine the core DNA/RNA sequences. The building blocks are then generated via transcription of the long strand DNA/RNA, or by chemical synthesis of the smaller sequences. Ultimately the individual subunits are mixed and assembled into the NP architecture. This generic scheme can be augmented by attaching the oligonucleotide modules to other functional units, such as metallic species, fluorescent probes, and others, either in the assembly stage, or after producing the desired nanoparticle structures.

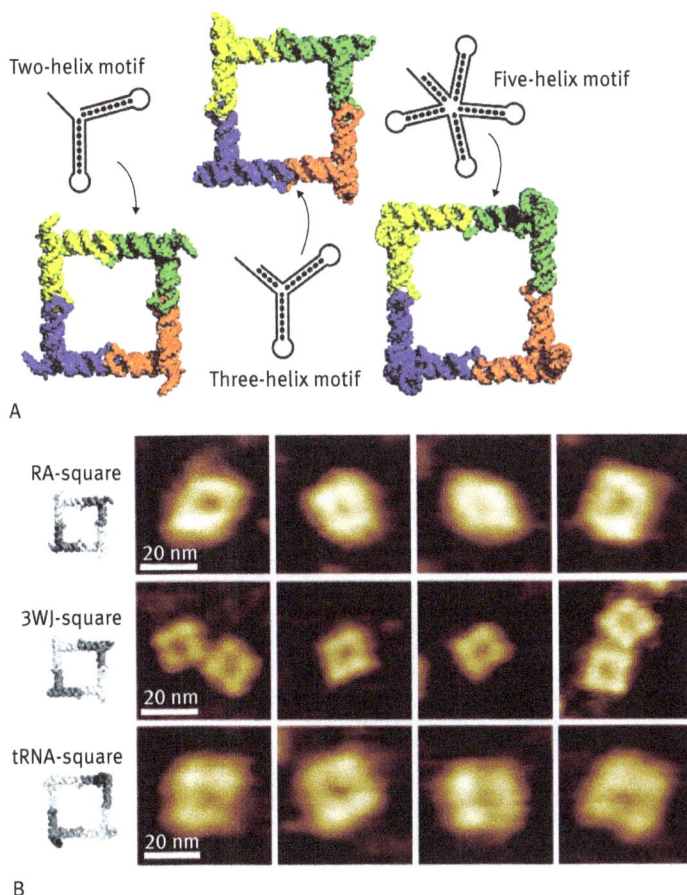

Fig. 5.12: RNA nanoparticles. **A:** Schematic drawing of three nanosquares assembled from distinct RNA motifs. The nanosquares comprise of L-shaped RNA modules (shown in different colors) which assemble by binding at the "stems". The RNA motifs determine the properties of the L-shaped components and the assembled nanosquares. **B:** Atomic force microscopy (AFM) images of the nanosquares. The schematic structures corresponding to the different RNA modules are shown on the left. Reprinted with permission from Severcan et al., *Nano Lett.* **9** (2009), 1270–1277, © 2009 American Chemical Society.

5.2 Organic and polymeric nanoparticles

Organic chemistry, polymers in particular, has contributed its "fair share" to nanoparticle science and technology. Advanced synthetic capabilities make production of sophisticated nanoparticles from organic building blocks possible. Similar to the biological NPs discussed above, several organic NP systems have been found to exhibit interesting properties as independent entities, while in some instances organic NPs were used as templates for assembly of (usually metallic) NPs.

Various techniques have been introduced to prepare organic and polymeric NPs. *Laser ablation*, mostly applied for generation of metallic NPs, is one of the "top-down" strategies successfully employed to produce organic NPs in solution. This method is based on intense laser irradiation of organic microstructures (e.g. powders) suspended in water; the "chipped" NPs can subsequently be dispersed in water. "Bottom-up" techniques for the production of polymer NPs have also been introduced. Such schemes generally utilize *internal cross-linking* within the polymer chain, generating condensed nanoparticles. Figure 5.13 illustrates a prototypical reaction scheme for generating polymer NPs. The process, developed by G. W. Coates and colleagues at Cornell University, employed a polymer chain with distinctively-spaced pendant *vinyl* residues (shown in red in Fig. 5.13). Reaction between the vinyl units resulted in intra-molecular crosslinking which stabilized condensed NPs. Many variations of this generic synthetic approach have been reported, endowing diverse physical, chemical, and biological properties to polymer-based NPs.

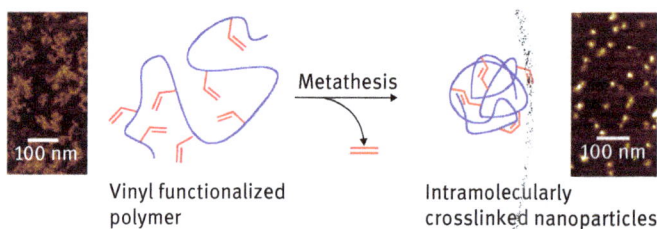

Metathesis

Vinyl functionalized
polymer

Intramolecularly
crosslinked nanoparticles

Fig. 5.13: Polymer nanoparticles assembled by internal crosslinking. Crosslinking is carried out through the pendant vinyl (i.e. ethylene) residues, shown in red. The atomic force microscopy (AFM) images depict the spread-out polymer chains (left) and the compact nanoparticles assembled by crosslinking (right). Reprinted with permission from Cherian et al., *JACS* **129** (2007), 11350–11351, © 2007 American Chemical Society.

Polymer quantum dots (or "*Pdots*") are a fascinating class of NPs with potentially broad applications. Pdots are constructed from π-conjugated semiconducting polymers exhibiting direct bandgaps due to overlap of π-electron energy levels (Fig. 5.14). Such conjugated polymers can assemble into bright, fluorescently-tunable NPs on precipitation in various solvents; Pdots have also been prepared via direct reaction of the monomers in emulsions. Similar to the inorganic semiconducting quantum dots (Qdots) discussed in detail in Chapter 2, Pdots exhibit significant potential as bioimaging agents. Pdots, in fact, offer advantages for imaging since they feature extremely bright fluorescence, as the NPs consist of a high concentration of the luminescent semiconducting polymers compacted in a small volume. In addition, Pdots can quite easily be chemically-modified via well-developed polymer chemistry routes, making display of various functional units on their surface possible. Furthermore, polymeric building blocks are likely to be less toxic to cells than inorganic Qdots.

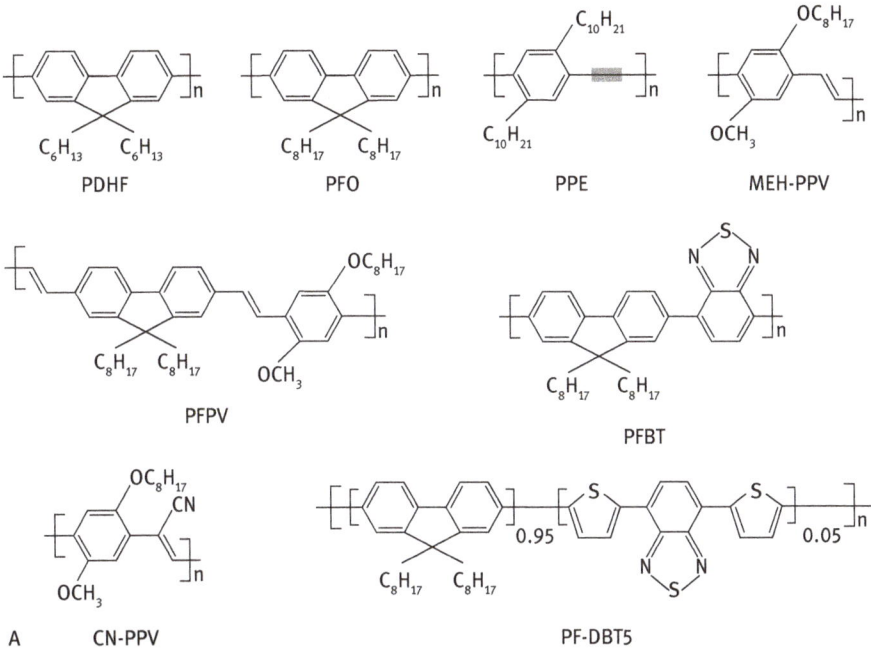

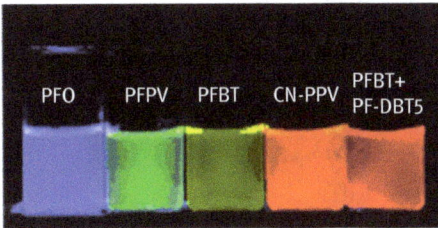

Fig. 5.14: Polymer quantum dots (Pdots). **A:** Conjugated semiconducting polymers used in construction of Pdots. **B:** Distinct colors (i.e. luminescence wavelengths) of Pdots comprising of different polymer materials. Reprinted with permission from Wu and Chu, *Angew. Chem. Int. Ed.* **52** (2013), 3086–3109, © 2013 John Wiley & Sons.

Pdots can be prepared either via condensation of the monomers followed by polymerization reactions, or (the more popular approach) by using the already-polymerized chains for assembly of the nanoparticles through crosslinking reactions. The latter technique is simpler for nonexperts to carry out, and obviates the use of specialized polymerization reagents which could interfere with biological applications of the NPs. Pdots have been produced from *hydrophilic, hydrophobic,* or *amphiphilic* polymer units, making these NPs a versatile platform. Furthermore, Pdots can contain other polymer (or biological) components in addition to the *luminescent* polymer species, and the semiconducting polymers themselves can easily be coupled with other functional units.

Figure 5.15 depicts a representative biologically-functionalized Pdot system and its application in bioimaging. The functionalized Pdots, developed by D. T. Chiu and colleagues at the University of Washington, were prepared by mixing two polymers: a semiconducting conjugated polymer (PFBT) and a polymer (polystyrene derivative)

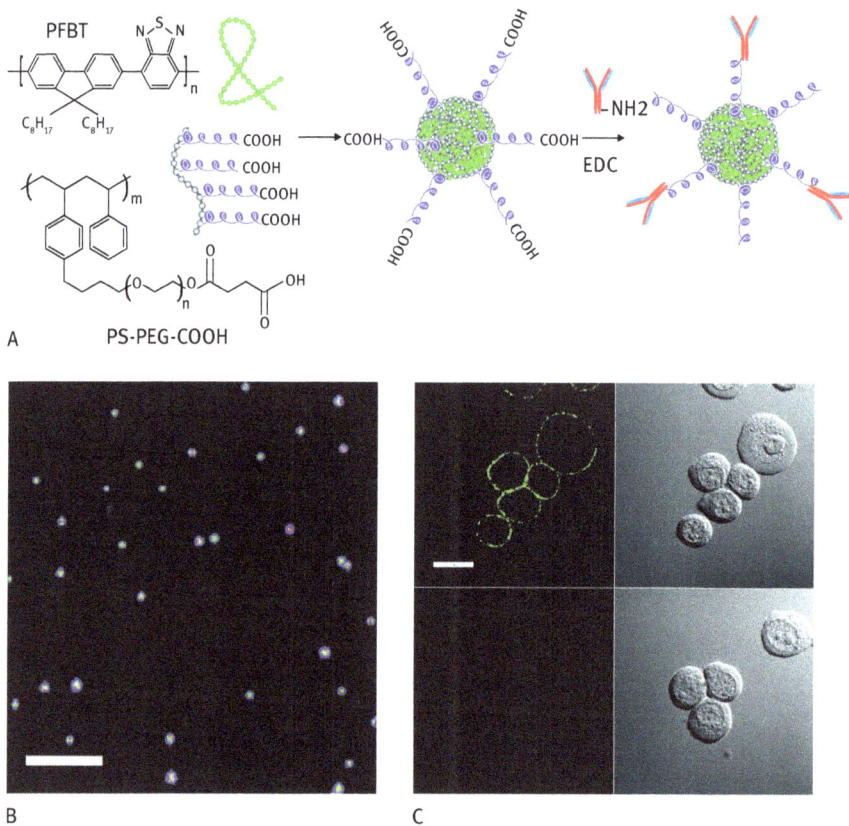

A PS-PEG-COOH

B   C

Fig. 5.15: Bio-conjugated polymer quantum dots (Pdots). **A:** Synthesis scheme: the Pdots are prepared by co-condensation of a semiconducting polymer (PFBT) and a polymer displaying carboxylic acid residues. The carboxylic residues are subsequently used for covalent coupling of antibody molecules (the "Y"-shaped residues). **B:** Single particle fluorescence microscopy depicting the bright Pdots. **C:** Cell imaging using the antibody-functionalized Pdots. The top left panel shows a fluorescence microscopy image of human breast cancer cells incubated with Pdots which were covalently-linked to antibodies recognizing a cell-surface receptor. No cell surface fluorescence was recorded when the cells were incubated with bare Pdots (not attached to antibodies, bottom panel). The panels on the right depict the bright field microscopy images indicating the cell positions. Reprinted with permission from Wu et al., *JACS* **132** (2010), 15410–15417, © 2010 American Chemical Society.

displaying abundant carboxylic moieties (Fig. 5.15A). The carboxylic units provided a facile chemical route for coupling the Pdots to biomolecules such as antibodies, which facilitated targeting of the particles to desired cellular locations. The single-particle fluorescence image in Figure 5.15B confirms the notable brightness of the Pdots; in fact, the fluorescence of the polymer NPs was notably more intense than many conventionally-used markers, such as organic fluorescent dyes or inorganic quantum dots. The high fluorescence and excellent photo-stability of the Pdots have made these NPs an excellent vehicle for cell imaging, highlighted in Figure 5.15C. The Pdots in the experiment were conjugated to antibodies directed against specific cell-surface receptors. The cell contours could be clearly deciphered when incubated with the Pdot-antibody constructs; no fluorescence was observed when bare Pdots (not coupled to the antibody) were used, confirming the occurrence of specific binding between the Pdot-antibody and the cell surface receptors.

While Pdots might serve as useful bioimaging agents, one should be aware of the limitations of their use, from both fundamental and technical points of view. Primarily, even though Pdots might exhibit minimal toxicity, they still constitute particles and as such, similar to conventional inorganic quantum dots, could have adverse effects when internalized within cells. It should also be emphasized that some semi-conductor polymers do not easily adopt NP structures, while others exhibit broad fluorescence emission peaks, making them less attractive for bioimaging applications compared to inorganic quantum dots, which exhibit narrow emission spectra (and color tunability).

Conductive polymers (CPs) are another broad and important class of conjugated polymers for which the formation of nanostructures gives rise to interesting properties and applications. The "claim-to-fame" of CPs has been their excellent, metallic conductivity and their adoption of various morphologies and three-dimensional configurations. Nanoparticles of *polypyrrole (PPy)*, a well-known CP, have attracted interest in light of their *optical absorption* profile rather than electrical conductivity. Figure 5.16 depicts an experimental scheme developed by Z. Dai and colleagues at Harbin Institute of Technology, China, for production of spherical PPy NPs and their use as vehicles for photothermal therapy. Photothermal therapy is based on local heat dissipation from NPs following irradiation with near infrared (NIR) light, which penetrates tissues. Such applications have mostly been reported in the context of metallic NPs (see, for example, discussion on photothermal effects of Au nanorods in Chap. 3). In the system presented in Figure 5.16, PPy NPs were prepared inside a polymeric "cage", serving as a stabilizing matrix. Importantly, the PPy NPs featured an absorbance peak at around 800 nm, well within the NIR spectral region. PPy NPs could be attractive for phototherapy applications as they should exhibit lower toxicity and greater long-term stability than Au and other metal NPs commonly studied as phototherapy platforms.

Similar to other nanoparticle systems, polymer NPs can be used as platforms for drug delivery. Contributing to such applications are synthetic methodologies developed to load therapeutic cargoes in polymer NPs, either by physical encapsulation (in

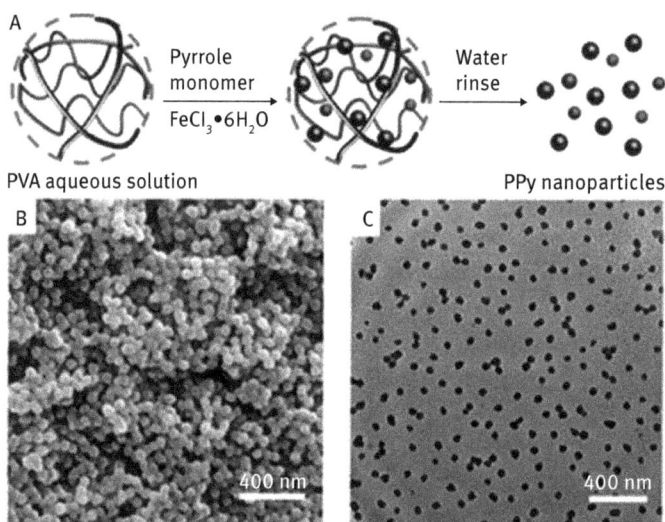

Fig. 5.16: Polypyrrole (PPy) nanoparticles. A: Synthesis scheme: pyrrole monomers embedded within a matrix of polyvinyl alcohol (PVA) acting as a stabilizer. Addition of the oxidizing agent ferric chloride (FeCl₃) produces reaction centers for generation of the PPy nanoparticles. The PVA cages are finally dissolved in an aqueous solution. B: Electron microscopy image of the PVA "cages" acting as reaction chambers. C: Electron microscopy image of the PPy nanoparticles after PVA dissolution. Reprinted with permission from Zha et al., *Adv. Mater.* **25** (2013), 777–783, © 2013 John Wiley & Sons.

porous polymer NP host matrixes), or by covalent attachment to the polymer chains. Moreover, the diversity of polymeric substances used as nanoparticle building blocks provides a means for controlled release of the therapeutic cargo, for example by chemical cleavage or physical swelling.

Figure 5.17 presents a strategy for drug delivery to tumors using polymer NPs. The particles, designed by J. Wang and colleagues at the University of Science and Technology of China, comprised of a *zwitterionic* polymer (displaying both positive and negative charges) encapsulating doxorubicin, a commonly-used anti-tumor compound. The NPs retained their zwitterionic properties in circulating blood due to the neutral pH, and the corresponding neutral charge upon the particle surface made them relatively "inert" to cell uptake. However, in tumor environments the blood vessels become "leaky", resulting in accumulation of NPs at the tumor site. Subsequently, since the micro-environment of cancer cells is highly acidic, the polymer surface became positive through an acid-promoted amide cleavage reaction. The positively-charged NPs were consequently attracted to the negatively-charged cell membrane and adsorbed by the cancer cells, releasing their therapeutic cargo (the doxorubicin molecules) inside the cells and destroying them.

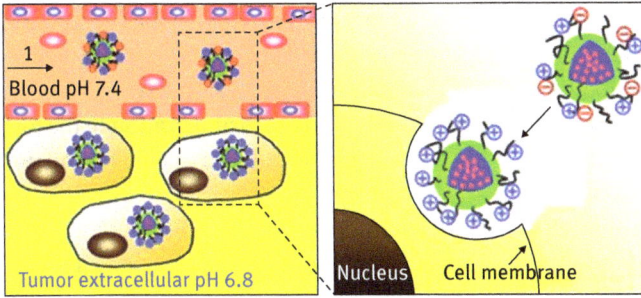

Fig. 5.17: pH-based release mechanism of therapeutic cargo in polymer nanoparticles. The nanoparticles containing a therapeutic molecule are constructed from zwitterionic polymer chains. The nanoparticles retain their neutral charge in the physiological pH of 7.4 in blood. In tumor extracellular environments the acidic pH results in cleavage of the negative residues on the nanoparticle surface; the positively-charged polymer nanoparticles get adsorbed by the cancer cells, releasing their therapeutic cargo. Reprinted with permission from Yuan et al., *Adv. Mater.* **24** (2012), 5476–5480, © 2012 John Wiley & Sons.

Figure 5.18 presents another NP polymer-based delivery system – a "nanocapsule" designed to deliver small interfering ribonucleic acid (siRNA) into cells. siRNA is a promising therapeutic agent as its short RNA sequences bind to target RNA strands and consequently block expression of specific proteins. The challenge, however, is to deliver siRNA into the cell; the NPs depicted in Figure 5.18 accomplish this goal

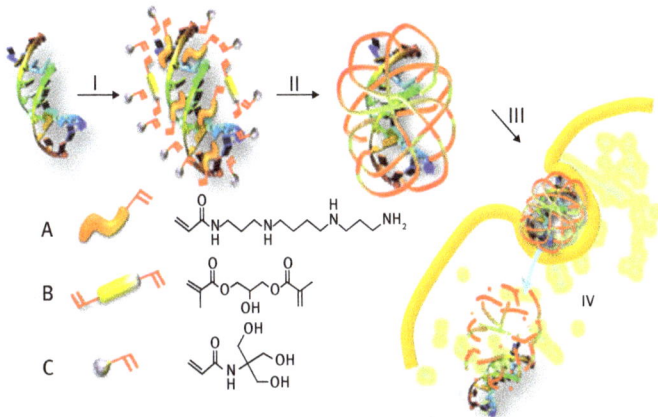

Fig. 5.18: Polymer nanocapsules for siRNA delivery. The nanocapsule construction process comprises of enrichment of the RNA fragment with the monomers and crosslinkers (fragments denoted by **A, B, C**) used for (**I**) assembly of the nanocapsule; (**II**) polymerization to create the nanocapsule embeding the siRNA cargo; (**III**) cell uptake via endocytosis; and (**IV**) disintegration of the polymer capsule and release of the siRNA cargo. Reprinted with permission from Yan et al., *JACS* **134** (2012), 13542–13545, © 2012 American Chemical Society.

by encapsulating the siRNA sequence within a biodegradable polymeric matrix. The polymer layer has several functions – maintaining intactness of the siRNA cargo while transporting through the bloodstream, protecting against degradation by serum enzymes, and eventually inducing intracellular release through disintegration in the low pH environments of the "endosomes", the cell compartments responsible for the uptake of extracellular particles.

While various polymer-based NPs for drug delivery feature particles prepared with the therapeutic cargo embedded within the polymer matrix comprising the nanoparticles, hollow NPs would have distinct advantages as they can be loaded with different molecular guests. Figure 5.19 depicts such hollow polymer NPs, prepared in different shapes and morphologies. Synthesis of the hollow NPs was carried out by simply drying micelle solutions of an amphiphilic block copolymer (polymer containing "blocks" of different monomer units). Rather surprisingly, R. K. O'Reilly and colleagues at the University of Warwick, UK, discovered that the drying conditions (specifically drying rate) had a profound effect on the NP morphology (spherical hollow NPs vs nanocylinders), internal topology, and volume. Importantly, the hollow

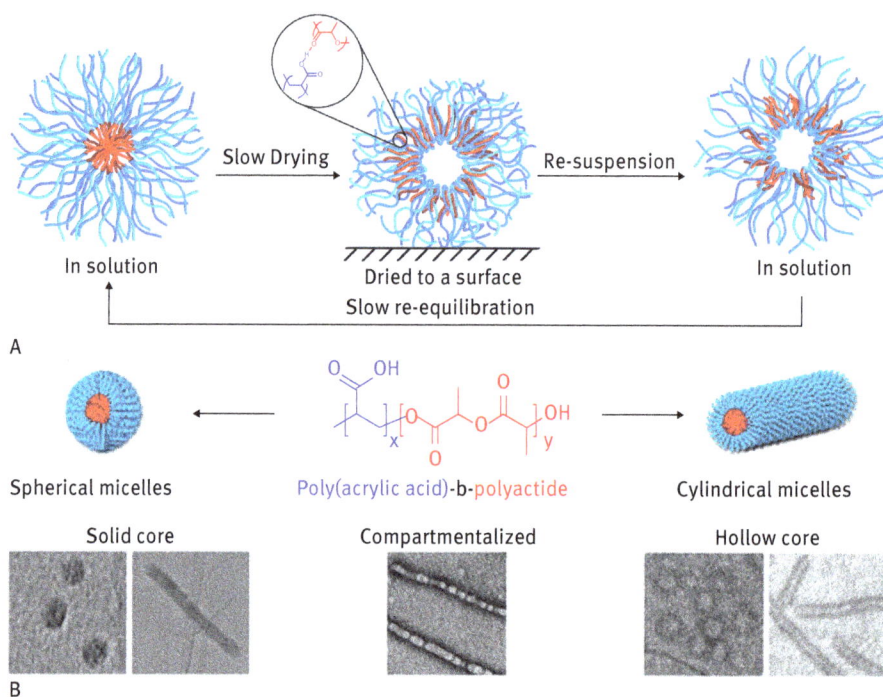

Fig. 5.19: Hollow polymer nanoparticles. **A:** Preparation of the hollow nanoparticles: upon slow drying, the polymer chains in the nanoparticle core rearrange and create a hollow space. **B:** Distinct polymer nanoparticle morphologies prepared. Reprinted with permission from Petzetakis et al., *ACS Nano* **7** (2013), 1120–1128, © 2013 American Chemical Society.

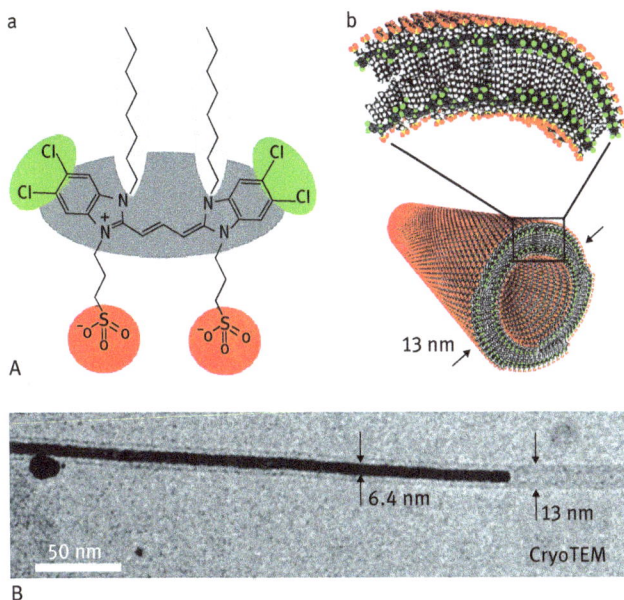

Fig. 5.20: Metal nanowire formation in cyanine nanotubular template. **A:** (**a**) Schematic diagram of nanotube assembly from cyanine; (**b**) the bilayer organization of the nanotube formed via interactions between the pendant hydrohphobic sidechains of cyanine. **B:** Electron microscopy image depicting the Ag nanowire formed within the cyanine nanotube. Reprinted with permission from Eisele et al., *JACS* **132** (2014), 2104-2105, © 2013 American Chemical Society.

NPs could be re-suspended after drying and loaded with relatively high concentrations of molecular cargoes, pointing to their potential use in drug delivery.

Similar to the peptide-templated nanoparticles described above, organic molecules have also been used as scaffolds for deposition of metal NPs (Fig. 5.20). In an experiment carried out by J. P. Rabe and colleagues at Humboldt-Universitaet zu Berlin, Germany, an amphiphilic cyanine dye was used to construct a template for silver nanowires. The dye molecule played a dual role. One function was structural in nature – adopting an extended nanotubular structure via interactions between the pendant amphipathic carbohydrate sidechains (Fig. 5.20A), in which both the inner and outer surfaces were available for metal deposition. The other role of the cyanine molecules was to promote adsorption of silver ions by electrostatic attraction to the dye's negatively-charged phosphate residues. The bound Ag^+ ions were subsequently reduced to metallic silver following light irradiation (i.e. "photo-initiated reduction"). Overall, the cyanine "superstructures" were shown to act as a combined chemical/physical platform for controlled deposition of uniform Ag nanowires.

6 Hybrid and composite nanoparticles

This chapter focuses on nanoparticles containing two or more atomic species belonging to different "families" – such as metal/semiconductor NPs or inorganic/organic NPs. The synthetic sophistication achieved in recent years has led to increasingly complex *multicomponent* NP systems. *Multifunctionality* is a common feature in hybrid NPs – characterized by the observation that in some cases the two (or more) components retain their individual properties. Yet in other scenarios, the properties of the hybrid NP are determined to a significant degree by the *synergy* between the distinct atomic or molecular components. Indeed, in some cases, hybrid NPs provide examples of how the "whole is greater than the sum of its parts" – specifically, composite NPs exhibit interesting, sometime novel properties. It should be noted, however, that many hybrid NPs exhibit *both* characteristics – displaying the fundamental properties of the individual atomic constituents, while these properties are significantly altered in the hybrid NP system.

In many respects, the remarkable diversity of hybrid NP structures mirrors that of hybrid metal NPs (discussed in Sect. 3.4) – including core-shell, Janus, dumbbells and other NP structures. This is, of course, not a coincidence since, conceptually and practically, hybrid NPs and hybrid *metal* NPs share many similarities, including generic synthetic routes and functionalities arising from the combination of more than a single chemical component. Numerous hybrid NPs have been reported in recent years, and this chapter presents just a sampling of the activity in this field, highlighting representative systems and NPs exhibiting unique structures and functions.

6.1 Metal/semiconductor nanoparticles

Integrating metals and semiconductors within hybrid NPs (i.e. metal-associated quantum dots and quantum rods) has spawned various research avenues. The distinct electronic and optical properties of the metal and semiconductor components, respectively, have been shown in many instances to be mutually modulated, affected via the interface between the two individual species. Synthesis of hybrid metal/semiconductor NPs has mostly been carried out via the three main generic routes outlined in Figure 6.1. One approach utilizes pre-assembled NPs (containing one of the two hybrid NP constituents) which function as a "seed" for growth of the second component (Fig. 6.1A). Using this approach, the two synthesis steps are essentially "orthogonal" – single-component nanoparticles are initially synthesized by conventional preparation routes, then placed in a solution containing precursors of the second component for which growth is favored to occur on the surface of the seed material rather than independently. Another strategy is to employ hybrid NPs themselves as starting materials (Fig. 6.1B). The NPs undergo further processing (usually

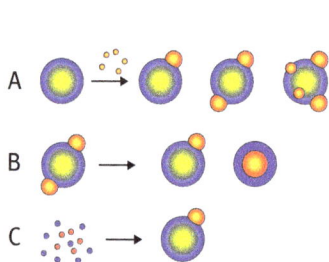

Fig. 6.1: Methods for the synthesis of hybrid nanoparticles. **A:** Initial growth of a single-component nanoparticle which subsequently functions as a "seed" for growth of the second component; **B:** modification of structure and/or organization of as-prepared hybrid nanoparticles; **C:** simultaneous nucleation and growth of the hybrid nanoparticle from precursors of the two components. Figure inspired by Costi et al., *Angew. Chem. Int. Ed.* **49** (2010), 4878–4897.

aided by elevated temperatures, resulting in melting and recrystallization), leading to phase separation and emergence of morphologies such as asymmetrical NPs and core-shell structures. A third route pursues simultaneous nucleation and growth of the two constituents into a single hybrid NP (Fig. 6.1C).

Asymmetric metal/semiconductor NPs (Fig. 6.2), sometimes referred to as hybrid *Janus NPs* or "nanodumbbells", have been prepared in various compositions (it should be noted that symmetrical dumbbell NPs have also been introduced and are discussed below). As a testament to the unexpected turns of scientific research, asymmetric metal/semiconductor NPs were initially fabricated in order to achieve better electrical connectivity (via the metallic component) between electrode surfaces and semiconductor nanorods (attaining electrical contact is still an important application of metal/semiconductor NPs). Subsequent research, however, has revealed that such hybrid NPs exhibit interesting properties even as independent entities.

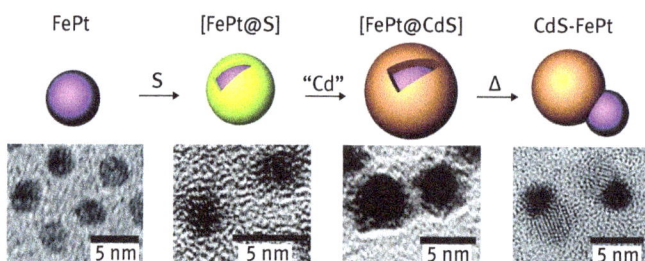

Fig. 6.2: Janus CdS-FePt hybrid nanoparticles. Synthesis stages and corresponding electron microscopy images of the nanoparticles. The metal (FePt) nanoparticle component is synthesized first, followed by growth of a semiconductor "shell". Thermal treatment results in formation of the Janus nanoparticle ("nanodumbbell"). Reprinted with permission from Gu et al., *JACS* **126** (2004), 5664–5665, © 2004 The American Chemical Society.

Procedures for synthesis of asymmetric metal/semiconductor NPs vary and are usually dependent on the atomic constituents. A common strategy is to first synthesize the metal NP component (by conventional metal NP synthesis), and subsequently grow the semiconductor part sintered onto the surface of the metal NP in solutions containing the semiconductor precursor molecules. Figure 6.2 depicts such a scheme, developed by B. Xu and colleagues at the Hong Kong University of Science and Technology. The synthesis procedure started with assembly of the magnetic metal core (iron-platinum). The NP core was coated with a layer of sulfur which was subsequently converted to CdS by addition of the Cd^{2+} precursor. Heating the core-shell NPs induced melting/recrystallization of the CdS shell in a position adjacent to (rather than around) the FePt core. Phase separation between the two NP components likely occurred because of the different crystalline organization of the core and shell materials (i.e. "lattice mismatch"). Importantly, the FePt/CdS dumbbell NPs in Figure 6.2 exhibited both magnetic and fluorescence properties associated with the respective metal and semiconductor building blocks.

Research in the past few years has generated hybrid metal/semiconductor NPs which did not exhibit clear separation between the metal and semiconductor components. Rather, varied NPs have been synthesized featuring random or semi-random distribution of one species (usually the metal) on the surface of the second constituent (the semiconductor NP, which acts as a nucleation "scaffolding"; this is different from the hybrid NP system shown in Fig. 6.2, in which the metal component served as a nucleating platform). Figure 6.3 presents examples of such NP morphologies showing PbSe nanowires decorated with Au NPs. The Au/PbSe hybrid nanowires were prepared by D. V. Talapin and colleagues at the IBM Research Division by initially synthesizing PbSe nanowires as scaffolds for subsequent nucleation and growth of the Au domains. As apparent in Figure 6.3, hybrid Au/PbSe NWs exhibiting diverse morphologies could be produced by this method.

Hybrid metal/semiconductor NPs in which the metal (Au) domains are arranged symmetrically rather than randomly around the semiconductor "core" have also been reported. An interesting example of such hybrid NP "ordering" is depicted in Figure 6.4, which shows Au/PbS NPs prepared by J. Y. Lee and colleagues at the National University of Singapore. The assembly mechanism of these ordered hybrid NPs was based on the use of the semiconductor PbS NPs as seeds for the metal "satelites". Specifically, the facets of the PbS nanocube-like NP cores provided reactive surfaces for nucleation and growth of the Au domains. Remarkably, the extent of gold deposition and resultant core-shell morphologies depended on the concentration of the Au precursors – providing a powerful means for controlling NP structure and three-dimensional organization.

The electronic and optical properties of hybrid metal/semiconductor NPs are shaped to a significant extent by the interactions between the plasmons (i.e. oscillations of the surface electrons) of the metallic components, and excitons (electron-hole pairs) created in the semiconductor parts (Fig. 6.5). A subtle interplay between these two fundamental physical properties occurs because the two distinct NP components are in close proximity in the hybrid NPs. Many optical and spectroscopic phenomena in hybrid NPs have been attributed to plasmon-exciton interactions, with energy transfer as a core process (Fig. 6.5). Indeed, tuning the plasmon-exciton interface

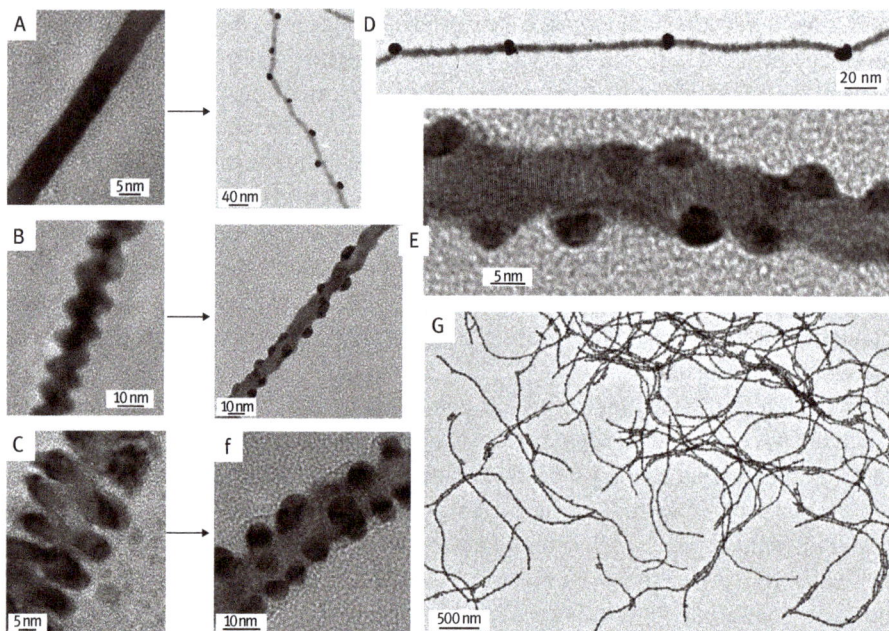

Fig. 6.3: Hybrid Au/PbSe nanowires. The hybrid systems were produced by first synthesizing the PbSe nanowires and subsequently growing Au nanoparticles on their surfaces. Reprinted with permission from Talapin et al., *J. Phys. Chem. C* **111** (2007), 14049–14054, © 2007 The American Chemical Society.

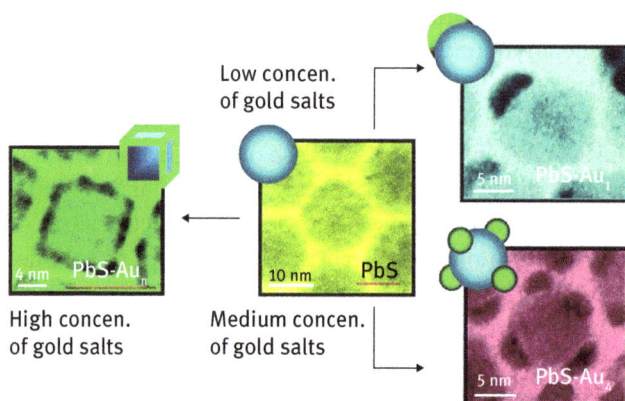

Fig. 6.4: Hybrid Au/PbS nanoparticles. The nanoparticle morphology is dependent on the concentration of the Au precursor (i.e. gold salt) in the nanoparticle growth solution. Reprinted with permission from Yang et al., *JACS* **128** (2006), 11921–11926, © 2006 The American Chemical Society.

has been a powerful tool for determining the properties and applications of hybrid metal/semiconductor NPs.

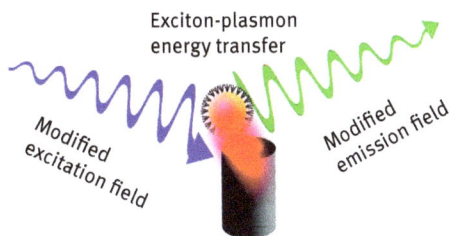

Fig. 6.5: Exciton-plasmon interactions in hybrid metal/semiconductor nanoparticles. Exciton-plasmon energy transfer might occur at the interface between the metal and semiconductor components, thereby modulating the electronic and optical properties of the individual nanoparticle species. Reprinted with permission from Achermann, *J. Phys. Chem. Lett.* **1** (2010), 2837–2843, © 2010 The American Chemical Society.

While a detailed discussion of plasmon-exciton interactions and the physical basis of the resultant opto-electronic effects are beyond the scope of this book, activity in this field attests to its scientific and technological potential. Researchers have recorded, for example, shifts in surface plasmon resonance (SPR) positions of metal NPs bound to semiconductor NPs, amplification of plasmon resonance through energy exchange, changes in exciton energies of the semiconductor NP constituents, both enhancement and attenuation of exciton emission (induced through quenching by the metal NP because of energy transfer from the semiconductor to the metal). Overall, the nature and consequences of plasmon-exciton interactions are closely related to the compositions and structures of the hybrid NPs. In addition to optical effects related to the semiconductor/metal interface, electronic properties have also been modulated. In particular, the "nanojunction" (i.e. the interface between the metal and semiconductor in the NPs) is responsible for shifts in the electronic energy levels and corresponding applications, such as field effect transistors.

Photocatalysis is a widely-encountered physical phenomenon in hybrid metal/semiconductor NPs and exploited for practical means. Semiconductor NPs have been recognized as promising platforms for harnessing sunlight to catalyze chemical reactions (see examples and discussion in Chapters 2 and 4). In general, the catalytic action of such NPs arises from utilization of the photo-excited electrons (and corresponding holes) for induction of chemical reactions. Importantly, the metal constituents in hybrid metal/semiconductor NPs have been found to stabilize light-induced excitons by "scavenging" the photo-excited electrons from the semiconductor conduction band, thereby extending the electron-hole charge separation lifetime. Figure 6.6A schematically depicts the catalytic action of hybrid nanoparticles. Specifically, an electron-hole pair is created by light irradiation. Crucially, the photo-excited electron at the conduction band of the semiconductor NP can be transferred to the metal through the relative proximity of the conduction band energy and the energy band of the metal NP (i.e. the "Fermi level"). From the metal NP the electrons can

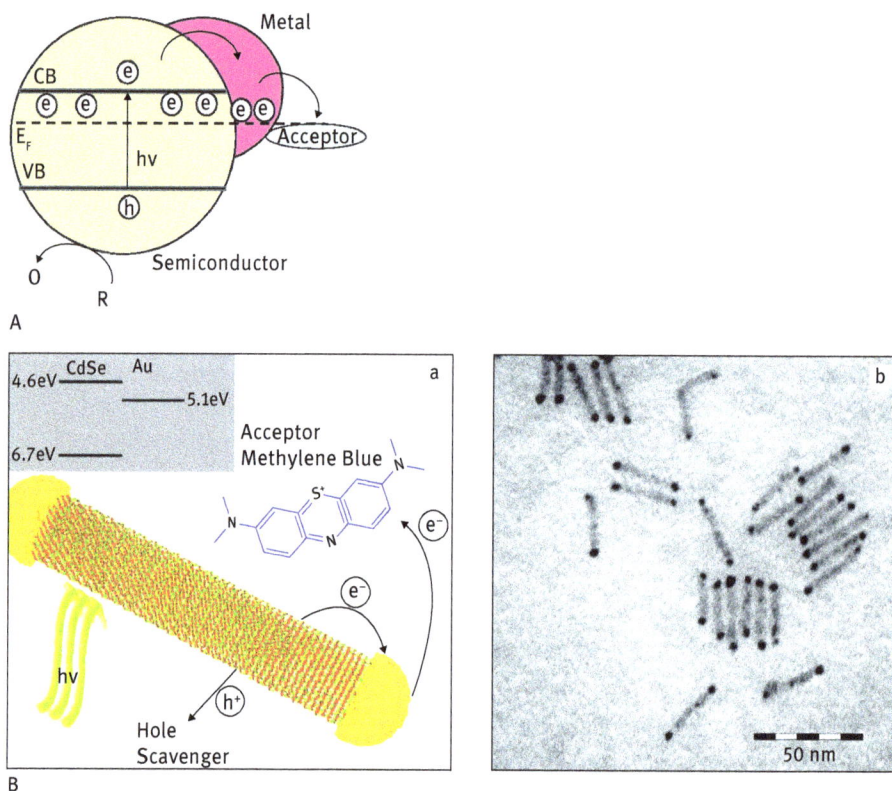

Fig. 6.6: Photocatalytic hybrid metal/semiconductor nanodumbbells. A: Energy level diagram accounting for the photocatalytic action of the hybrid nanoparticles; photo-excited electrons are scavenged from the conducting band of the semiconductor to the Fermi level of the metal nanoparticle component, thereupon transferred to acceptor molecules. The holes likewise can catalyze oxidation reactions. Reprinted with permission from Subramaniam et al., *JACS* **126** (2004), 4943–4950, © 2004 The American Chemical Society. B: Hybrid Au/CdSe nanodumbbells and their use in photocatalysis; (a) schematic depiction of the nanodumbbell and the energy level diagram; (b) electron microscopy image showing the nanodumbbells. Reprinted with permission from Costi et al., *Nano Lett.* **8** (2008), 637–641, © 2008 The American Chemical Society.

be further transferred to atomic/molecular "acceptors", and thus utilized as catalytic reducing agents. Similarly, the holes in the semiconducting valence band can catalyze oxidation processes (Fig. 6.6A).

Figures 6.6B,C present an example of dumbbell metal/semiconductor NPs and their application in organic catalysis. The system, designed by U. Banin and colleagues at the Hebrew University, Israel, consisted of CdSe nanorods capped at both ends with gold tips (synthesis of the Au/CdSe NPs was carried out by first synthesizing the semiconductor nanorods, which were then placed in an aqueous solution containing the precursors for Au NPs). The hybrid NPs efficiently reduced a model organic molecule (methylene blue) upon light irradiation. Notably, charge separation occurring within the metal/semiconductor NPs even allowed delayed photocatalysis – first irradiating the hybrid NPs, and subsequently transferring the NPs to a dark environment in which reduction of the organic molecule occurred.

Other configurations of hybrid metal/semiconductor NPs have been employed to enhance catalytic activity through different mechanistic pathways. An interesting Au@TiO$_2$ core-shell NP structure was designed by F. Zaera and colleagues at the University of California, Riverside, facilitating enhanced Au-catalyzed organic reactions (Fig. 6.7). Au NPs bound to TiO$_2$ surfaces have been shown to act as good catalysts in a variety of organic reactions (for example the environmentally-important reaction in which harmful carbon monoxide is oxidized to CO$_2$), however Au NPs tend to lose their catalytic action over time due to low stability and aggregation. The "yolk-shell" nanoparticle morphology shown in Figure 6.7, comprising a porous TiO$_2$ shell encapsulating multitude of small Au NPs, exhibited good catalytic activity. In particular, the TiO$_2$ shell appeared to both aid long-term preservation of the embedded Au NPs, and enable rapid and efficient diffusion of the reaction substrates from the external solution inward through the pores.

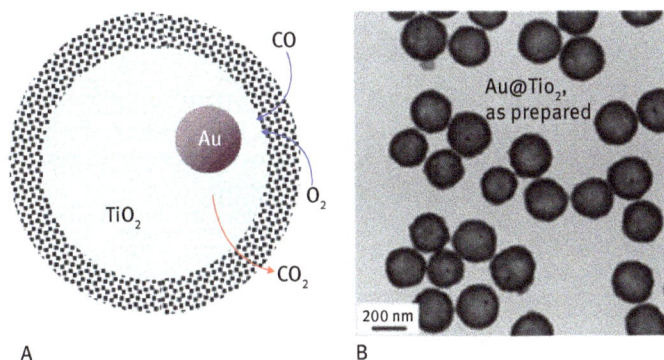

A   B

Fig. 6.7: Core-shell Au@TiO$_2$ nanoparticles. **A:** Schematic diagram of the catalytic action carried out by the Au nanoparticles embedded within the porous TiO$_2$ shells. **B:** Electron microscopy image showing the core-shell nanoparticle morphology. Reprinted with permission from Li et al., *Angew. Chem. Int. Ed.* **50** (2011), 10208–10211, © 2011 John Wiley & Sons.

Other hybrid metal/semiconductor core-shell nanoparticles have been reported. Figure 6.8 depicts extraordinary morphologies of Au@Cu$_2$O core-shell NPs synthesized by M. H. Huang and colleagues at the National Tsing Hua University, Taiwan. Control of the NP structures was achieved by using Au NPs in various shapes acting as "structure-directing" cores upon which the Cu$_2$O layer was assembled. As expected, interactions between the metal and semiconductor constituents exerted significant

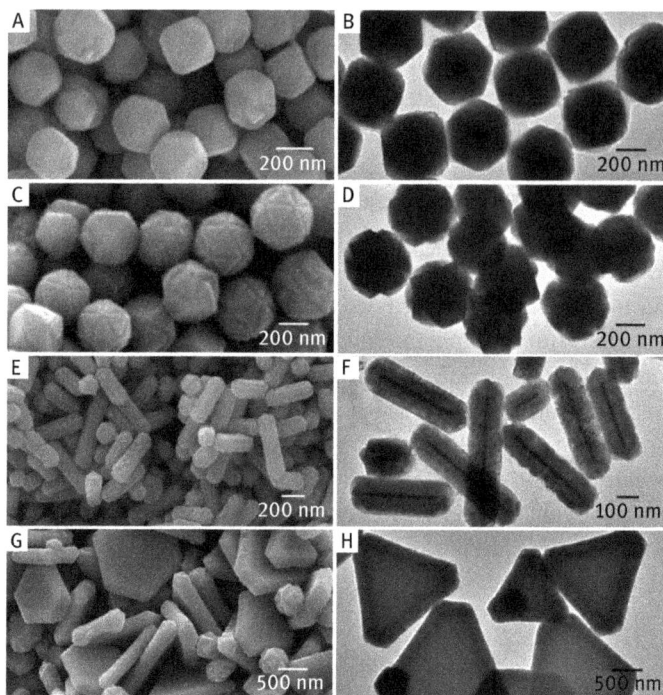

Fig. 6.8: Au@Cu$_2$O core-shell nanoparticles. Different nanoparticle morphologies were obtained by using distinctly-shaped Au nanoparticles as cores. **Left panels:** Scanning electron microscopy (SEM) images. **Right panels:** Transmission electron microscopy images. Reprinted with permission from Kuo et al., *JACS* **131** (2009), 17871–17878, © 2009 American Chemical Society.

effects on their respective physical properties; the plasmon resonance of the Au NP cores, for example, was quenched by the Cu$_2$O coating in these hybrid NPs.

Figure 6.9 depicts Au@Cu$_2$O core-shell NPs and highlights the close relationship between particle structure and optical properties. These hybrid NPs, synthesized by H. Wang and colleagues at the University of South Carolina, featured Au cores encapsulated by highly porous Cu$_2$O shells of varying thicknesses. In contrast to the Au@Cu$_2$O NP system discussed above (e.g. Fig. 6.8) and most core-shell metal/semiconductor NPs reported overall, in this case the Cu$_2$O shell was polycrystalline. Indeed, this unique shell organization gave rise to significant red-shifts of the Au NP cores' surface plasmon resonance (Fig. 6.9). This observation was ascribed to the "dampening" effects of the semiconducting shell on the gold plasmons, and further underscores the importance of geometrical parameters in modulating the physical properties of the metallic (Au) and semiconductor (Cu$_2$O) components in hybrid NPs.

Applications of hybrid metal/semiconductor core-shell NPs in catalysis (and other uses) have faced challenges, primarily due to difficulties in achieving uniform shell coatings. Such hurdles are generally traced to the "lattice mismatch" between the

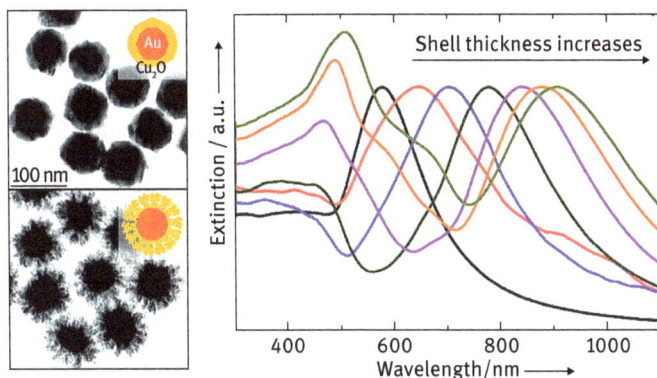

Fig. 6.9: Optical modulation in Au@Cu$_2$O core-shell nanoparticles of different shell thickness.
Left: Electron microscopy images showing nanoparticles with different Cu$_2$O shell thickness. **Right:**
Pronounced red shifts of the Au nanoparticle cores' surface plasmon resonance (SPR) on increasing shell thickness. Reprinted with permission from Zhang et al., *ACS Nano* **6** (2012), 3514–3527,
© 2012 American Chemical Society.

metal and semiconductor crystalline components. Indeed, "Janus-type" hybrid organizations appear more favorable, since Janus NPs have smaller interface areas between the two nanoparticle components than core-shell configurations, thus reducing the lattice mismatch.

6.2 Hybrid SiO$_2$-containing nanoparticles

Coupling silica (SiO$_2$) and inorganic materials such as metals (usually gold) and semiconductors in hybrid NPs has been a useful strategy designed to modulate and improve the properties of inorganic NPs and broaden their potential uses, particularly in biological imaging and catalysis. As discussed in more detail in Chapter 4.4, SiO$_2$ exhibits favorable properties including optical transparency and chemical stability. Moreover, SiO$_2$ is inert towards many biological and chemical molecules, although it can be chemically-coupled to various functional groups.

Core-shell metal@SiO$_2$ NPs have likely been the most widely-studied hybrid metal/SiO$_2$ NPs. SiO$_2$ shells are effective in preventing aggregation of metal NPs, a common problem which has limited the applicability of metal NPs in applications requiring long-term stability such as catalysis . Furthermore, SiO$_2$ coating of metal NPs usually makes the resultant NPs more stable in the harsh conditions often encountered in organic catalysis processes, such as prolonged exposure to high temperatures. SiO$_2$ is generally transparent, thus it does not significantly affect the optical properties of the metal NP cores. However, in some cases (particularly involving anisotropic core-shell structures such as nanorods) the SiO$_2$ shell did modify the plasmon resonances of the embedded NPs. An important contribution of SiO$_2$ coating of metal NPs is the availability of various functionalization avenues for the SiO$_2$ shell, enabling dissolution of the hybrid NPs into desired solvents, attachment of residues which minimize adverse biological effects (i.e. cytotoxicity), and/or providing the means for targeting the NPs through surface display of recognition elements.

A generic route for synthesis of core-shell metal@SiO$_2$ NPs is outlined in Figure 6.10. The metal NP core is first synthesized and subsequently coated with a molecular layer designed to attract (and/or react with) SiO$_2$ monomers such as triethoxysilane (TEOS). Coatings utilized in such reaction schemes have included silanol (e.g. alcohol-displaying silicon) units, amphiphilic polymers which allow transfer of the particles into nonaqueous solutions, and surfactants. Figure 6.10B portrays core-shell Au@SiO$_2$ and Ag@SiO$_2$ NPs prepared by C. Graf and colleagues at Utrecht University, Holland, by first coating the respective metal NP cores with the well-known amphiphilic polymer poly(vinylpyrrolidone), PVP. The PVP shell subsequently reacted with SiO$_2$ monomers (TEOS) which underwent polymerization to form the SiO$_2$ shell.

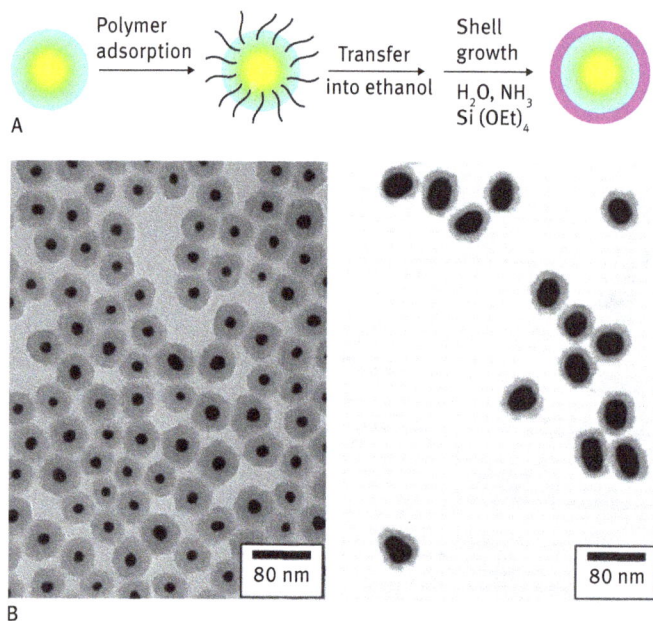

Fig. 6.10: Metal@SiO$_2$ core-shell nanoparticles. **A:** Generic synthesis scheme: the metal nanoparticle core is first synthesized and coated with amphiphilic polymers or surfactants. Shell growth can be carried out in organic solvent via reaction with SiO$_2$ monomers such as Si(OEt)$_4$ (TEOS). **B:** Electron microscopy images of Au@SiO$_2$ core-shell nanoparticles of different core and shell thickness, prepared via the procedure described in **A**. Reprinted with permission from Graf et al., *Langmuir* **19** (2003), 6693–6700, © 2003 American Chemical Society.

The synthesis technique outlined in Figure 6.10 has intrinsic capabilities for controlling the shell thickness by tuning concentrations and incubation times for monomer adsorption onto the surface of the metal NP cores.

SiO$_2$-coated Au nanorods (NRs) are considered promising candidates for "multifunctional" biological applications – potentially encompassing imaging, photothermal therapy, and drug delivery in a single hybrid particle. Figure 6.11 illustrates this concept as envisioned by C. Chen and colleagues at the National Center for Nanoscience and Technology, China. Specifically, the researchers constructed Au NRs encased within mesoporous SiO$_2$, a shell matrix which can further embed additional molecular cargo, in this case doxorubicin, a known anti-cancer compound. The Au NR cores exhibit two important functions: optical imaging via the localized surface plasmon resonance (LSPR) arising from the nanorod configuration (see Chap. 3 for discussion of LSPR in Au NRs), and the possibility of localized heating and destruction of target cells (i.e. hyperthermia therapy) via irradiation with infrared (IR) light. Importantly, the latter effect could also be employed for laser-induced, controlled release of the SiO$_2$-loaded drug cargo via localized heating and partial melting of the SiO$_2$ polymer network.

The SiO$_2$ shell in the Au NR@SiO$_2$ system illustrated in Figure 6.11, in fact SiO$_2$-coated core-shell NPs in general, often leads to reduced toxicity. This effect might be related to eliminating the direct interface between the metal NP and physiological/cell environments, thus preventing adverse processes such as electron transfer or formation of harmful reactive oxygen species (ROS). Lower toxicity is also explained by blocking release of metal ions and metallic fragments from the NP cores (shielded by SiO$_2$ in the core-shell nanoparticles). Nanoparticle aggregation – a common problem when using bare metal NPs in physiological environments – can also be minimized by SiO$_2$ coating. Finally, in the context of drug delivery applications, the mesoporous SiO$_2$ matrix constitutes an excellent platform for encapsulation of drug compounds (and their subsequent release through light irradiation in desired tissue targets).

Semiconductor NPs (i.e. quantum dots) have also been coated with SiO$_2$, yielding in some instances core-shell hybrid NPs exhibiting more favorable properties than bare quantum dots. Similar to metal NP core systems, SiO$_2$ coating reduced toxicity of semiconductor NPs – a critical issue since quantum dots have attracted intense interest as a bio-imaging tool. Indeed, in such applications the optical transparency of SiO$_2$ shells is essential since practical utilization of the hybrid NPs depends on retaining the luminescence properties and brightness of quantum dots. Another notable issue echoing hybrid metal@SiO$_2$ NPs discussed above is the feasibility of displaying functional units and recognition elements on the SiO$_2$ shell surface – thus providing means of molecular targeting and imaging of specific tissues, cells, and intracellular environments.

A typical synthesis scheme for creating surface-functionalized quantum dot@SiO$_2$ core-shell NPs is outlined in Figure 6.12. The chemical pathway, developed by D. Gerion and colleagues at the University of California, Berkeley, started with

A

50 nm

B

50 nm

C

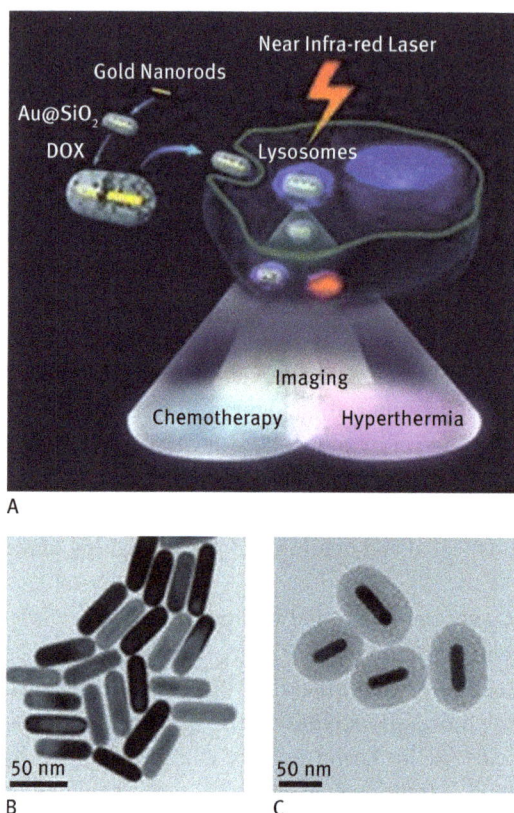

Fig. 6.11: Au@SiO$_2$ core-shell nanorods for cancer therapy. **A:** Schematic illustration of the technique. Au nanorods are coated with mesoporous SiO$_2$ shells which also host doxorubicin (DOX), an anti-cancer agent. The Au@SiO$_2$ nanorods are taken up by cells and can be employed for imaging (through surface plasmon resonance) and chemotherapy (release of drug cargo through hyperthermia – localized heating induced by infrared irradiation). **B:** Electron microscopy image of noncoated Au nanorods. **C:** Electron microscopy image of Au nanorods coated with the porous SiO$_2$ shell. Reprinted with permission from Zhang et al., *Adv. Mater.* **24** (2012), 1418–1423, © 2012 American Chemical Society.

quantum dots capped with a common stabilization surfactant (trioctylphosphine-oxide, TOPO). TOPO coating was then substituted by a layer of thiolated silane monomers, converted to silanol, and subsequently hydrolyzed to form a SiO$_2$ layer. The pendant hydroxyl moieties on the SiO$_2$ shell could be further covalently linked to silane monomers displaying tailored functional units – such as phosphonates, amines and thiols – for coupling of additional molecular units.

SiO$_2$ coating of *magnetic NPs* has also been demonstrated. These hybrid NPs have been designed mostly for biological and therapeutic applications, combining the advantages of the SiO$_2$ shell with the functionalities of magnetic NPs (see Chap. 4.1). In particular, the magnetic component in such core-shell NPs has important functions, including positional targeting onto desired body locations (through application of an external magnetic field), serving as a contrast agent in magnetic resonance imaging, and providing a mechanism for drug release through localized heating via application of alternating magnetic fields.

While the majority of hybrid core-shell SiO$_2$ NPs studied have been directed towards production of SiO$_2$ shells, inverse configurations were reported in which metals or substances other than SiO$_2$ formed the shells. Such NPs, in fact, display useful

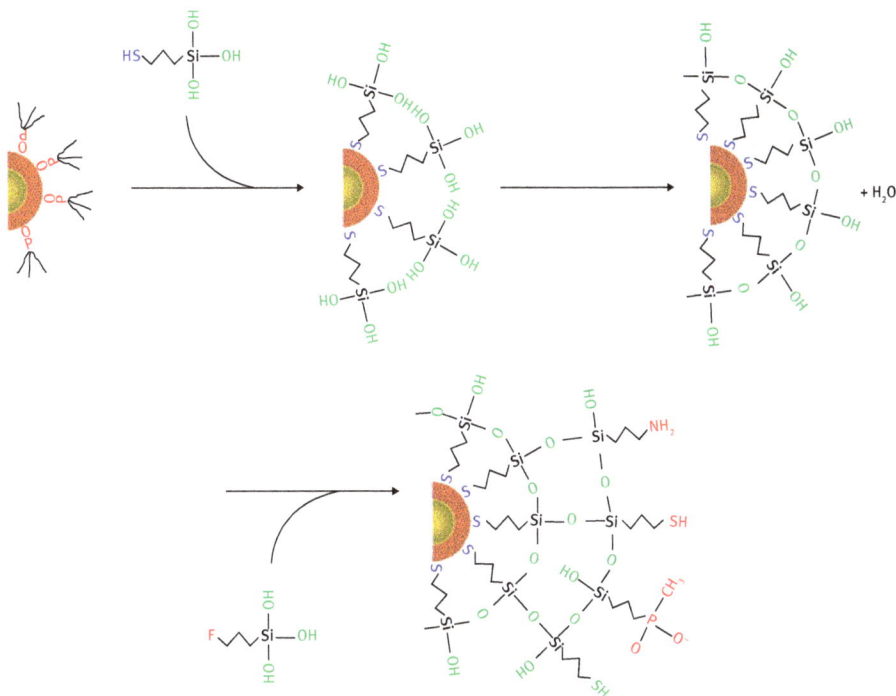

Fig. 6.12: Synthesis of quantum dot@SiO$_2$ nanoparticles. The method is based on substitution of the as-synthesized protective layer of the quantum dots with silanol monomers which are subsequently polymerized, producing a SiO$_2$ shell. Reprinted with permission from Gerion et al., *J. Phys. Chem. B* **105** (2001), 8861–8871, © 2001 American Chemical Society.

properties, for example in the field of heterogeneous catalysis. Figure 6.13 illustrates such a hybrid NP comprising of cobalt hydroxide [Co(OH)$_2$] "nanopatches" attached to the surface of SiO$_2$ NPs, which in this case constitute the particle "cores". Co(OH)$_2$ has been known as a catalyst of water oxidation reactions (i.e. generation of oxygen gas), however it is usually chemically unstable and tends to aggregate and lose its catalytic capacity. Interestingly, the hybrid NPs shown in Figure 6.13, synthesized by S. V. Lymar and colleagues at Brookhaven National Laboratory, were found to exhibit good and long-lasting catalytic activity. This observation has been ascribed to the role of the SiO$_2$ NP core in providing high surface area scaffold for immobilization of the Co(OH)$_2$ "satellite" NPs acting as catalytic centers, additionally preventing their aggregation and deactivation. The SiO$_2$ core also probably enabled enhanced adsorption of water molecules which were subsequently oxidized. Furthermore, the transparency of the SiO$_2$ NPs facilitates application of spectroscopic techniques for characterization and analysis of the catalytic reaction pathways, aiding research into improving the performance of hybrid NPs.

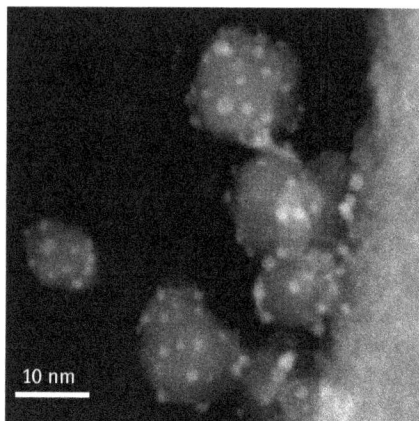

Fig. 6.13: Hybrid $Co(OH)_2/SiO_2$ nanoparticles. Electron microscopy image depicting the SiO_2 cores with $Co(OH)_2$ nanodomains (shown as whitish spots) adsorbed onto the SiO_2 surface. Reprinted with permission from Zidki et al., *JACS* **134** (2012), 14275–14278, © 2012 American Chemical Society.

Fully spherical SiO_2@Au core-shell NPs exhibit interesting optical and photothermal properties, specifically associated with the Au nanoshells. Likely the most important phenomenon of core-shell SiO_2@Au NPs, which could potentially be employed for practical applications, is their tunable surface plasmon resonance (SPR). Specifically, the position and intensity of SPR signals arising from the thin Au nanoshell depend on the inner and outer diameters of the shell, and thus can be readily tuned by varying the NP dimensionalities (Fig. 6.14).

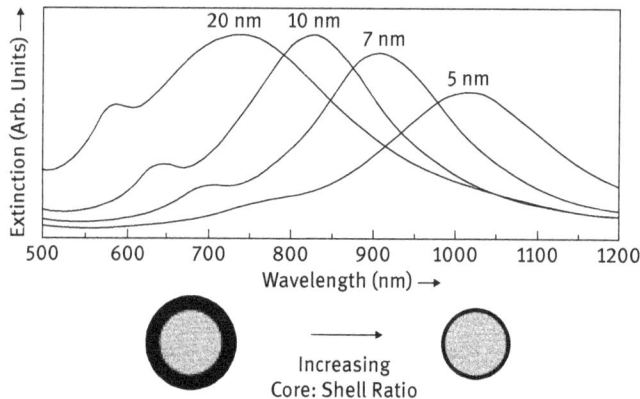

Fig. 6.14: Surface plasmon resonance (SPR) of SiO_2@Au core-shell nanoparticles depends on thickness of the Au shell. Calculated SPR spectra of core-shell nanoparticles of different Au shell thickness (shell thickness values are indicated above the spectra). Reprinted from *Chemical Physics Letters*, vol. 288, S. J. Oldenburg et al., *Nanoengineering of optical resonances*, pp. 243–247, © 1998, with permission from Elsevier.

Importantly, Au nanoshells which generally adsorb significant amounts of energy over their relatively wide cross-section cannot emit all the adsorbed energy as light, instead releasing heat at the NP surface. An intriguing application of heat-releasing SiO$_2$@Au core-shell NPs is illustrated in Figure 6.15. The SiO$_2$ core in this particle configuration served as a platform for growth and stabilization of the Au shell, which functioned as a "solar steam generator". Specifically, when illuminated, the Au nanoshells absorbed energy through their surface plasmon resonance, subsequently emitting part of this energy as heat. Exploiting this phenomenon, N. J. Halas and colleagues at Rice University have shown that continuous illumination of the SiO$_2$@Au NP suspension resulted in generation of steam around the NPs – arising from the significant heat dissipated at the Au nanoshell surface. Steam was subsequently released to the surrounding environment through transport of the core-shell NPs to the air/liquid interface.

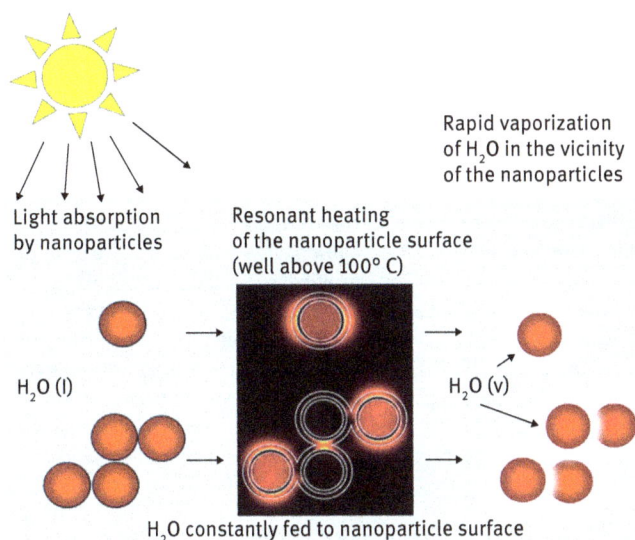

Rapid vaporization of H$_2$O in the vicinity of the nanoparticles

Light absorption by nanoparticles

Resonant heating of the nanoparticle surface (well above 100° C)

H$_2$O (l)

H$_2$O (v)

H$_2$O constantly fed to nanoparticle surface

Fig. 6.15: SiO$_2$@Au core shell nanoparticles as "solar steam generators". The nanoparticles absorb energy through the surface plasmon resonance of the Au shells. The absorbed energy is partly released as heat at the nanoparticles' surface, producing steam. Reprinted with permission from Neumann et al., *ACS Nano* **7** (2013), 42–49, © 2013 American Chemical Society.

Light-induced heating of SiO$_2$@Au core-shell NPs, combined with readily available chemical functionalization routes, makes these hybrid particles amenable for various biological applications. Figure 6.16 presents an example of a useful cell imaging application utilizing DNA-functionalized SiO$_2$@Au core-shell NPs. In the experiment, N. J. Halas and colleagues at Rice University covalently linked a single-strand DNA to the surface of the Au nanoshell (covalent bonding was facilitated by using DNA strands displaying thiol moieties). The single strand DNA was then hybridized with

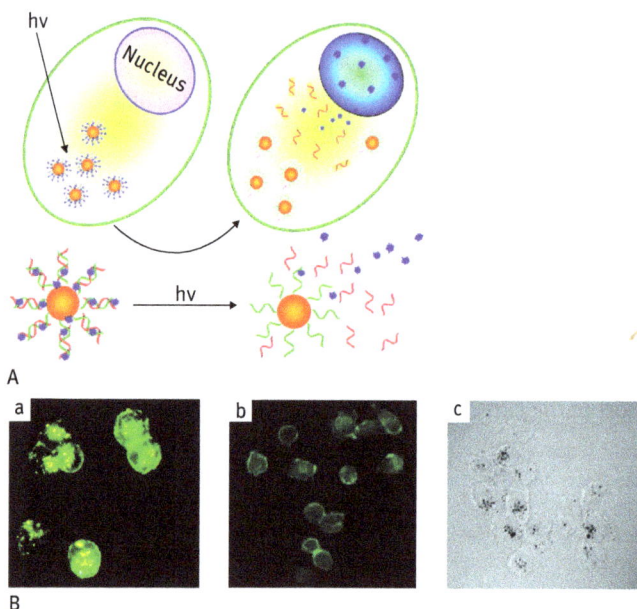

Fig. 6.16: Cell imaging using bio-functionalized SiO_2@Au core-shell nanoparticles. **A:** Schematic diagram of the experiment: SiO_2@Au core-shell nanoparticles were surface-functionalized with double-strand DNA which further hosted a fluorescent dye. Following internalization of the nanoparticles into cells, near infrared (NIR) irradiation was applied, resulting in localized heating of the Au nanoshells and consequent DNA de-hybridization leading to release of the fluorescent dye. **B:** Microscopy images demonstrating cell labeling: **(a)** highly-fluorescent cells subjected to the experimental procedure outlined in **A**; **(b)** control cells, not incubated with the SiO_2@Au core-shell nanoparticles; **(c)** optical microscopy image showing internalization of the SiO_2@Au core-shell nanoparticles into the cells (the nanoparticles correspond to the black dots). Reprinted with permission from Huschka et al., *Nano Lett.* **10** (2010), 4117–4123, © 2010 American Chemical Society.

the complementary strand, and the double-stranded DNA further hosted (via inter-strand intercalation) additional molecular guests (fluorescent dyes in the experiment depicted in Figure 6.16, although other guest molecules can be intercalated as well). Following uptake of the hybrid NPs into cells, light irradiation in the near infrared (NIR) spectral region (which is not absorbed by water or physiological tissues) resulted in heating of the Au nanoshell, consequently resulting in de-hybridization of the DNA double strand and release of the fluorescent cargo.

An elegant variation of the localized heating effect in SiO_2@Au core-shell NPs and its application for biological targeting and therapeutics is illustrated in Figure 6.17. In this hybrid NP system, developed by D. S. Kohane and colleagues at MIT, the SiO_2@Au NPs were chemically conjugated to a specific recognition element (peptide ligand), which was further embedded within a protective layer of a heat-sensitive hydrophobic polymer. The polymer layer effectively shielded both the peptide ligand and the

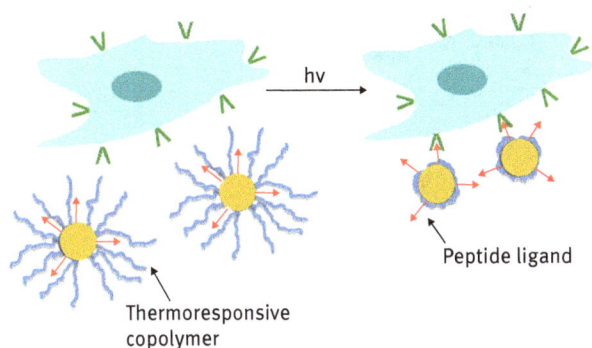

Fig. 6.17: Light-induced targeting of functionalized SiO_2@Au core-shell nanoparticles. The core-shell nanoparticles are coated with a recognition element (red arrows) and a thermo-responsive polymer layer. Upon irradiation of the nanoparticles with near-infrared (NIR) light, localized heating at the vicinity of the Au nanoshell gives rise to disintegration of the polymer layer and consequent exposure of the peptide recognition element – making cell targeting by the nanoparticles possible. Reprinted with permission from Barhoumi et al., *Nano Lett.* **14** (2014), 3697–3701, © 2014 American Chemical Society.

nanoparticle itself from degradation in physiological solutions, making the hybrid NPs biologically insert. However, irradiation with NIR light gave rise to localized heating of the Au nanoshell, resulting in "collapse" of the copolymer layer and consequent exposure of the peptide recognition units, making specific interactions of the NPs with their cellular destinations possible. This strategy might be used for targeted photothermal therapy, drug delivery (for example through encapsulating therapeutic molecules within the polymer layer), and imaging.

Core-shell SiO_2 nanoparticles not containing metal components have been synthesized as well. Figure 6.18 portrays bright fluorescent SiO_2@SiO_2 core-shell NPs. The synthesis scheme, developed by U. Wiesner and colleagues at Cornell University, utilized SiO_2 monomers covalently attached to a fluorescent dye which formed the NP core. A transparent outer SiO2 shell was then grown around the fluorescent core, stabilizing the NPs and additionally providing a means for controlling the size of the core-shell NPs. The remarkable brightness of these core-shell NPs (which was close to that of semiconductor quantum dots widely used in biological imaging applications) has been ascribed to compacting of the fluorescent dyes within the NP cores. Furthermore, SiO_2 NPs are likely to be less toxic than semiconductor-based QDs, and thus might constitute a useful bio-imaging platform.

6.3 Hybrid polymer-metal nanoparticles

Creating nanoparticles which integrate polymers and metals has been a powerful strategy for modulating and improving the properties of purely metallic NPs. Even

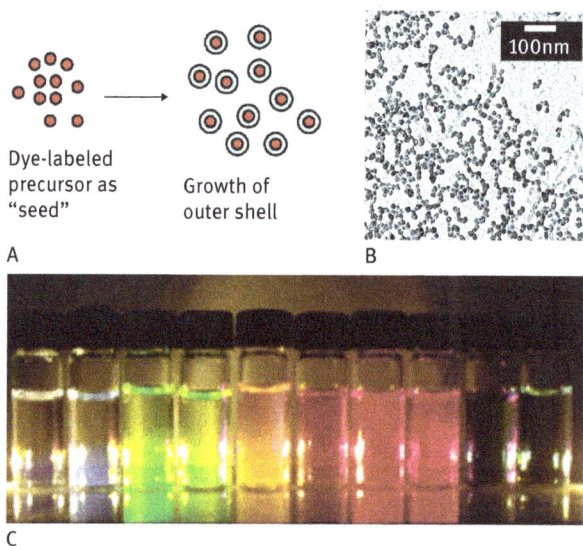

Fig. 6.18: Fluorescent $SiO_2@SiO_2$ core-shell nanoparticles. **A:** Nanoparticle preparation procedure: SiO_2 labeled with a fluorescent dye is used to construct the core, subsequently coated with
a transparent SiO_2 nanoshell. **B:** Microscopy image showing the homogeneous distribution of the
nanoparticles. **C:** Photograph of $SiO_2@SiO_2$ core-shell nanoparticles with different fluorescent dyes
attached to the SiO_2 cores. Reprinted with permission from Ow et al., *Nano Lett.* **5** (2005), 113–117,
© 2005 American Chemical Society.

more pronounced than the SiO_2-containing hybrid NPs discussed in the previous
sections, the huge variety of polymeric materials has spawned new applications and
research avenues focusing on hybrid polymer/metal NPs. The main morphologies
reported for these NP species echo other hybrid systems summarized in this chapter – mostly Janus and core-shell configurations. In particular, the polymer shells in
metal@polymer core-shell NP organizations constitute a platform for biological and
chemical surface functionalization, and also provide protection and stabilization layers to the metal cores. Moreover, the choice of polymer material facilitates control of
the thickness, rigidity, and porosity of the shells.

Multifunctional polymer-containing hybrid NP systems have attracted particular
interest in this field, because in many instances each component contributes distinct
characteristics which affect the overall functionalities of the composite nanoparticle. As an example, B. H. Chung and colleagues at the Korea Institute of Bioscience
and Biotechnology synthesized core-shell NPs in which the cores comprised of interspersed fluorescent polymer and magnetic iron oxide NPs, while the shell material
was SiO_2 (Fig. 6.19). These hybrid NPs were proposed as potential agents for biological imaging and therapeutic applications, building on the complementary functions
of the different building blocks. Specifically, the iron oxide was a vehicle both for
magnetic resonance imaging and localized tissue destruction by alternating magnetic

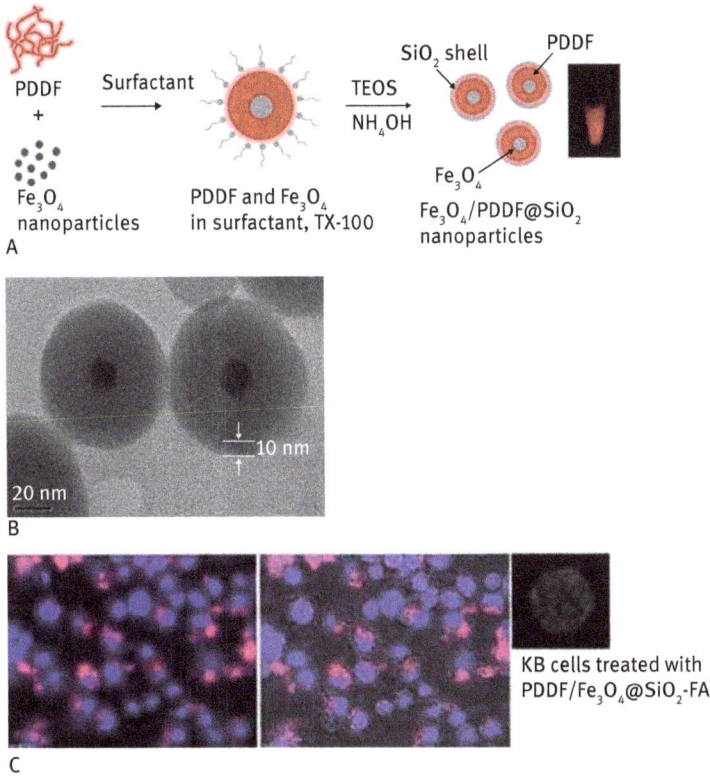

Fig. 6.19: Multifunctional hybrid nanoparticles for biological applications. **A:** Synthesis scheme: a fluorescent polymer (PDDF) is interspersed with Fe_3O_4 nanoparticles in the presence of a surfactant stabilizer, yielding core-shell Fe_3O_4@PDDF nanoparticles. These nanoparticles are subsequently incubated with silica monomers (triethylorthosilicate, TEOS), ultimately forming SiO_2 nanoshells. **B:** Electron microscopy image showing the iron oxide core, the internal polymer shell, and external (thinner) SiO_2 shell. **C:** Fluorescence microscopy images (left and middle) and magnetic resonance image (MRI, right) obtained following incubation of cancer cells with the Fe_3O_4@PDDF@SiO$_2$ core-double-shell nanoparticles coupled to the folate (FA) ligand. Insertion of the nanoparticles into the cells is evident by the cell-associated purple-pink fluorescent domains (cell positions are traced with the aid of a blue fluorescent dye labelling the cell nuclei). The picture on the left corresponds to fluorescence confocal microscopy image while the middle image is a superposition of both fluorescence and optical microscopy images. The magnetic resonance image on the right demonstrates signal attenuation (darker domains) due to magnetic relaxation induced by the cell-associated Fe_3O_4 nanoparticle cores. Reprinted with permission from Lee et al., *Macromol. Biosci.* **13** (2013), 321–331, © 2013 John Wiley & Sons.

fields generating heat (e.g. hyperthermia effect; Sect. 4.1). The fluorescent polymer in the hybrid NP system enabled cell imaging, and the SiO_2 shell endowed biocompatibility, low cellular toxicity, and biological targeting through covalent display of biological recognition elements. A practical application of the polymer/Fe_3O_4@SiO$_2$ core shell NPs is presented in Figure 6.19C. The SiO_2 surface of the NPs was functionalized with folate, a residue recognized by receptors over-expressed on the surface of cancer cells; binding the NPs to the cell surface resulted in their efficient uptake into the cells which made both fluorescence microscopy visualization of the cells and enhanced contrast of the MRI image possible.

Fluorescent metal NPs represent another hybrid NP system utilizing fluorescent polymers. As discussed in Chapter 3, a well-known feature of Au NPs (and other metal NPs like Ag NPs) is the quenching of fluorescence associated with species in close proximity to the NP surface, occurring via transfer of the fluorescence energy from the fluorescence-emitter onto the metal nanoparticle. However, in cases where the fluorescent dye and metal maintain large and constant distances, fluorescent enhancement effects are actually recorded, termed *metal-enhanced fluorescence (MEF)*. This phenomenon has not yet been fully explained, however it is believed to arise from electronic energy radiating from the NPs and transferred to the fluorescent dyes (which accordingly constitute "fluorescence energy acceptors"). Strategies have been devised to exploit MEF for designing fluorescent metal NPs. The key requirement in such systems is to prevent direct contact between the metal core and the fluorescent dye/residue, thereby disabling fluorescence quenching by the metal. Instead, the Au cores in such NP species (or other noble metals in some cases) can be coated with nonmetallic layers in order to achieve separation of the fluorophores and the metal surface, thereby enabling MEF.

Figure 6.20 depicts an example of an MEF-generating NP system. The NPs, synthesized by L. Li and colleagues at the University of Science and Technology, China, comprised of core-shell Au "nanostars". The nanostar configuration had been selected due to the enhanced MEF, associated with the anisotropic NP shape and surface discontinuities – believed to give rise to more pronounced electronic energy transfer from the Au core. The Au nanostars were further coated with two polymeric layers (Fig. 6.20A). The first shell comprised of gelatin which provided a dense barrier around the gold nanostar core. A molecular cross-linker (glutaraldehyde) was also added to further stabilize the gelatin layer, preventing its disruption and dissolution in physiological solutions. A conjugated fluorescent polymer (PFVCN) was subsequently docked onto the surface of the core-shell NPs by electrostatic attraction. The experimental data confirmed that these composite NPs exhibited significantly enhanced fluorescence, believed to occur through MEF. Furthermore, the positive charge on the hybrid Au@polymer core-shell NP surface and the relatively low toxicity of the NPs enabled their use as cell imaging agents (Fig. 6.20B).

Asymmetric Au/polystyrene "nanocorals", of interest for biological imaging and analysis, are depicted in Figure 6.21. This hybrid NP design, reported by L. P. Lee and colleagues at the University of California, Berkeley, combined Au "nanoshells" and *nanohemispheres* comprising polystyrene (PS), a broadly-used and readily-functionalized polymer. The researchers showed that the "roughened" Au nanocoral exhibited excellent sensing capabilities by surface enhanced Raman spectroscopy (SERS)

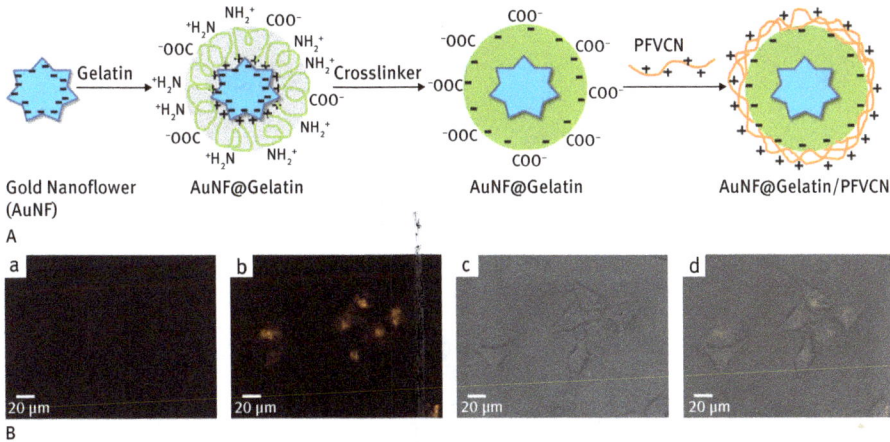

Fig. 6.20: Fluorescent Au/polymer core-shell nanoparticles. **A:** Preparation scheme: Au "nanoflower" (AuNF) synthesized and coated with cross-linked gelatin layer. The AuNF@gelatin core-shell nanoparticles are further coated with a fluorescent polymer (PFVCN) overall yielding nanoparticles exhibiting enhanced fluorescence through the metal-enhanced fluorescence (MEF) effect. **B:** Cell imaging with the fluorescent AuNF@gelatin/PFVCN nanoparticles; (**a**) fluorescence microscopy image of cells not treated with the nanoparticles; (**b**) cells incubated with the AuNF@gelatin/PFVCN nanoparticles – fluorescent domains are apparent; (**c**) bright field microscopy image of the cells; (**d**) combined bright field/fluorescence microscopy image confirming insertion of nanoparticles into the cells. Reprinted with permission from Cui et al., *ACS Appl. Mater. Interfaces* **5** (2013), 213–219, © 2013 American Chemical Society.

(see Chapter 3 for discussion of SERS phenomena). The hybrid NPs could be directed to specific targets via functionalization of the PS surface, for example with antibodies directed against cell surface-displayed residues (Fig. 6.21B); the fluorescence microscopy experiment in Figure 6.21B employed fluorescently-labeled PS to highlight cell binding of the NPs.

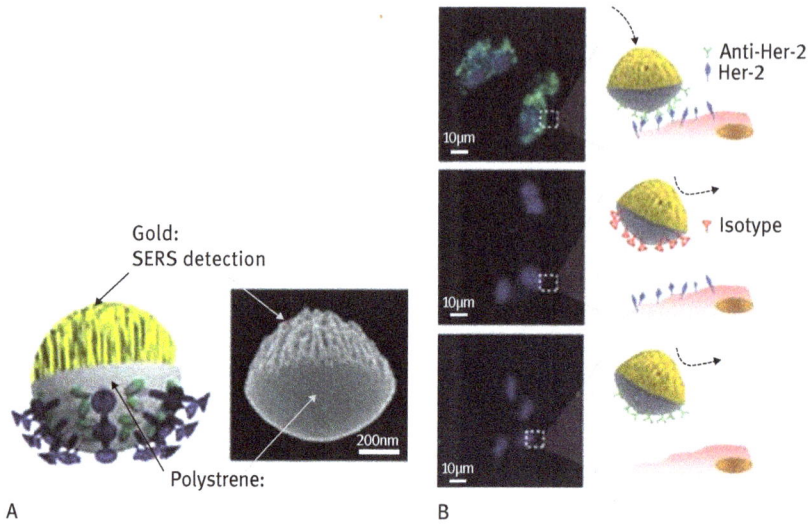

Gold:
SERS detection

Polystrene:

A

Anti-Her-2
Her-2

Isotype

B

Fig. 6.21: Au/polymer "nanocorals". **A:** Schematic depiction of the nanoparticles, comprising of Au nanoshell attached to a polystyrene nanohemisphere. The polystyrene surface can be function-alized with recognition elements. **B:** Cell imaging using the nanocorals functionalized with cell-specific antibodies. Cell labeling is apparent only when the nanocorals are functionalized with an-tibodies recognizing a cell displayed protein, HER-2 (green fluorescent domains, top row). No cell surface fluorescence labeling is recorded when the nanocorals are functionalized with a different antibody (middle row), or when the nanocorals are incubated with cells not displaying the HER-2 protein (bottom row). The blue fluorescence corresponds to cell nuclei labeled with a dye employed for cell tracking. Reprinted with permission from Wu et al., *ACS Appl. Mater. Interfaces* **5** (2013), 213–219, © 2013 American Chemical Society.

7 Nanoparticle interactions with biomolecules and cells

The expanding applications of nanoparticles in biology and biomedicine are the backdrop for extensive research efforts aimed at elucidating the effects of NPs upon biological entities. This issue is particularly pertinent in numerous studies analyzing nanoparticle *toxicity*. In a broader context, the similar dimensionalities of NPs and many biological molecules, the diverse chemical reactions involving NPs, and the many examples of biomolecular display on NP surfaces have been a powerful driving force for investigating the interface between NPs and biological systems. This chapter discusses several aspects of this burgeoning field; this topic has been also examined in other chapters as it is naturally central to biological applications of many NP families.

A prominent aspect of NPs concerns their small, sub-cellular size regime. The implication of this fundamental physical feature is that NPs might easily penetrate into cells and transverse cellular and tissue barriers in the body. Figure 7.1 summarizes the prevalent transport mechanisms by which NPs can pass through cellular

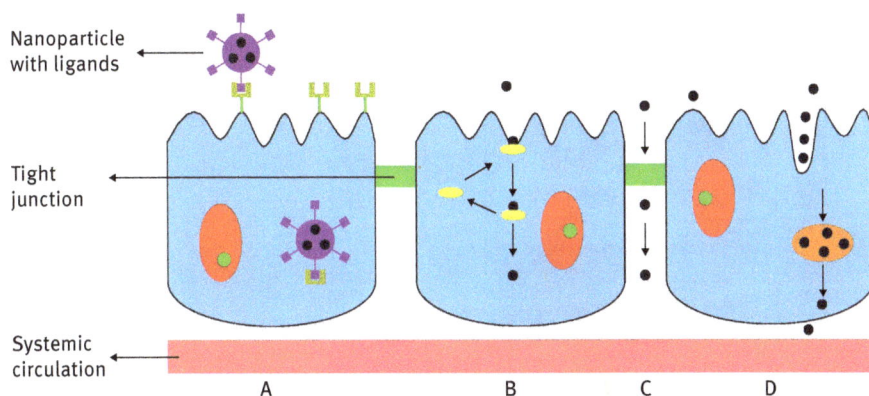

Fig. 7.1: Potential mechanisms of nanoparticle transport across cell barriers. Schematic diagram depicting common pathways by which nanoparticles (black dots) can traverse cell barriers and reach the circulatory system. **A:** Receptor-mediated transport: nanoparticles displaying biological ligands bind to specific cell surface receptors which subsequently induce cell internalization of the nanoparticles and release into the circulatory system. **B:** Carrier-mediated transport: a process similar to **A**, in which nanoparticle uptake is mediated by carrier molecules (usually proteins) which help overcome the membrane barrier. **C:** Paracellular transport: passive diffusion through the space between cells (usually through the "tight junctions"). **D:** Transcellular transport: generic mechanism in which nanoparticles (and typical small molecules) diffuse through the cell membrane into epithelial cells and are transported to the circulatory system. Drawing inspired by Yun et al., *Adv. Drug Deliv. Rev.* **65** (2013), 822–832.

barriers (membranes) and reach the circulatory system. Since NPs mimic biological molecules they can, in principle, traverse epithelial cell barriers via common pathways such as *receptor-mediated* or *carrier-mediated transport* (in which NPs coupled to specific ligands bind onto cell surface receptors, subsequently internalized and transferred through the cells to blood vessels; Fig. 7.1A,B); *paracellular transport* of NPs through the spaces between adjacent cells (i.e. the "tight junctions"; Fig. 7C); and *endocytic transcellular transport*, a generic means by which external particles are taken up into cells within enclosed endosome "compartments" (Fig. 7.1D). While the possibility of seamless NP transport would be beneficial in certain biomedical applications, particularly drug delivery and cellular targeting, it poses clear risks, as NPs could interfere with and/or disrupt physiological and cellular processes in undesired locations.

Nanoparticle size has been found to closely affect their transport into and out of blood vessels. This aspect is particularly important in *anti-tumor* drug delivery applications in which the therapeutic action is dependent upon transport of the bioactive particles within the nanoporous vasculature surrounding tumors. Figure 7.2 schematically depicts an NP "filtering" mechanism enabling particle passage from the circulation network to target tissues (i.e. tumors) through "leaky" blood vessels. Specifically, only smaller NPs can penetrate the nanometer-size gaps between the epithelial cells coating the vessel walls.

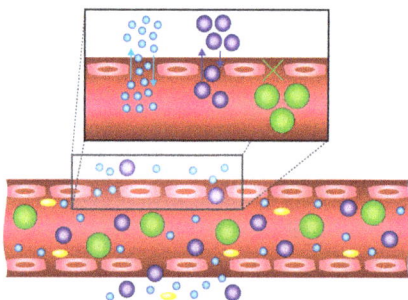

Fig. 7.2: Size-dependent nanoparticle penetration through blood vessel walls. Movement of nanoparticles to/from tissues through intercellular spaces closely depends on size. Reprinted with permission from Farokhzad and Langer, *ACS Nano.* **3** (2009), 16–20, © 2009, American Chemical Society.

Shape has a profound effect on the interactions of NPs with biological entities, particularly cells. Many studies have addressed the relationship between NP geometry and cell entry. It is generally accepted that NP shape effects are mostly manifested in processes occurring at interfaces, specifically the cell membrane. However, while NP shape seems to significantly affect the efficiency and rate of cell uptake, there is no consensus yet on the existence of specific shape/morphology parameters which either promote or prevent NP penetration through cell membranes. Part of the reason for this uncertainty is the realization that NP uptake into cells follows diverse routes, not only those dictated by receptor-mediated processes which are expected to depend on NP shape (e.g. Fig. 7.1A). NP shape also affects cell physiology after internalization. Some studies have pointed to distinct intracellular signaling pathways induced *after* cell uptake of NPs of different shapes. Specifically, different cytokines (a prominent class of proteins released by cells on triggering an immune response) were secreted by cells treated with NPs of various shapes.

Figure 7.3 outlines a representative experiment demonstrating shape-dependent mechanism of NP cell entry. The system illustrated in Figure 7.3, developed by B. Yan and colleagues at Shandong University, China, comprised of polystyrene NPs with the same diameters and surface chemistry, albeit exhibiting different morphologies (nano*spheres* vs nano*discs*). Indeed, transformation from *three-dimensional* morphology (spheres) into *two-dimensional* structures (discs) significantly reduced cell uptake. These results were ascribed to the more extensive interface between the disc-shaped NPs and the cell membrane, which presumably led to attachment and accumulation of the NPs at the membrane. On the other hand, the spherical NPs were more amenable to internalization via endocytic pathways as they exhibit a smaller direct contact area with the membrane. Notably, the study outlined in Figure 7.3 represents a typical conundrum encountered in cell/NP studies: reduced cell entry via NP shape manipulation might on the one hand minimize the adverse toxic effects incurred following cell entry by NPs, but on the other hand could also limit the therapeutic potential of the NPs.

In addition to cell entry, numerous studies have attempted to decipher the effects of NPs on cellular processes such as motility (cell motion), cell shape changes, and cell division. In many instances, NPs were shown to adversely affect those funda-

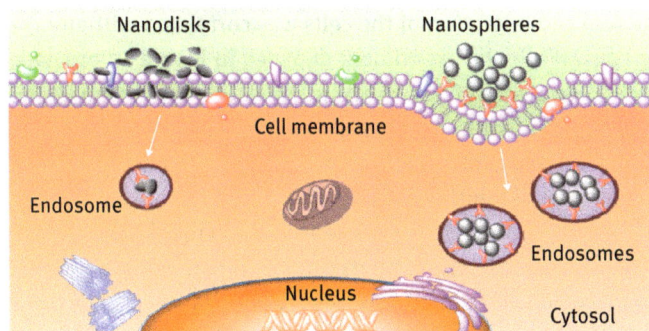

Fig. 7.3: Effect of nanoparticle shape on cell entry. Polymer nanodiscs accumulate at the cell membrane while nanospheres are internalized more as they exhibit a smaller interface with the membrane surface. Reprinted with permission from Zhang et al., *ACS Appl. Mater. Interfaces* **4** (2012), 4099–4105, © 2012, American Chemical Society.

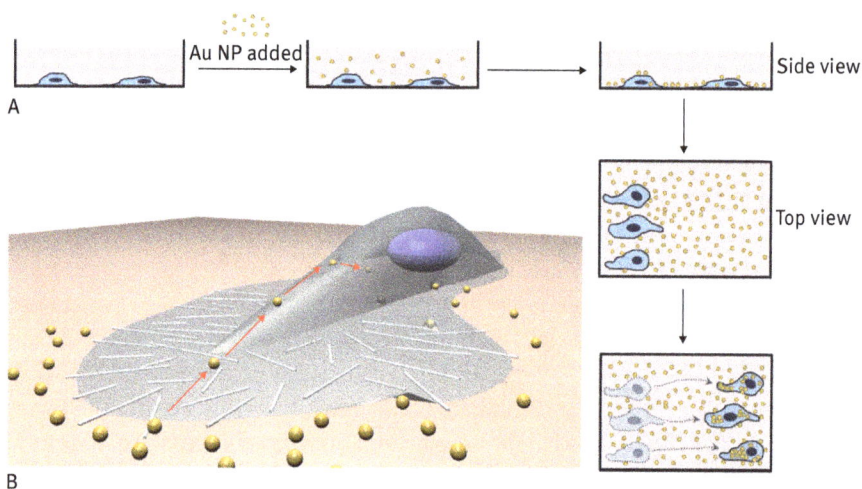

Fig. 7.4: Surface-deposited nanoparticles reduce cell mobility. **A:** Experimental scheme: cells are plated on a surface followed by deposition of Au nanoparticles. Migration of the cells is monitored, revealing that the Au NPs are internalized by the cells while simultaneously retarding their motion upon the surface. **B:** Diagram showing a moving cell absorbing nanoparticles in its vicinity. Reprinted with permission from Yang et al., *Nano Lett.* **13** (2013), 2295–2302, copyright (2013), American Chemical Society.

mental cell events, although experimental data are highly dependent on cell types and NP properties. Figure 7.4 illustrates an elegant experiment highlighting the impact of Au NPs on cell motility. The system devised by C. J. Murphy and colleagues at the University of Illinois comprised of cells attached to a surface on which Au NPs were co-deposited. The researchers recorded two intriguing and inter-related phenomena. Specifically, cell movement on the surface was clearly retarded in the presence of the NPs. Furthermore, the reduced migration of the cells was correlated with uptake of the NPs by the moving cells. While the experiment depicted in Figure 7.4 provides compelling evidence for the interference of NPs in a fundamental process such as cell movement on a solid substrate, that work did not shed light on the exact molecular-mechanistic aspects, underscoring the formidable difficulties in elucidating the precise parameters at play in the NP/cell interface.

Cells respond to NPs in their environment even when no actual uptake of particles occurs. Studies have shown, for example, that cells experience lower motility and impeded proliferation when placed on surfaces containing an abundance of nanowires and nanorods. Such effects have been ascribed to the putative distortion of the cell membrane in contact with these nanometer-sized species. NP-induced morphological changes of the cell membrane are believed to also indirectly affect intracellular organelles, resulting in adverse consequences for cellular processes. Overall, these observations underlie the fact that cells recognize and respond to nanoparticles, likely through interactions at the cell surface interface.

Nanoparticle *toxicity* is an acute concern in light of the expanding applications of NPs in biology and biomedicine. Awareness of this aspect was raised early on, since NPs exhibit sizes similar to many proteins and biological particles like small viruses. Numerous studies have indeed produced evidence of NP interference in intracellular signaling and metabolic processes. Other than size and shape, the surface chemistry of NPs is a predominant factor in inducing cytotoxicity. Factors such as charge and re-activity of functional groups displayed on NP surfaces exert significant biological ef-fects. *Surfactant molecules* (such as cetyltrimethylammonium bromide, CTAB, widely used for synthesis of metallic NPs) which are commonly used as NP capping agents are particularly toxic, likely because of (electrostatic) interactions with cellular mem-branes and consequent disruption of the membrane framework. Accordingly, since most cellular membranes exhibit net negative charge (due to excess of anionic lipids), it is believed that NPs coated with cationic surfactants are generally more toxic than their anionic or uncharged counterparts.

The central roles of the capping layers in shaping NP toxicity – specifically their molecular composition and surface charge – have led to development of diverse strate-gies to mitigate the adverse biological impact by modulating NP surface architecture. The most effective approach in that regard has been to substitute the surfactant molecules with more biocompatible layers. Various chemical treatments have been designed to replace reactive surfactant residues with inert substances such as silica, biocompatible polymers, hydrophobic coatings and others. In many instances, sur-face modifications produce added benefits such as enhanced NP stability (i.e. less disintegration in physiological solutions), which also contributes to reduced toxicity. Indeed, maintaining physical and chemical integrity of NPs in actual physiological conditions is considered a major challenge that has not been effectively addressed yet. The physiological "soup" contains a multitude of ions, biomolecules, and cells, which could interact with NPs, adsorb onto their surface, and result in dissolution, size/shape alteration, or in many cases aggregation of NPs. All of these processes could significantly alter the biological activity of NPs.

While size, shape, and surface chemistry are properties that can be studied and tuned, other complex factors probably contribute to the toxicity of NPs. Figure 7.5 il-lustrates substances present in NP solutions which might adversely affect cells and tissues. Among the species that might contribute to toxicity are organic molecules and reaction side-products from NP synthesis procedures. Particularly acute are metal species originating from NP cores; cadmium ions, for example, released from semi-conductors quantum dots (widely used as bioimaging agents) have well-known toxic profiles. Other potentially harmful substances are proteins and charged species inad-vertently released from the NP surface which induce physiological damage.

Experiments have been carried out to assess the relative contributions to toxicity of NPs themselves in comparison to substances released to solution from NPs (such

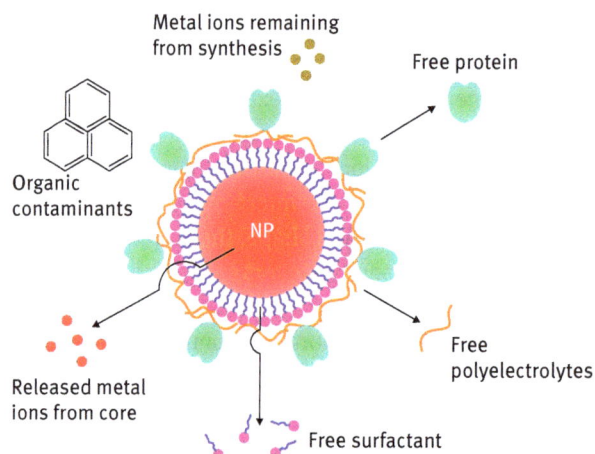

Fig. 7.5: Substances associated with nanoparticles which have possible adverse biological effects.

as those depicted in Fig. 7.5). Figure 7.6, for example, summarizes the results of a simple experiment designed by C. J. Murphy and colleagues at the University of Illinois to address this issue. The researchers exposed cells to two suspensions: Au nanorods

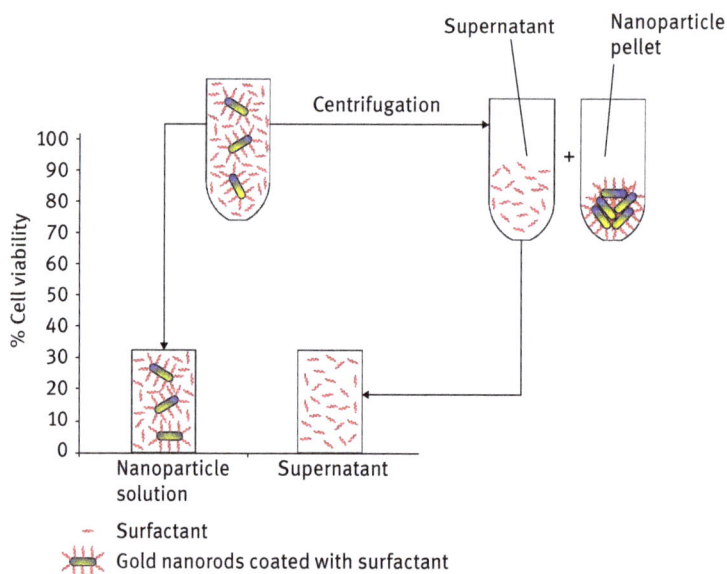

Fig. 7.6: Assessing toxic effects of soluble small molecules and ions in nanoparticle solutions. Similar cell viabilities were recorded when cells were incubated with whole solutions of gold nanorods, and the same solutions centrifuged to remove the nanorods. This result suggests that soluble species – small molecules and ions – were largely responsible for the toxicity. Figure inspired by Alkilany and Murphy, *J. Nanoparticle Res.* **12** (2010), 2313–2333.

capped by surfactant layers and the same solution after removal of the NRs via centrifugation. Importantly, the extent of cell death was almost identical in the two cases, indicating that soluble substances were responsible for cell toxicity (probably surfactant molecules released from the NR coating) rather than the NRs themselves. Overall, while organic reagents, metal ions, and surfactants present in NP solutions (highlighted in Fig. 7.5) are not actual constituents of the NPs themselves, from a practical standpoint such species could nevertheless have significant biological effects and need to be considered in potential therapeutic applications of NPs.

Another point to consider on evaluating NP cell uptake and toxicity is the fact that surface coating of the particles cannot be made totally inert in physiological environments. Indeed, the physiological milieu probably gives rise to attachment of proteins (and other biomolecules) to the NP surface. Such adsorption processes are dynamic in nature and depend on factors related both to NP properties, such as shape and surface functional groups, as well as solution conditions including concentration and affinity of molecules to the NPs. However, binding of different biomolecules onto NPs could significantly alter their surface architecture, charge, and reactivity, and consequently influence their biological/toxic profiles. Some researchers, in fact, have asserted that formation of protein "corona" on NP surfaces is the most important factor determining their "biological identity". Such coronas are highly complex, however, and according to some estimates they are constructed from hundreds of different molecules, making exact analysis of their impact challenging.

While NP toxicity has been widely reported, it should be noted that most work in this field has been conducted in cell culture models (e.g. in vitro) rather than in actual body environments (e.g. in vivo: functional conditions maintained in tissues or whole animals). Notably, in some cases NPs which displayed high cytotoxicity in cell culture models were, unexpectedly, much less toxic in animal models. These observations underscore the fact that the biological interface and impact of NPs are closely dependent on their physiological environments. Furthermore, the well-known distinctions between cell culture models (two-dimensional cell systems which grow in controlled media) and actual human (or animal) systems, which constitute three dimensional matrixes interfacing with multicomponent complex solutions interacting with both cells and NPs, likely have a pivotal effect on NPs and their functionalities.

Characterization of the biological and toxic profiles of NPs has not been limited to cells, tissues, or whole animals. In fact, an expanding body of work has focused on NP interactions with biological molecules, primarily proteins. Such investigations are particularly pertinent because in many instances the dimensionalities of NPs are close to those of biological macromolecules (and vice versa). The diverse and reactive coating layers of NPs further contribute to the variety of interactions occurring between NPs and biomolecules. Figure 7.7, for example, presents an Au NP system designed to probe interactions between a protein (α-chymotrypsin, a well-known protease, or protein-degrading enzyme) and different functional residues coating the NP. The diagram in Figure 7.7 illustrates the NP structural features. Specifically, V. Rotello

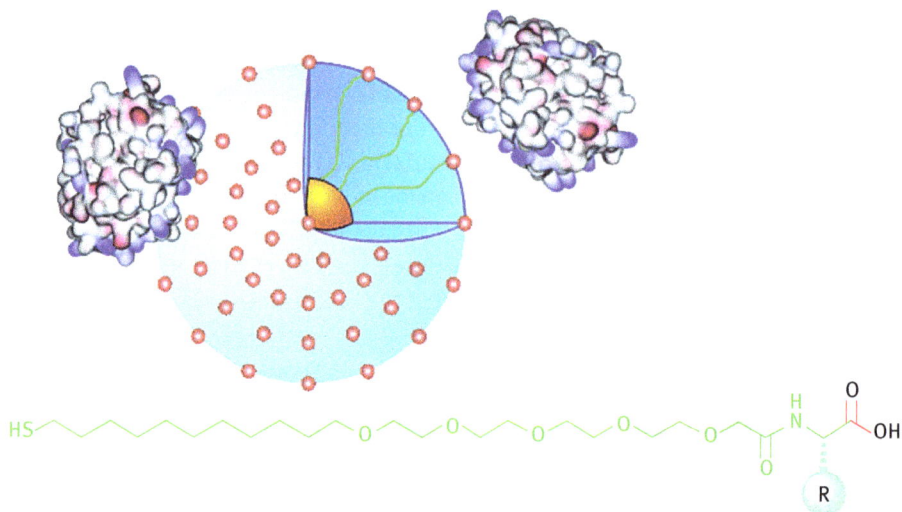

Fig. 7.7: Interactions between proteins and functionalized gold nanoparticles. The Au nanoparticles were functionalized with thiolated hydrocarbon chains (chemical structure shown below the nanoparticle drawing) terminated by different amino acids (indicated as "R" units). Display of charged amino acids induced binding and subsequent denaturation of α-chymotrypsin (two representative molecules are shown); electrostatic binding occurred between the nanoparticles and charged domains on the protein surface, indicated in blue and red. Reprinted with permission from You et al., *JACS* **127** (2005, 12873–12881, © 2005, American Chemical Society.

and colleagues at the University of Massachusetts bound elongated chains terminated with different functional groups to the Au NP. Indeed, displaying amino acids with charged terminal groups on the NP surface resulted in tight binding between the NPs and the protein via electrostatic attraction with negatively- or positively-charged domains on the protein surface. Significantly, the interactions between chymotrypsin and the functionalized NPs led to denaturation of the protein, underscoring the significant effects of NP binding on protein properties.

Figure 7.8 portrays an interesting set of experiments designed to probe nanoparticle size effects on protein-NP interactions. The hypothesis examined by R. F. Minchin and colleagues at the University of Queensland, Australia, was that the unfolding of fibrinogen – a prominent blood-plasma protein involved in clot formation – can be induced by binding to NPs, and that the NP-induced unfolding process is dependent on particle size. Fibrinogen is a cylindrical protein and could bind to NPs coated with a negatively-charged polymer via its two far ends (Fig. 7.8). The researchers discovered, however, that configurations of the bound protein were determined by the size of the NPs: smaller NPs could bind only a single fibrinogen molecule, resulting in linking two NPs through the protein. Large-diameter NPs, on the other hand, enabled multiprotein binding on the particle surface. Importantly, "stretching" the fibrinogen between two small NPs (i.e. Fig. 7.8A) resulted in partial unfolding and exposure of protein domains

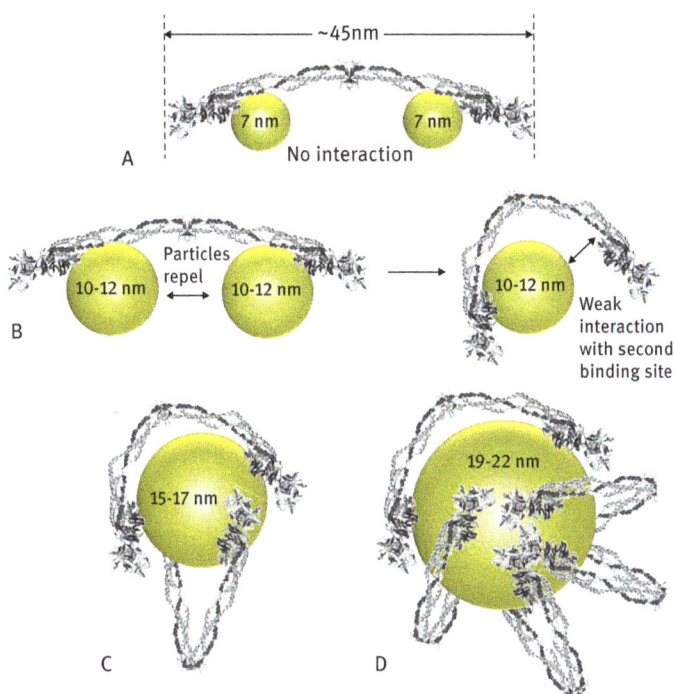

Fig. 7.8: Effect of nanoparticle size on protein-nanoparticle interactions. Different modes of binding between fibrinogen and Au nanoparticles coated with poly(acrylic acid; PAA). **A:** A single fibrinogen molecule binds two small nanoparticles at the two ends of the protein fiber. **B:** Larger nanoparticles induce detachment of one end of the protein because of electrostatic repulsion between the two adjacent nanoparticles. **C,D:** Binding of a single or several proteins on the surface of larger nanoparticles. Reprinted with permission from Deng et al., *ACS Nano.* **6** (2012), 8962–8969, © 2012, American Chemical Society.

which could elicit an immune response (known as "epitopes"). In contrast, crowding of the fibrinogen molecules at the surface of the large NPs (Fig. 7.8C,D) inhibited access to the epitope domains, reducing immunogenic activity of the protein. The model in Figure 7.8 might explain the size-dependent inflammatory activity encountered in many NP systems.

The effects of NPs on protein folding and unfolding have nurtured research aimed to shed light on NP-induced *protein aggregation* (or *fibrillation*). This emphasis is related to the predominance of peptide aggregation phenomena in many devastating pathologies such as Alzheimer's and Parkinson's diseases. It should be noted that experiments analyzing NP-associated peptide aggregation have produced diverging results – some reports indicated catalysis of protein aggregation through interactions with NPs, while others demonstrated the opposite effect of NP-induced inhibition of fibrillation processes. These seemingly contrasting outcomes underscore the complexity of this field and the need to consider the fact that a broad range of parameters,

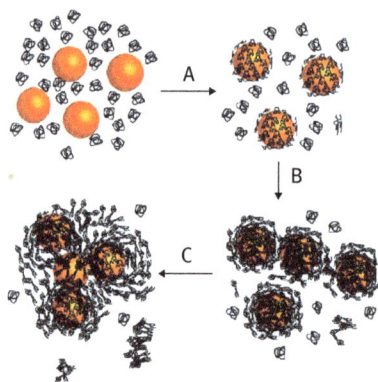

Fig. 7.9: Aggregation of lysozyme induced by gold nanoparticles. **A:** Lysozyme monomers are adsorbed onto the nanoparticle surface undergoing particle unfolding when attached to the NP surface. **B:** The unfolded proteins serve as nucleating sites for protein aggregation. **C:** Fragments of the aggregated protein are released and constitute aggregation seeds in solution. Reprinted with permission from Zhang et al., *Nano Lett.* **9** (2009), 666–671, © 2009, American Chemical Society.

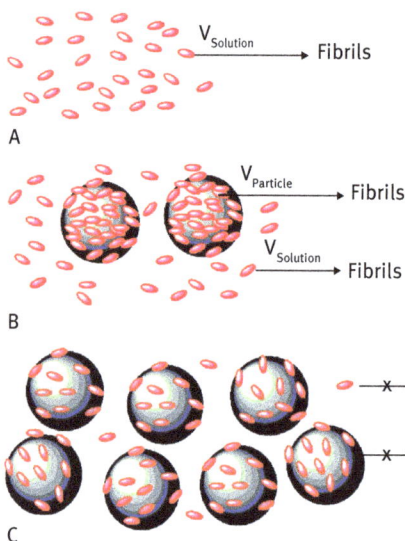

Fig. 7.10: Effect of nanoparticle concentration on fibrillation of the beta-amyloid peptide. **A:** In solution, the peptide monomers (small red ovals) spontaneously aggregate forming fibrils. **B:** Low concentration of positively-charged polymer nanoparticles. The peptide monomers are adsorbed onto the nanoparticle surface, and their high concentration there induces fibrillation. Fibrillation also occurs in solution. **C:** High concentration of the nanoparticles inhibits fibrillation because of the greater dilution of adsorbed peptides on the overall larger surface area of the nanoparticles. Reprinted with permission from Cabaleiro-Lago et al., *ACS Chem. Neurosci.* **1** (2010) 279–287, © 2010, American Chemical Society.

including protein and NP species, NP properties, and overall solution conditions, all contribute to modulating the aggregation pathways of proteins by NPs.

Figures 7.9 and 7.10 present examples of this experimental divergence and the implications for the relationship between NPs and protein aggregation phenomena. Figure 7.9 depicts a scheme outlined by N. J. Halas and colleagues at Rice University accounting for the effect of Au NPs on aggregation of *lysozyme*, a ubiquitous protein which self-assembles into fibrils in certain conditions. Specifically, the researchers observed clear enhancement of lysozyme fibril formation in the presence of the NPs. They proposed a mechanism in which individual protein molecules partially unfold upon adsorption onto the NP surface; these unfolded proteins constitute nucleating seeds for protein aggregation, yielding extended NP-protein aggregate networks. The protein aggregates can also be released from the NPs, promoting further fibrillation in solution, independent of the NPs.

A different model for NP association with fibril-forming peptides is illustrated in Figure 7.10. In the experiment S. Linse and colleagues at Lund University, Sweden, analyzed fibrillation of the *beta-amyloid peptide* – the peptide comprising the amyloid plaques observed in brains of patients with Alzheimer's disease – in the presence of varying concentrations of cationic polymeric NPs. The researchers surmised that amyloid fibrils form via self-assembly of the beta-amyloid monomers in solution (Fig. 7.10A). In scenarios of low NP concentrations (Fig. 7.10B), fibrillation was faster since the dense coverage of peptides attached to the NP surface more easily gave rise to peptide co-assembly and fibril formation. However, in a scenario of high NP concentration (Fig. 7.10C), the peptides attached to the NP surface were largely dispersed, resulting in inhibition of fibril formation.

Nanoparticles might also exert cell toxicity by their interaction with the genetic materials – DNA and the chromosomes packaging the DNA molecules. Such *genotoxic effects* are of particular concern, since gene modifications are associated with the onset of cancer and other diseases and can be propagated by cell division. Genotoxic profiling of NPs is a broad research area. Indeed, studies have shown that diverse NP classes (metal, metal oxides, semiconductor, others) induced chromosome fragmentations, DNA point mutations and breakage, and overall modulated gene expression. Figure 7.11 shows computer simulations of Au NP interactions with DNA, based on experiments which measured changes in distances within DNA chains, induced by NP

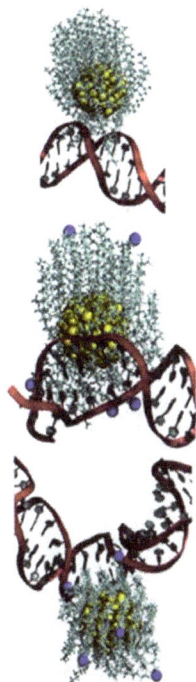

Fig. 7.11: Effect of electrostatic interactions between DNA double strands and nanoparticles. Computer simulations showing distortions of a double-strand DNA induced by electrostatic interactions between the negatively-charged oligonucleotide chains and positively-charged residues within the coating layer of gold nanoparticles. **Top:** Uncharged Au nanoparticle does not bind to the DNA double strand. **Middle:** Opening of the double strand in the DNA minor groove induced by electrostatic attraction between the phosphate groups within the DNA strands and amine residues upon the nanoparticle surface (blue spheres). **Bottom:** Electrostatically-induced DNA double strand separation in the DNA major groove. Reprinted with permission from Railsback et al., *Adv. Mater.* **24** (2012), 4261–4265, © 2012, John Wiley & Sons.

addition. The experiments and computer simulations, carried out by A. V. Melechko and colleagues at North Carolina State University, confirmed the significance of electrostatic interactions between the negatively-charged DNA and positive moieties at the surface of NPs, which could lead to pronounced structural transformations of DNA including bending and strand separation.

While parameters such as size, shape, and surface charge of NPs have been linked to genotoxic effects, it should be noted that few reports provided experimental evidence for actual insertion of NPs into the cell nucleus, in which it could interact with DNA. It is, in fact, conceivable that indirect processes such as NP-induced oxidative stress or interference with intracellular signaling and gene expression pathways are the real culprits in genotoxicity, and further work is needed to resolve this issue.

The interface between NPs and biological entities will likely continue to attract considerable interest, in light of the increasing use of NPs as platforms for displaying biologically-active molecules and delivering these ligands to their physiological sites of activity. A crucial question guiding research in this field is whether NPs function as "innocent bystanders" (serving, for example, as surfaces to immobilize biological recognition residues), or whether the NPs are actively and intimately involved in modulating the biological processes targeted. Experimental evidence supporting either scenario will continue to be assessed on a case-by-case basis.

8 Nanoparticle assemblies

While the preceding chapters in this book discuss the unique physical, chemical, and biological properties of nanoparticles and their relationship to NP compositions, dimensions, shapes, and atomic structures, in many cases it is longer-range assemblies, or "superstructures", of NPs which are responsible for interesting phenomena and applications. Indeed, organized two- and three-dimensional NP assemblies often exhibit novel collective and synergistic properties. Furthermore, large-scale NP assemblies, rather than individual NPs, usually enable fabrication of practical devices and constitute the basis for NP-based applications. In some respects, NPs can be perceived as forming a "new Periodic Table" – providing distinct building blocks which can be organized in diverse architectures and functional assemblies. This chapter summarizes materials in which NPs have been arranged in ordered assemblies, the unique properties of such systems, and their potential uses.

Methodologies for long-range organization of NPs can be divided into "top-down" techniques involving manipulation by external instrumentation (primarily lithography), and "bottom-up" approaches, relying on self-assembly processes. Consistent with the overall focus of this book, the discussion here centers on bottom-up NP organization strategies. Self-assembly of NPs can be brought about both via physical means (i.e. not involving modification of the nanoparticle structure or chemical properties but rather using varied templates) and chemical strategies (modulation of NP properties, usually by tuning the coating layers and surface-displayed residues). In many instances, both strategies are combined to create organized NP assemblies of varied spatial configurations.

8.1 Gold nanoparticle assemblies

The impetus for creating organized Au NP architectures emanates to a large degree from well-known phenomena such as electrical conductivity and surface plasmon resonance (SPR) which are highly sensitive to macroscale organization. For example, the plasmon resonance of Au NPs depends on the aggregation state – i.e. Au NPs held in close proximity exhibit different optical properties compared to individual Au NPs which are spatially separated (see discussion in Chap. 3). Interest in organized Au NP systems also stems from the expectation that linking Au NPs in macroscale configurations could open the way for utilizing such assemblies in nano- and microelectronic devices, taking advantage of their intrinsic electrical conductivity. Indeed, most practical applications of Au NP-based systems (such as in sensors, electro-optic devices, catalysts, and others) are generally based on macroscale NP assemblies rather than individual particles.

Creating organized assemblies of Au NPs has often been pursued with the use of molecular scaffolds or templates which constrain and/or direct NP organization. This approach, referred to as "template-directed" self-assembly, is versatile as the structural features of the NP assemblies formed are determined by the templates used. The templates employed have mostly included inorganic or organic materials, but also biological molecules such as peptides and oligonucleotides. The challenges in this line of research mostly arise from the need to develop the means for chemical or physical confinement of the NPs inside the template without adversely affecting the integrity and physical properties of the NPs themselves.

Confinement of Au NPs within porous templates is a broad-based strategy for creating two-dimensional and three-dimensional NP assemblies. In such systems, pore structures within the host matrix are exploited as the means for ordering (and/or growing) NPs. In general, following synthesis of the embedded Au NPs, the host material is dissolved, leaving behind the organized NP assemblies. Templates used in such schemes have included porous inorganic and polymer frameworks, biological materials, and even microorganisms. Figure 8.1 portrays an experimental scheme for deposition of Au NPs inside the aligned channel network of a porous alumina matrix. In the procedure developed by V. V. Tsukruk and colleagues at Georgia Institute of Technology, the internal cylindrical surface of alumina was first functionalized with a polyelectrolyte containing positively-charged amine groups. The polyelectrolyte layer facilitated immobilization of Au NPs which passed through the porous matrix as the positive polyelectrolyte moieties substituted the surfactant coating of the NPs. The alumina-encapsulated Au NP assembly was found to be useful for sensing gaseous analytes through application of surface-enhanced Raman spectroscopy (SERS). Indeed, SERS effects were specifically ascribed to Au NP aggregates within the alumina chan-

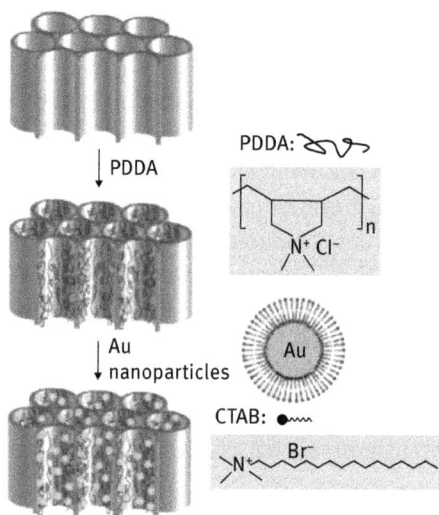

Fig. 8.1: Gold nanoparticle assemblies within the channels of porous alumina. The alumina template is first functionalized with a positively-charged electrolyte (PDDA); the electrolyte captures and immobilizes Au nanoparticles within the channels through substitution with the positive surfactant molecules used to coat the Au nanoparticle. Reprinted with permission from Ko and Tsukruk, *Small.* **4** (2008), 1980–1984, © 2008, John Wiley & Sons.

nels. Sensing was further aided by the fact that negatively-charged analytes could be co-adsorbed inside the amine-coated channels, generating SERS signals.

In contrast to the porous systems discussed above, many types of templates enable construction of organized NP assemblies through external deposition of the particles on rigid scaffoldings. Polymer chains, for example, have been employed as oriented scaffolds to create ordered Au NP structures. Figure 8.2 illustrates the use of cylindrical micelles comprising block copolymer (e.g. polymers containing different monomeric building blocks) as templates for selective adsorption of Au NPs. Importantly, the cylinders, synthesized by I. Manners and colleagues at the University of Toronto, were made of "A-B-A" blocks, of which the "B" (i.e. central) block displayed a positive surface charge. Consequently, negatively-charged Au NPs were selectively attached only to the central region of the cylinder by electrostatic attraction, producing localized coating of the Au NPs.

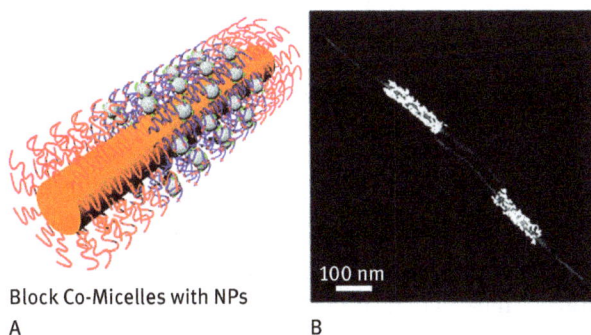

Block Co-Micelles with NPs

A   B

Fig. 8.2: Cylindrical block copolymers as templates for selective deposition of gold nanoparticles. **A:** Diagram showing the two polymer blocks (the red and blue chains); the blue polymer is positively-charged. Au nanoparticles coated with a negatively-charged layer (grey spheres) are attached to the positively-charged polymer in the middle of the cylinder via electrostatic attraction. **B:** Electron microscopy image demonstrating the selective deposition of the Au nanoparticles on the cylinders. Reprinted with permission from Ko and Tsukruk, *Small.* **4** (2008), 1980–1984, © 2008, John Wiley & Sons.

Template-directed assembly of Au NPs has also been demonstrated using biological or biomimetic host matrixes. Figure 8.3 illustrates an elegant experiment in which hollow tubular structures built from lipid molecules facilitated assembly of Au NPs within the internal space of the tubes. Interestingly, the synthetic route developed for production of the cylindrical tubes by T. Shimizu and colleagues at AIST, Japan, appeared to induce migration of the Au NPs from the aqueous solution into the tubes via "suction" affected by capillary forces. Subsequent high-temperature treatment of the Au NP/tubes eliminated the lipid scaffolding, further inducing sintering of the NPs to produce long Au nanowires.

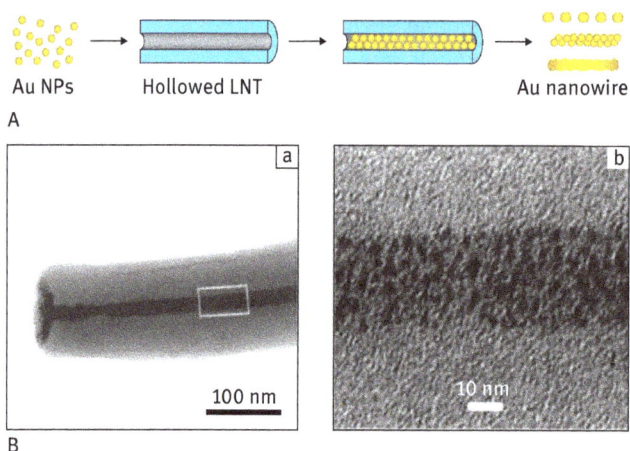

Fig. 8.3: Gold nanowires produced via assembly of Au nanoparticles within the channels of lipid nanotubes. **A:** Experimental scheme: Au nanoparticles are inserted into the channels of hollow lipid nanotubes (LNTs); following high-temperature sintering, the nanotube template is removed and the nanoparticles fuse, producing Au nanowires. **B:** Electron microscopy image showing the Au nanowire formed inside the LNT channel. Reprinted with permission from Yang et al., *Chem. Mater.* **16** (2004), 2826–2831, © 2004, American Chemical Society.

Peptides have also been used as physical scaffolds for NP assemblies. Peptides and proteins offer important advantages as conduits for creating organized NP architectures since their structural features (in particular secondary structure elements such as alpha helix, beta sheet, etc.), and overall macroscale morphologies can be tuned via modification of their amino acid sequence. Moreover, the amino acid residues in a peptide sequence provide diverse reactive sites for docking and binding of Au NPs (and other NP species). In particular, thiol-containing cysteine residues have been widely used for attachment and subsequent organization of Au NPs. While peptides have served as scaffoldings for assembly of already-synthesized nanoparticles, some studies have demonstrated the use of peptide templates as platforms for actual NP synthesis (usually employing peptide fibers; Chap. 5).

Figure 8.4 shows an Au NP double helix, spontaneously formed on the skeleton of an amphiphilic peptide. The single-step synthesis scheme, developed by N. L. Rosi and colleagues at the University of Pittsburgh, combined the process of peptide self-assembly with nucleation and growth of the Au NPs. The researchers pre-selected a 12-residue peptide exhibiting a strong affinity to gold. A long-chain fatty acid was covalently attached to the peptide, inducing adoption of a ribbon-like helical morphology. Remarkably, co-incubation of the amphiphilic peptide with the molecular precursor of Au NPs (HAuCl$_4$) resulted in formation of an organized array of Au NPs within the helical clefts. Nucleation and growth of the Au NPs was ascribed to the presence

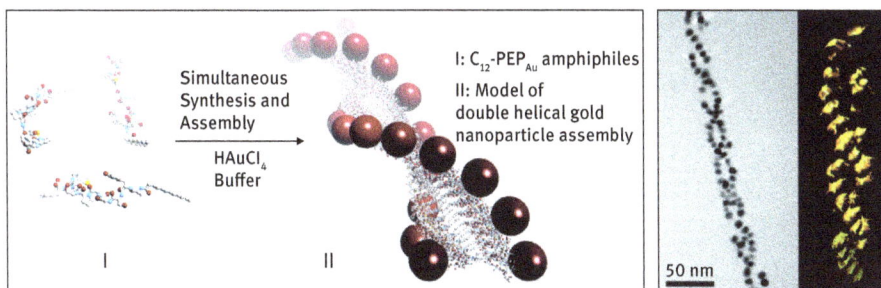

Fig. 8.4: Double helix assembly of gold nanoparticles on a peptide template. A short amphiphilic peptide is co-incubated with the gold precursor salt. Au nanoparticles are slowly formed within the clefts of the peptide double helix. The electron microscopy images on the right show the Au nanoparticle double helix. Reprinted with permission from Chen et al., *JACS.* **130** (2008), 13555–13557, © 2008, American Chemical Society.

of tyrosine in the peptide sequence, believed to provide electrons for gold reduction through oxidation of its phenolic ring.

Deoxyribonucleic acid (DNA) is one of the most diverse biological templates for creating organized NP assemblies. DNA-based NP patterning is aided by the modular, programmable nature of DNA (intrinsic to base-pair complementarity) and readily-available chemical routes for linking Au NPs and DNA. Figure 8.5 presents an imaginative system designed by J. Kim and colleagues at the University of Arkansas. The researchers bound single strand DNA units to Au NPs. Depending on the number of DNA strands linked to the central Au NP, discrete "NP complexes" were produced through base-pair complementarity, in which symmetries of the NP modules (linear, square planar, octahedral, etc.) were determined by the minimal steric hindrance of the ligands surrounding the central Au NP. Overall, the geometry of the Au NP assemblies could be programmed simply via the number of DNA strands linked to the Au NPs, and double strand DNA complementarity.

Figure 8.6 illustrates a DNA template which enabled assembly of two-dimensional Au NP patterns (or arrays). The DNA scaffolding, designed by R. A. Kiehl and colleagues at the University of Minnesota, comprised of elongated DNA "tiles", or strips, built by mixing several pairs of complementary sequences. While the DNA "quilt" was mostly constructed from double-stranded DNA, some of the tiles were designed to display "free" strands functioning as "anchors" for Au NPs displaying the complementary DNA sequences. Hybrid NP patterns (containing, for example, NPs of different diameters) could be achieved using this strategy, since tiles could be prepared displaying different sequences – one row displaying free strands complementary to small NPs, while another row displayed strands complementary to large NPs.

Optical phenomena which are closely linked to the plasmon properties of Au NPs and Au nanorods (NRs) are among the attractive physical properties of long-range Au NP assemblies. DNA complementarity has been used for construction of *plasmonic*

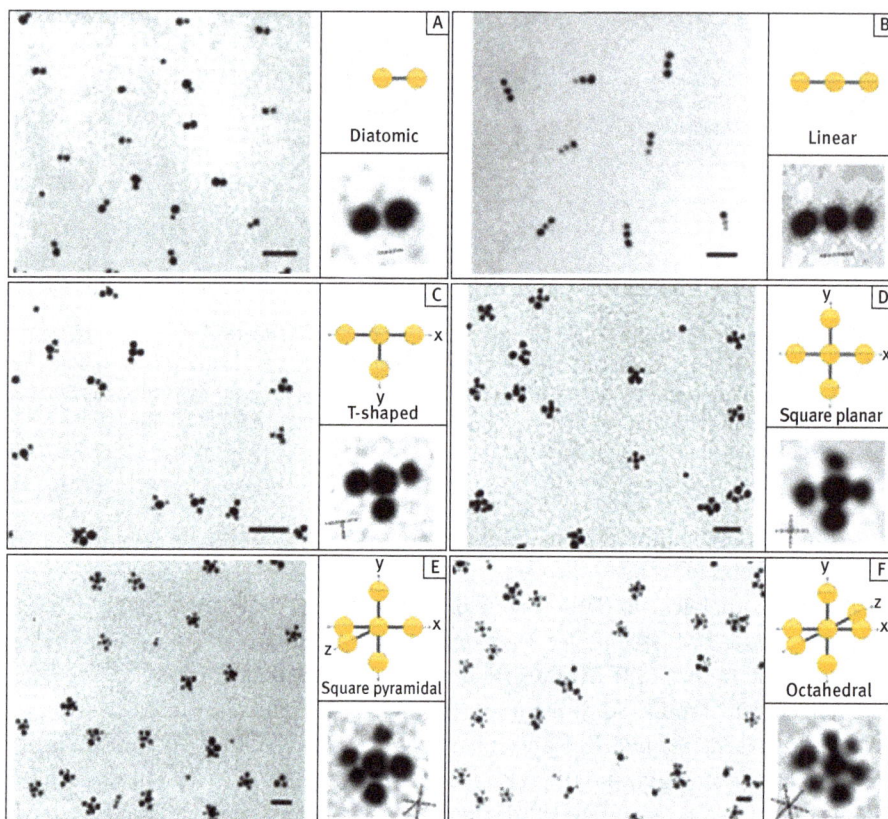

Fig. 8.5: DNA-assembled gold nanoparticle coordination complexes. Electron microscopy images and corresponding geometries of Au nanoparticle/DNA complexes. The Au nanoparticles are functionalized with single strand DNA and linked to each other through base-pair complementarity. Au NPs (yellow spheres) with different symmetries are produced through successive attachment of DNA fragments with thiol ends. The relative positions of the DNA strands reflect the configurations which minimize the steric hindrance around the nanoparticle. The bold lines in the schematic diagrams indicate double-stranded DNA. Scale bars correspond to 20 nm. Reprinted with permission from Kim, J. et al., *Angew. Chem.* **50** (2011), 9185–9190, © 2011 John Wiley and Sons.

super-structures consisting of several Au NPs bound together through DNA strands attached to the NP surface. Figure 8.7 depicts such a system, in which Au NRs were conjugated with spherical Au NPs. Construction of the Au NP/Au NR assemblies, designed by N. A. Kotov and colleagues at the University of Michigan, was based on the observation that surfactant molecules bound at the two ends of Au NRs exhibit less affinity to the NR surface, and thus can be substituted by thiolated DNA molecules in a rather low concentration. In comparison, higher DNA concentration is required to substitute the surfactants encasing the cylindrical surface (e.g. the "sides" of the NRs).

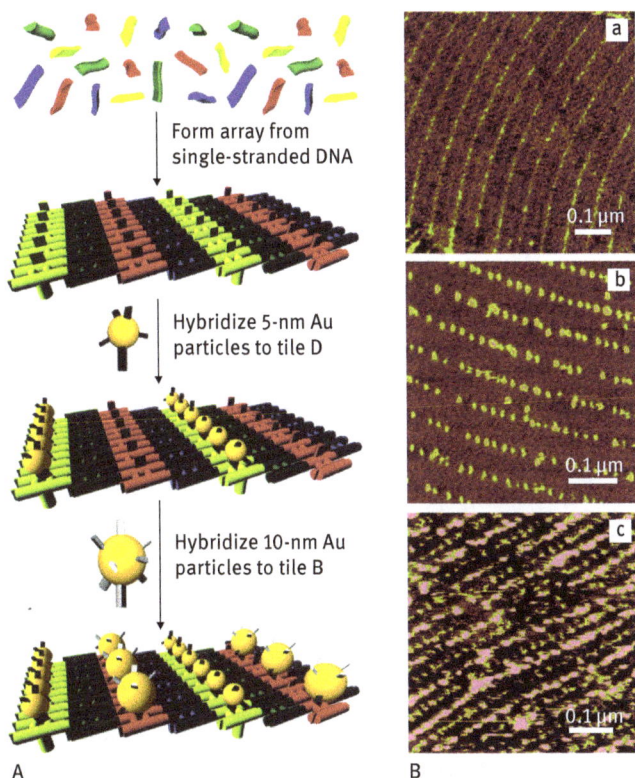

Fig. 8.6: Surface patterns of gold nanoparticles via DNA "tiles". **A:** Experimental scheme: two-dimensional tile structure is prepared by mixing complementary DNA strands. Some of the strands display free ends which are complementary to single strand DNA covalently attached to Au nanoparticles of different sizes. Addition of the Au nanoparticles results in formation of particle arrays through double-strand complementarity. **B:** Atomic force microscopy (AFM) images showing Au nanoparticles arrays formed using this strategy: (**a**) small Au nanoparticles added; (**b**) large Au nanoparticles added; (**c**) both small and large Au nanoparticles added; alternating rows can be discerned in the image. Reprinted with permission from Pinto et al., *Nano Lett.* **5** (2005), 2399–2402, © 2005 American Chemical Society.

This difference in affinity to the NR surface by the surfactant layer allows attachment of specific DNA strands onto the ends or sides of the NRs by controlling the concentration of the thiolated DNA (Fig. 8.7). Accordingly, DNA sequences which were complementary to strands attached to spherical Au NPs were selectively placed either at the two ends (resulting in attachment of the NPs at the ends of the NR), on the cylinder sides (leading to "side attachment" of the NPs), or in both ends and sides. The distinct "plasmon assemblies" formed (in which the spherical Au particles were either attached at the two ends or sides of the NRs, or completely surrounded them) gave rise to dramatic shifts in the spectral positions of the plasmon resonance.

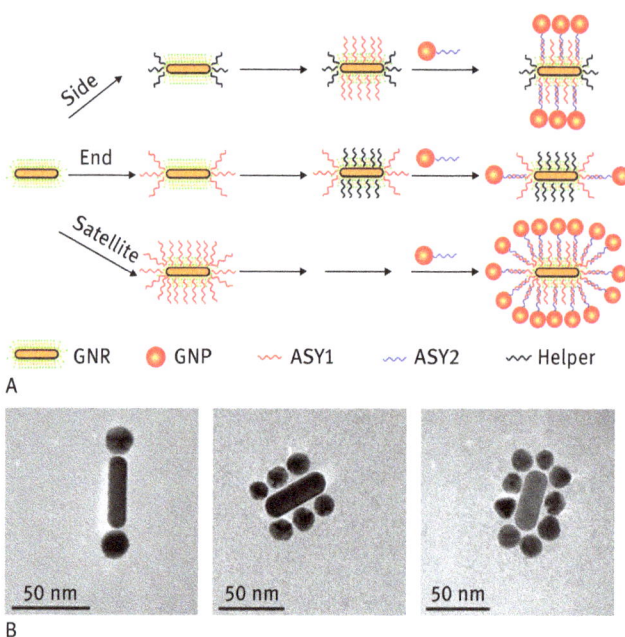

Fig. 8.7: "Region-specific" gold nanoparticle assemblies. The diagram shows the construction of different Au nanoparticle/nanorod configurations by substituting the surfactant layer of the nanorods in different positions. **Side conformation:** A "helper" DNA strand (not complementary to DNA strands attached to Au nanoparticles) is added in low concentration, thus replacing only surfactant molecules at the two ends. A second DNA strand (denoted ASY1) is then added in high concentration, thereby substituting the surfactants at the "sides" of the nanorod. ASY1 binds to its complementary strand (denoted ASY2), which is displayed on the spherical Au nanoparticles, generating the "side" conformation. **End Conformation:** The ASY1 strand is first added in low concentration, substituting the surfactants at the two ends. The helper DNA strand is subsequently added, replacing the surfactants at the nanorod's sides. Addition of Au nanoparticles displaying the ASY2 strand results in their binding at the two ends. **Satellite conformation:** Only ASY1 DNA strand is added in both concentration regimes, resulting in attachment of the Au nanoparticles both at the ends and at the sides. **Bottom images:** Electron microscopy images showing the three configurations. Reprinted with permission from Xu et al., *JACS*, **134** (2012), 1699–1709, © 2012 American Chemical Society.

Controlled "core/satellite" Au NP assemblies have been created using ligands other than DNA. Figure 8.8 presents such a system in which bonding between the Au NP core and NP satellites was achieved by chemical means. Specifically, S. Singamaneni and colleagues at Washington University coated the Au core with a bi-functional linker (p-aminothiophenol; p-ATP). This residue was covalently attached to the NP core surface through a thiol unit on the one end, displaying an amine residue on the opposite end and thus attracting Au NPs serving as satellites (the linker concentration and solution pH were carefully adjusted to prevent formation of dipolar or chain structures rather than core-shell NP configuration). By tuning the solution pH (which, in turn, affected the surface charge of both core and satellite NPs) and ratio between the concentrations of the core and satellite NPs, the researchers constructed remarkable assemblies exhibiting differing numbers of satellites (Fig. 8.8B). Notably, the number of satellite NPs affected the position of localized surface plasmon resonance (LSPR) making color tuning of the NP assemblies possible (Fig. 8.8B).

Au NPs coated with polymeric moieties have been shown to adopt interesting *amphiphilic micellar* superstructures (i.e. spheres, cylinders, "nanoworms", and others) either individually or in conjunction with other amphiphilic molecules. Figure 8.9 outlines a strategy for self-assembly of micellar superstructures constructed from Au NPs coated with a layer of block-copolymers. The copolymers employed by Z. Nie and colleagues at the University of Maryland comprised of a hydrophobic block tethered to the NP surface by a thiol unit, and an amphiphilic chain extended towards the solvent. The polymer-coated NPs were initially dissolved in an organic solvent. Subsequent addition of water to the NPs resulted in assembly of the polymer-coated NPs into micelles, which are more energetically-favorable due to the repulsion forces between the hydrophobic polymer blocks on the surface of the Au NPs in aqueous environments.

Remarkably, changing the lengths of the block copolymers had a profound effect on the morphologies of the NP assemblies – giving rise to single particle micelles, NP clusters, and unilamellar vesicles (Fig. 8.9B); formation of the different superstructures was ascribed to minimizing exposure of the hydrophobic chains to the aqueous environment. Formation of the Au NP micellar assemblies was accompanied by redshift of the plasmon resonance due to the coupling of surface electrons as the distances between the NPs became smaller in the aggregates. Indeed, light absorption by these Au NP aggregates in the near-infrared (NIR) spectral region (which is not absorbed by tissues) might open the way to biological imaging and photothermal therapy applications using the polymer-coated Au NPs.

A self-assembly approach similarly driven by amphiphilic block-copolymer layers tethered to Au NPs is presented in Figure 8.10. In that study, E. R. Zubarev and colleagues at Rice University constructed long polymer chains comprising a hydrophobic component (polystyrene) and a hydrophilic arm [poly(ethylene-oxide)]. The researchers correctly surmised that addition of water to a solution of the block copolymer would result in micelle formation, and placement of the Au NPs at the intersection between the hydrophobic and hydrophilic constituents of the polymer would result in localization of the NPs at the external micelle surface. Careful tuning of the lengths of the two polymer arms yielded highly stable "worm-like" NP aggregates. While the results presented in Figures 8.9 and 8.10 underscore the diversity of micellar Au NP assemblies, it should be emphasized that generating such structures is not trivial, and control of NP organization intimately depends on particle size, composition and properties of polymer coatings, and ratios between water and organic solvents.

Coating Au NPs with pH-sensitive polymers has been employed to construct Au NP assemblies by tuning the solution acidity (Fig. 8.11). In the experiment, reported by H. Xia and colleagues at Shandong University, China, pH-responsive polymer "brushes"

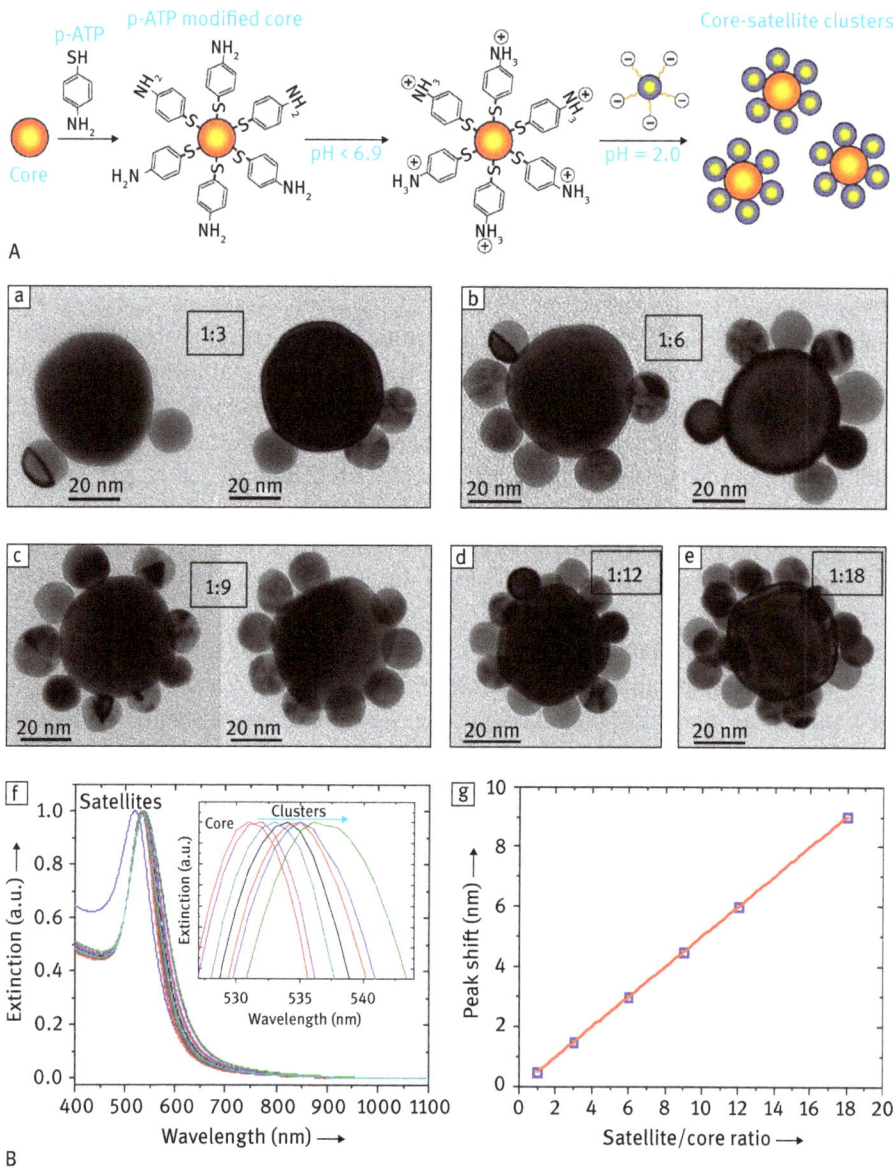

Fig. 8.8: "Core/satellite" gold nanoparticles. **A:** Synthetic diagram. The Au nanoparticle core is functionalized with p-aminothiophenol (p-ATP). In an acidic pH the amine residues are positively-charged, thus attracting surfactant-coated satellite NPs which are negatively-charged in low pH. **B:** Electron microscopy images of core/satellite NP assemblies produced in solutions with different ratios between the core and satellite NPs. Panels f and g depict the shift in localized surface plasmon resonance (LSPR) on changing the satellite/core ratio. Reprinted with permission from Gandra et al., *Nano Lett.* **12** (2012), 2645–2651, © 2012 American Chemical Society.

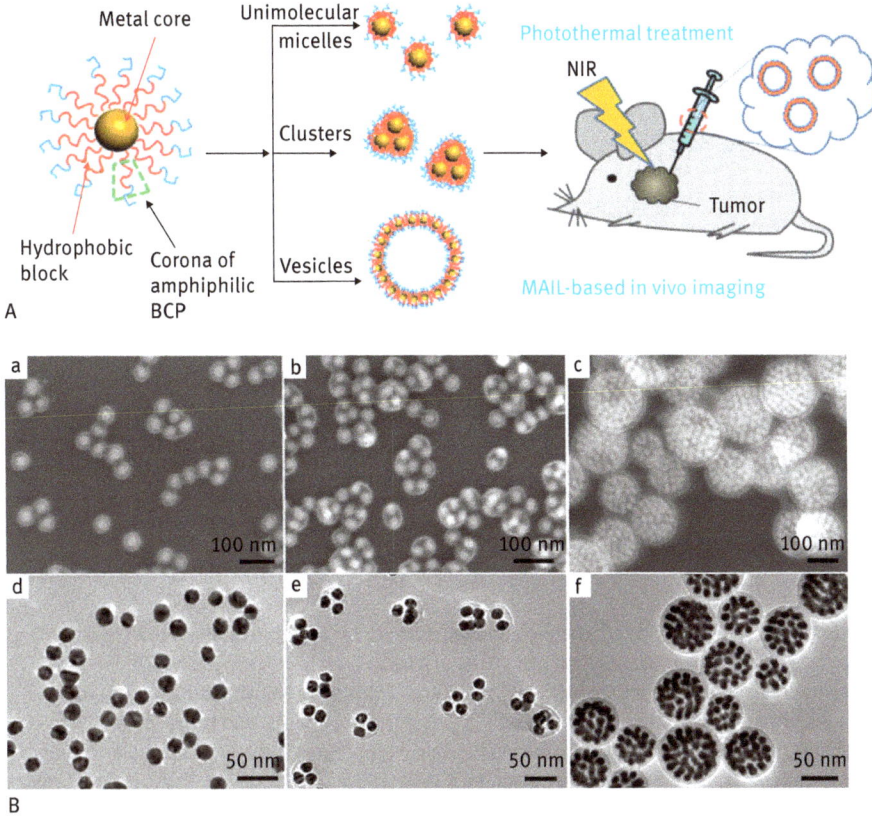

Fig. 8.9: Spherical assemblies of copolymer-coated gold nanoparticles. **A:** Schematic structure of the Au nanoparticles and their potential biological application. The Au core is covalently linked to block copolymer (BCP) chains comprising an inner hydrophobic block and an outer amphiphilic unit; the length of the block copolymer, particularly the amphiphilic corona, determines the morphology of nanoparticle assemblies (single-particle micelles, clusters, or vesicles). Aggregation of the nanoparticles induces shifts of plasmon resonance absorbance to the near infrared, making imaging and therapeutic applications possible. **B:** Scanning electron microscopy (top) and transmission electron microscopy (bottom) images showing aggregates of Au nanoparticles coated with block copolymers of different chain-lengths; larger clusters are generated on increasing the chain length (right image). Reprinted with permission from He et al., *JACS.* **135** (2013), 7974–7984, © 2013 American Chemical Society.

were deposited on NP surfaces. Specifically, one coating featured polymer brushes which were deprotonated (e.g. H^+ ions removed) in basic pH with the result that the NPs were hydrophobic and separated from each other. However, at pH values lower than 7.0, partial protonation occurred, leading to positive electrostatic charge on the NP surface. A different NP coating comprised of a pH-sensitive polymer coupled to an anionic species at slightly acidic environments (i.e. 4 < pH < 7). Consequently,

A

B

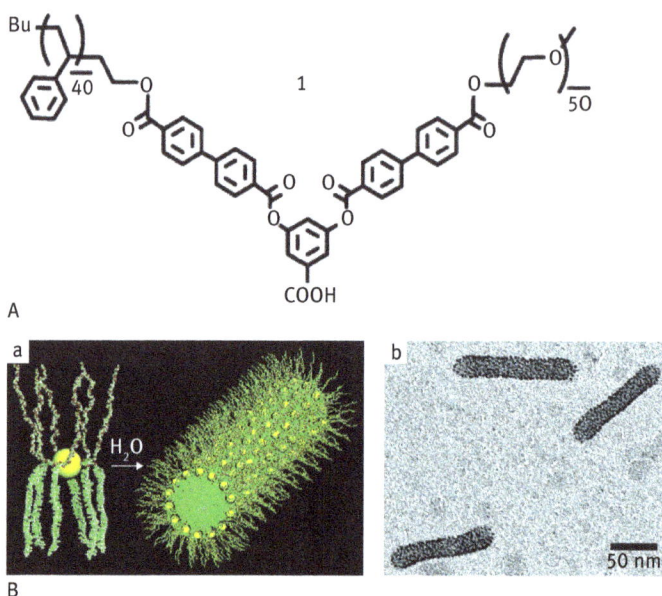

Fig. 8.10: "Worm-like" gold nanoparticle-containing micelles. **A:** Structure of the block copolymer used for micelle assembly; the polymer comprises of a hydrophobic block (polystyrene, left arm) and a hydrophilic block, poly(ethylene-oxide), right arm. **B:** Diagram showing the Au nanoparticle located between the hydrophobic and hydrophilic polymer components, and the resultant worm-like micellar structure formed in water. The microscopy image shows the Au nanoparticle assembly. Reprinted with permission from Zubarev et al., *JACS.* **128** (2006), 15098–15099, © 2006 American Chemical Society.

tuning the pH of the medium had a pronounced effect on association of the NPs into long-range structures (affected by modulating the electrostatic attraction between the negative and positive residues on NP surface). As apparent in Figure 8.11, tuning the solution pH led to formation of NP "necklaces" whose lengths were intimately dependent on the degree of acidity. Notably, the experiments revealed that inter-particle repulsion was smaller at the tips of the growing NP chains than at the sides, giving rise to necklace elongation rather than aggregation.

While spherical Au NPs have commonly been used in self-assembled superstructures, Au nanorods (NRs) have had distinctive roles in this research field. In particular, Au NR assemblies could be used for tuning of both the longitudinal and transverse localized surface plasmon resonance (LSPR) associated with the surface electrons on the nanorod tips and cylindrical surface, respectively (see Chap. 3 for a detailed discussion of these two types of plasmon resonance in Au NRs). In this context, distinct shifts of LSPR spectra were recorded in chains of Au NRs linked at their tips (Fig. 8.12). An elegant procedure for assembly of end-to-end Au NRs was developed by E. Kumacheva and colleagues at the University of Toronto by substitution of the surfac-

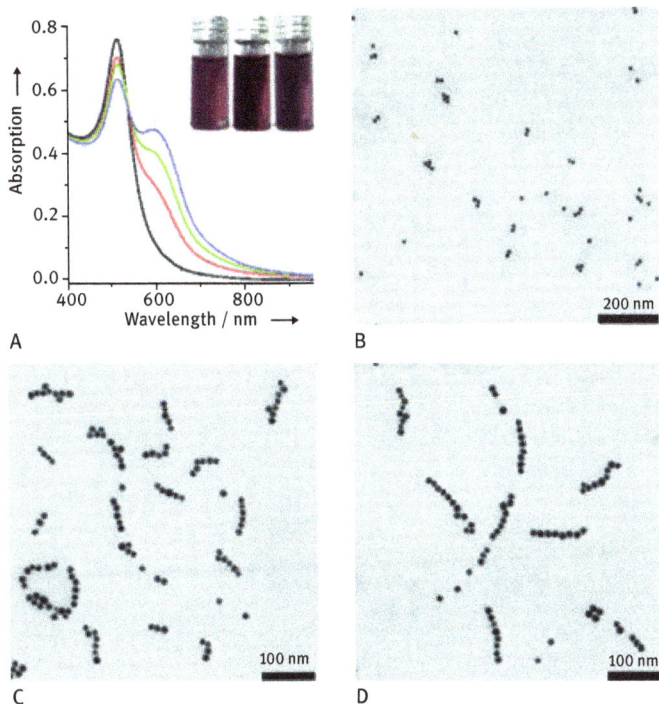

Fig. 8.11: pH-sensitive polymer-coated gold nanoparticles used for controlled nanoparticle assemblies. **A:** Evolution of the surface plasmon resonance on formation of long-range Au nanoparticle assemblies, induced by pH modulation. **B–D:** electron microscopy images showing formation of elongated necklaces on lowering the solution pH: (**B**) 3.5, (**C**) 3.0, and (**D**) 2.5. Reprinted with permission from Xia et al., *Angew. Chem. Int. Ed.* **52** (2013), 3726–3730, © 2013 John Wiley & Sons.

tants at the NR tips with an amphiphilic ligand (thiol-terminated polystyrene). Water was then added to the Au NRs (which were initially dissolved in an organic solvent). The addition of water resulted in self-association of the hydrophobic polystyrene moieties, consequently producing Au NR chains formed by the head-to-tail linkages of the NRs (Fig. 8.12A). Note the gradual shift of the longitudinal plasmon resonance (similar to the well-studied spectral shifts occurring on spherical Au NP aggregation). Importantly, end-to-end Au NR chains also produced "hot spots" in surface enhanced Raman scattering (SERS) sensing applications (hot spots refer to physical regions within periodic arrays of plasmonic nanoparticles, in which proximity of adjacent Au nanorods leads to enhanced electrical fields and corresponding greater signal sensitivity upon adsorption of analytes).

Optical modulation phenomena in Au NR assemblies form the basis for other sensing schemes. The objective of the experimental design outlined in Figure 8.13, developed by N. A. Kotov and colleagues at the University of Michigan, was the detection of a microcystin-LR (MC-LR), a water-soluble toxin. To achieve this goal, Au NRs were

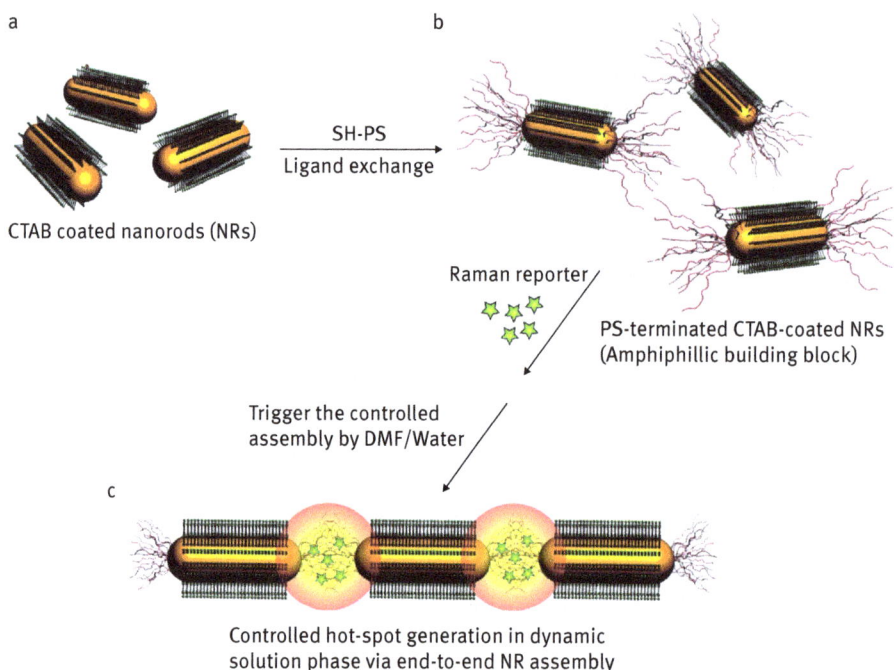

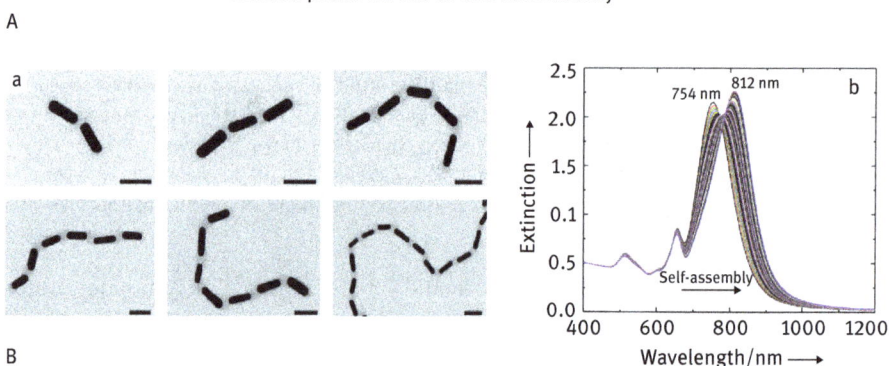

Fig. 8.12: "Head-to-tail" assembly of surface-modified gold nanorods. **A:** Nanorod assembly scheme: **(a)** preparation of conventional surfactant-coated Au nanorods; **(b)** substitution of the surfactants at the two tips with hydrophobic ligands; **(c)** induction of the head-to-tail assembly by dissolving the modified nanorods in a water/organic solvent mixture – water promotes attraction between the hydrophobically-coated nanorod tips. **B:** Characterization of nanorod linear assemblies: (a) electron microscopy images; scale bar is 40 nm; (b) shift of the longitudinal localized surface plasmon resonance (LSPR) on increasing the nanorod chain lengths. Reprinted with permission from Lee et al., *JACS.* **133** (2011), 7563–7570, © 2011 American Chemical Society.

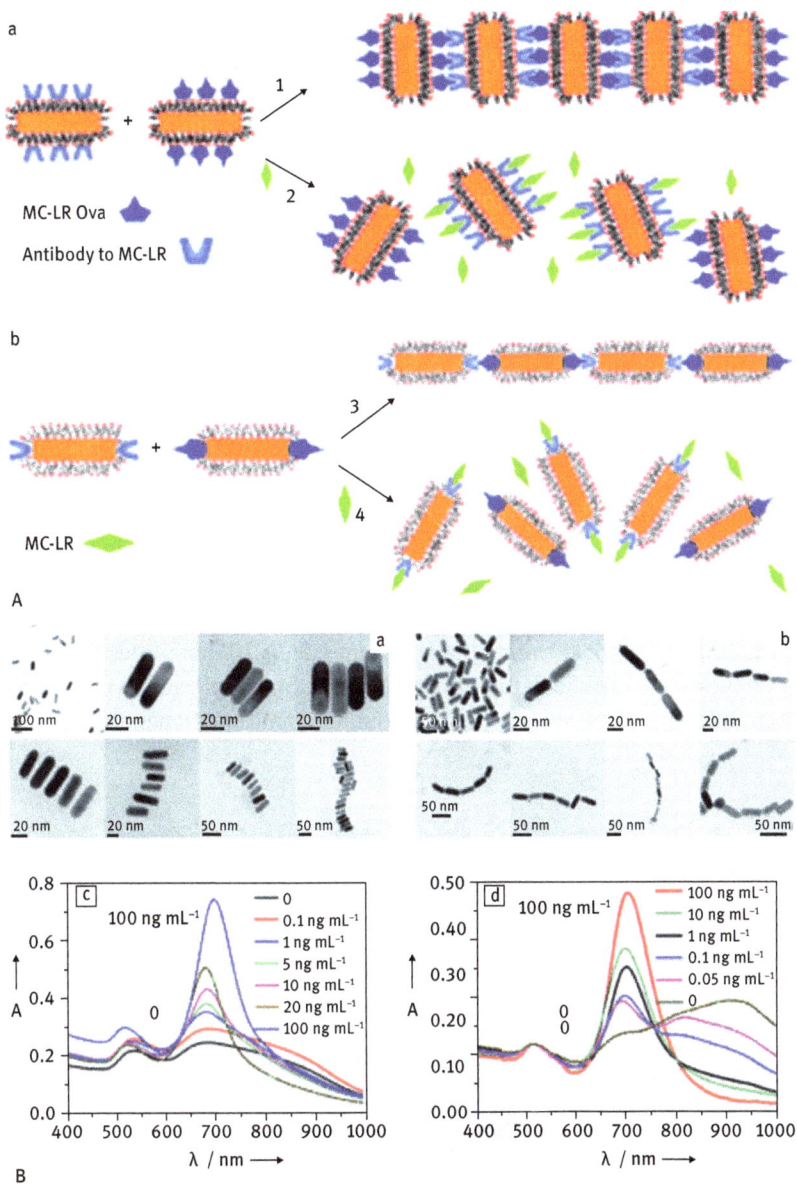

Fig. 8.13: Biomolecular sensing through assembly/disassembly of gold nanorods. **A:** Experimental strategy: two types of surface functionalized Au nanorods are prepared: (a) nanorods displaying antibodies directed against the target toxin (MC-LR) and their complementary synthetic antigens (MC-LR Ova) on the sides, and (b) antibodies displayed at the two tips. Au nanorod chains are generated via antigen/antibody complementarity. However, the nanorod assemblies are disintegrated in the presence of the target toxin – MC-LR – which has higher affinity to the antibody than the synthetic antigen. **B:** a-b): Electron microscopy images depicting the side-by-side and end-to-end Au nanorod assemblies; c-d): changes in the transverse localized surface plasmon resonance (panel c) and longitudinal plasmon resonance (panel d) following addition of the target toxin. Note that the toxin-induced spectral changes occur in different regions, corresponding to the transverse and longitudinal plasmon resonances, respectively. Reprinted with permission from Wang et al., *Angew. Chem. Int. Ed.* **49** (2010), 5472–5475, © 2010 John Wiley & Sons.

coated with antibodies which recognized MC-LR. The antigens were deposited on the nanorod sides via electrostatic attraction (as the NR side surfaces are more accessible for electrostatic binding), while placement of the antibody at the tips was carried out via covalent bonds (using thiolated antibodies). These functionalized NRs were induced to assemble either sideways or end-to-end by incubation with Au NRs displaying synthetic antigens which were complementary to the antibodies (Fig. 8.13A). Both architectures were used for toxin detection via displacement of the antibody-bound synthetic antigens by the toxin molecules, resulting in NR disassembly and the corresponding shifts in plasmon resonances. Notably, however, distinct spectral shifts were induced when the toxin was added to the side-by-side configuration or the end-to-end motif. As expected, disassembly of the end-to-end architecture affected only the longitudinal plasmon resonance (due to the altered length of the NR chains), while disassembly of the side-by-side NR assembly resulted in shifts of the transverse plasmon resonance associated with reduced thickness of the NR aggregates. Consequently, the two disassembly events provided two distinct sensing modes which could be used.

As the examples above attest, the coating of Au NRs plays a crucial role in generating organized assemblies. Figure 8.14 illustrates an interesting approach for arranging Au NRs in two dimensions, accomplished by modifying the surface properties of Au NRs. In the experiment, H. Nakashima and colleagues at NTT laboratories in Japan substituted the surfactant layer with thiol-linked phospholipid molecules which were covalently bound to the Au NR surface (through the thiol moiety). Depending on the

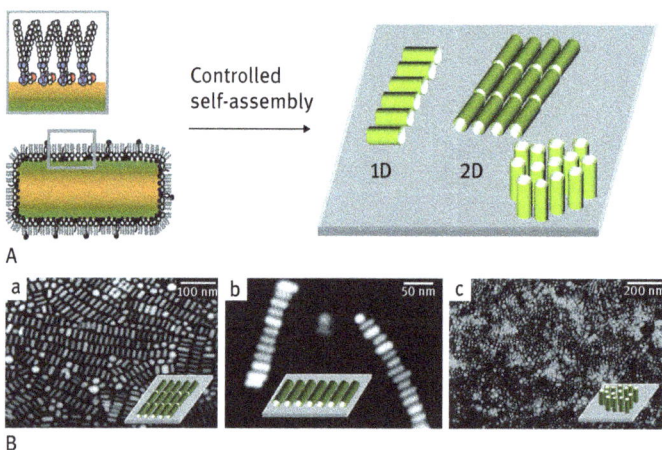

Fig. 8.14: Coating layer-mediated self-assembly of gold nanorods. **A:** Schematic depiction of the experimental approach: the nanorods are coated with phospholipid molecules; different drying conditions result in alignment of the nanorods on the surface. Ordering of the nanorods is maintained via hydrophobic interactions between the lipid-coated nanorods. **B:** Electron microscopy images and models of the aligned nanorod configurations. Reprinted with permission from Nakashima et al., *Langmuir* **24** (2008), 5654–5658, © 2008 American Chemical Society.

drying rate of the organic solvent, the NRs could be assembled in different configurations on the surface, including a side-by-side linear arrangement, "stacked sheet" configuration, or even "standing up" NR arrays. In all these scenarios, the hydrophobic attraction between the phospholipid layers was the primary factor pulling the NRs together and determining their organization.

Macroscale organization of Au NPs has been achieved not only by chemical means (i.e. manipulation of NP surfaces) or confinement within molecular templates, but also via physical manipulation of the particles in interfaces (such as solution/air or solution/solid interfaces). Figure 8.15 illustrates a three-dimensional "super-crystal" of Au NRs assembled via immersion of a rigid template within a solution of NRs, followed by slow evaporation. Specifically, U. Bach and colleagues at Monash University, Australia, demonstrated that the evaporation process resulted in deposition of vertically aligned Au NRs on the patterned surface. This remarkable long-range Au NR organization was ascribed to the combined action of both capillary forces exerted between the template surface and the NRs, and interactions between adjacent NRs. Accordingly, NR organization was highly dependent on the nature of the molecular layers coating the Au NRs. The researchers realized, for example, that displaying hydrophilic ligands (specifically carboxylic moieties) on the Au NR surfaces played a critical role in achieving macroscale organization, ascribed to affinity between the hydrophilic residues and rigid surface. The organized Au NR assembly constituted an excellent platform for SERS-based sensing. SERS is highly dependent on the geometry and spatial organization of Au NPs (specifically nanorods). In particular, SERS amplification has been closely linked to NP periodicity and gaps between the plasmonic NPs. The researchers surmised that particle separation in the Au NR assembly shown in Figure 8.15 corre-

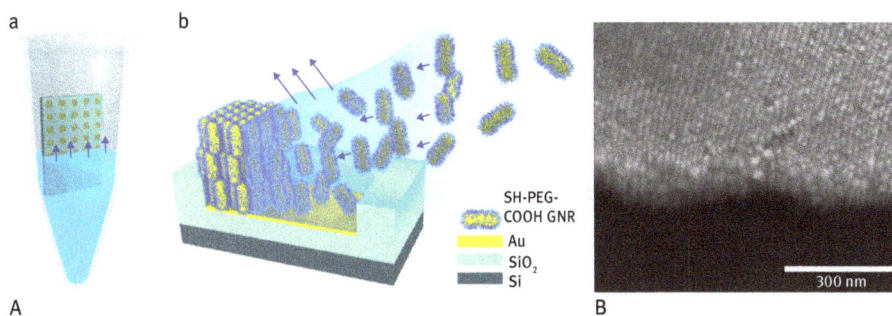

Fig. 8.15: Vertical alignment of gold nanorods through evaporation-induced templating. **A:** Experimental strategy: a patterned surface is dipped into a solution containing Au nanorods coated with carboxylic acid residues. Following liftoff, the liquid slowly evaporates resulting in alignment of the nanorods. **B:** Electron microscopy image showing a close-up of the vertically aligned Au nanorods within the rigid template. Reprinted with permission from Thai et al., *Angew. Chem. Int. Ed.* **51** (2012), 8732–8735, © 2012 John Wiley & Sons.

sponded to an optimal distance for generating a strong SERS effect, referred to as a "hot spot" distance.

Organization of hydrophobic Au NPs in two-dimensional film systems at the air/water interface has recently been demonstrated (Fig. 8.16). The technology relies on dissolution of alkyl chain-coated Au NPs within two-dimensional templates of surfactant molecules placed on the water surface. The film template comprised of two surfactants: one surfactant forming condensed domains in the film, while the other surfactant adopted a fluid phase in which the Au NPs were solubilized. Gradual compression of the tri-component film (by reducing the film area between two moving barriers) resulted in accumulation of the Au NPs between the condensed surfactant domains, ultimately producing extended wires comprising Au NPs. Important from

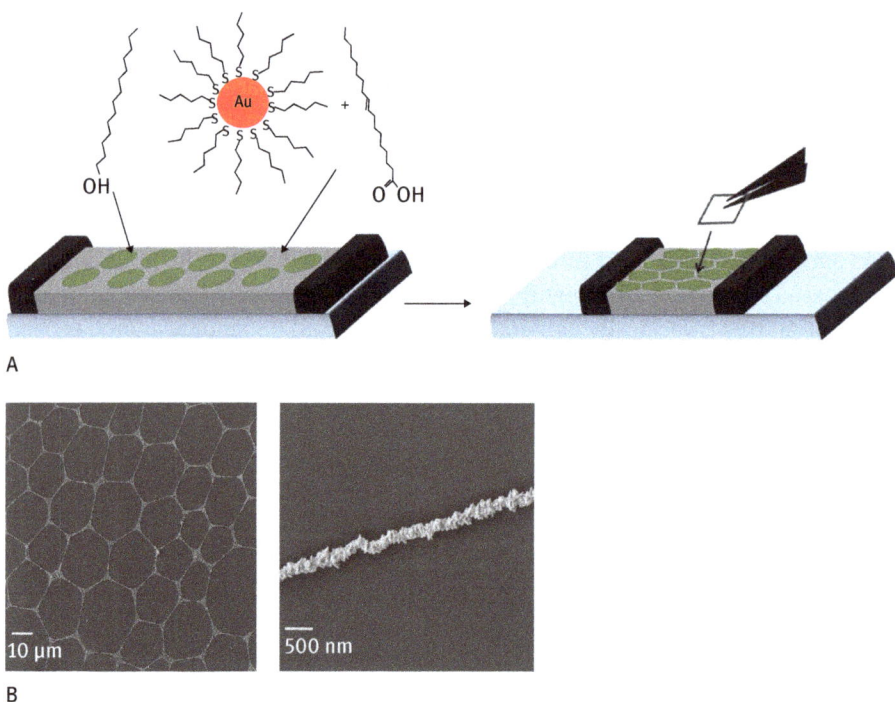

Fig. 8.16: Gold nanoparticle assemblies in surfactant template films. **A:** Schematic depiction of the technology: the surfactant template film, deposited on water, comprises of two surfactants: alkyl hydroxide forming condensed domains (darker circles), and a fatty acid adopting a fluid phase (lighter green background). Hydrophobically-coated Au nanoparticles added to the film are solubilized only in the fluid surfactant phase. Following film compression (by reducing the surface area between the two barriers), the Au nanoparticles are confined between the condensed film domains, ultimately forming a two-dimensional nanowire network. **B:** Electron microscopy images showing the Au nanoparticle assembly following film compression and transfer onto a solid substrate. Reprinted from Morag et al., *Adv. Mater.* **23** (2011), 4327–4331.

a practical standpoint was the observation that the Au NP wires could be further annealed, yielding continuous Au nanowires which were electrically conductive.

8.2 Assemblies of other metal nanoparticles

Organized assemblies of NPs comprising metals other than gold have also been reported and employed in various applications. Similar to the Au NP aggregates discussed above, such systems often exploit the plasmon shifts induced when metallic NPs are in close proximity. Silver NPs are among the most studied NPs in this context, based in large part upon the close likeness of silver plasmonic properties to those of Au NPs. Ag NPs also resemble Au NPs in the chemical routes available for functionalization and self-assembly, and similar to Au NP assemblies, macroscale organization of Ag NPs has often been achieved by chemical modifications of the nanoparticle surfaces.

A surface functionalization process designed to induce silver NP assembly into "microleaves" is portrayed in Figure 8.17. The experiment, carried out by Q. Qang and colleagues at the Chinese Academy of Sciences, was based on using a phenolic compound, p-aminothiophenol (PATP), as chemical linker and structural mediator of the micrometer-scale leaves. PATP was covalently bound to the surface of one Ag NP by the thiol moiety, while another NP was captured via the amine residue at the opposite position of the phenol ring (Fig. 8.17B). Intriguingly, interactions between the phenol rings (referred to as "$\pi-\pi$ stacking") produced highly uniform "microleaves" comprising lamellar organization of the PATP-bound NPs.

Another seemingly simple route for creating Ag NP assemblies is based on the "coffee-ring" effect – the familiar "ring" left behind after a drop of coffee (or other solute/solvent mixture) evaporates. This evaporation-based technique relies on the propensity of dissolved NPs to aggregate at the edge of the receding evaporating liquid via capillary and inter-particle forces. The coffee-ring effect has been utilized, for example, by S. Magdassi and colleagues at the Hebrew University, Israel, to create electrically-conductive patterns, attained by exploiting the physical contact between adjacent NPs following solvent evaporation (Fig. 8.18).

Template-directed assembly has been a useful strategy for achieving macroscale ordering of Ag NPs. Similar to Au NP arrays produced using this approach, the technique relies on physical matrixes serving as external scaffolding for organization of the NPs. Figure 8.19 presents an interesting example of Ag NPs which were actually grown inside a porous template. The template-based strategy developed by A. P. H. J. Schenning and colleagues at the Technical University of Eindhoven, Netherlands, comprised of a liquid crystalline polymer forming an oriented porous network. Ag^+ ions were then embedded between the liquid crystalline sheaths and subsequently reduced, forming aligned Ag NP "wires".

A

AgNPs   AgNPs   AgNLs

B   C

D

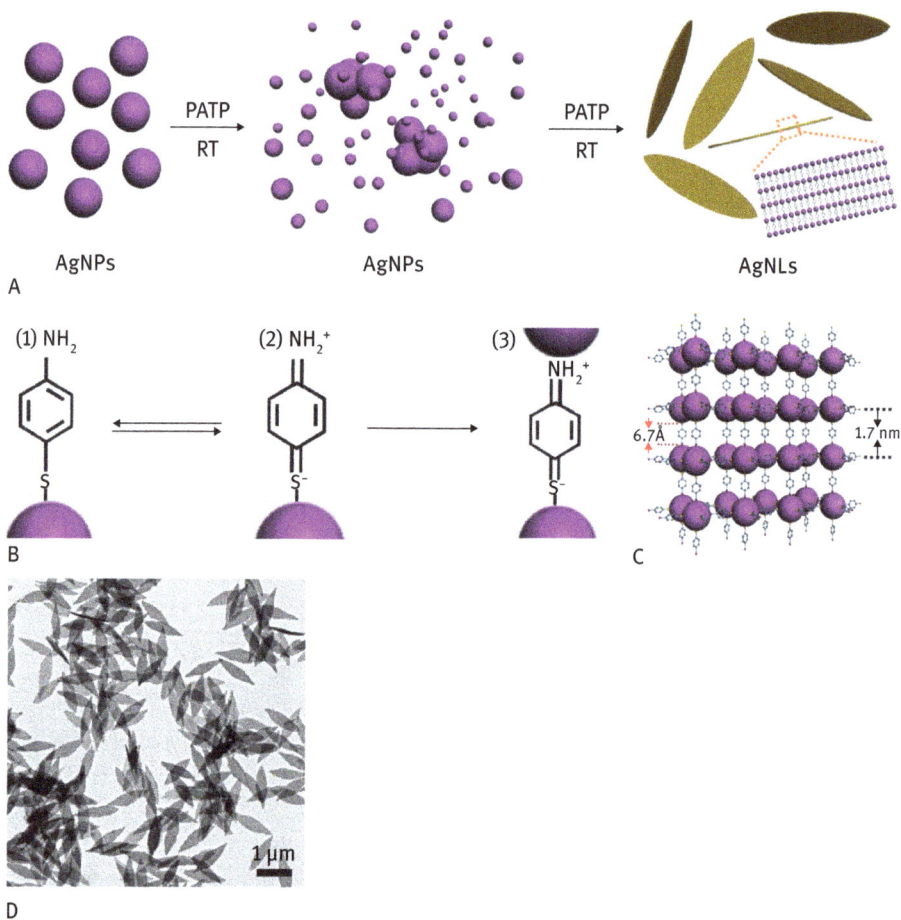

Fig. 8.17: "Microleaves" from self-assembled silver nanoparticles. **A:** Schematic illustration of the assembly process, mediated by the linker molecule PATP. **B:** Linking two adjacent Ag nanoparticles by the thiol and amine residues of PATP. **C:** Ordered lamellar structure of the Ag nanoparticles induced by PATP. **D:** Electron microscopy image of the Ag nanoparticle microleaves. Reprinted with permission from Li and Wang, *ACS Nano.* **7** (2013), 3053–3060, © 2013 American Chemical Society.

Figure 8.20 depicts fabrication of ordered core-shell *aluminum@aluminum-oxide NP* arrays using patterned surfaces as a physical template. Organized aluminum NP architectures have potential practical applications since Al NPs exhibit intense localized surface plasmon resonance (LSPR) in the deep ultraviolet range, which can be exploited in diverse applications such as sensing, optical devices, antennas, and others. Aluminum LSPR, however, is highly sensitive to geometrical parameters, particularly particle size and spatial organization of the NPs, and even slight variations in these parameters result in resonance shifts and spectral features which are too broad. The experimental strategy developed by F. Bisio and colleagues at CNR-SPIN, Italy,

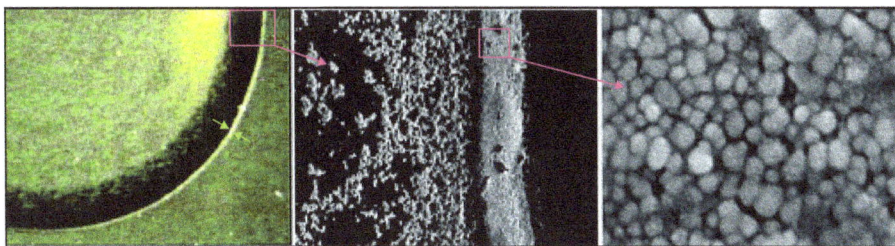

Fig. 8.18: The "coffee-ring" effect. Silver nanoparticles aggregate at the edge of an evaporating solution. The physical contact between the nanoparticles enables electrical conductivity. Sizes of individual silver nanoparticles on the right image were approximately 20 nm. Reprinted with permission from Magdassi et al., *Langmuir.* **21** (2005), 10264–10267, © 2005 American Chemical Society.

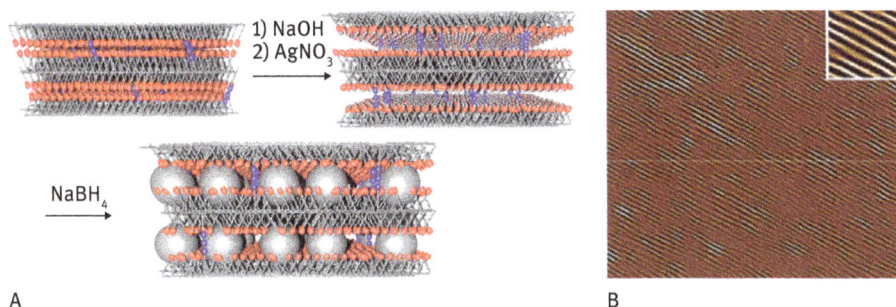

A   B

Fig. 8.19: Aligned silver nanoparticle assemblies via a liquid-crystal template. **A:** Preparation scheme: Ag^+ ions are inserted into the inter-sheath spaces within a pre-synthesized liquid crystalline polymer network. Subsequent reduction by $NaBH_4$ generates Ag nanoparticles, aligned within the liquid crystalline framework. **B:** Atomic force microscopy (AFM) image showing the aligned Ag nanoparticle assemblies. Image size is 500 nm x 500 nm. Reprinted with permission from Dasgupta et al., *JACS*, **135** (2013), 10922–10925, © 2013 American Chemical Society.

addressed these limitations via direct, on-site nanoparticle synthesis on a patterned template. As depicted in Figure 8.20, a thin Al film was initially deposited on one edge of each groove within the corrugated surface. Subsequent de-wetting and oxidation of the Al film resulted in formation of aligned arrays of $Al@Al_2O_3$ core-shell NPs. Importantly, this NP organization gave rise to sharp and intense deep ultra-violet LSPR, which was ascribed to the ordered aluminum metal cores.

In addition to the use of artificial and inorganic templates to create metallic NP arrays, biological macromolecules and even whole microorganisms have been employed as scaffolds. Biomolecules contain various chemical groups, such as amino acids, which could function as docking sites for metal NPs, and as such provide programmable surfaces for the construction of complex architectures. Figure 8.21, for example, depicts the use of a rod-like virus as a template for deposition of Ag and Au NPs. The *tobacco mosaic virus (TMV)*, a thoroughly studied virus having cylindrical

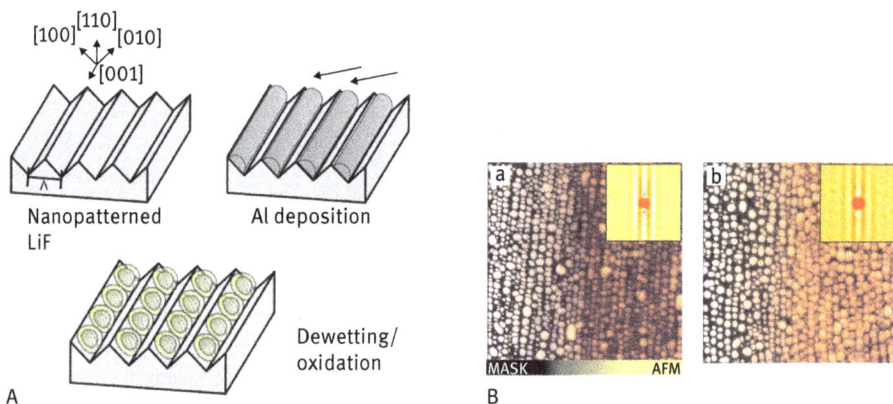

Fig. 8.20: Aligned aluminum nanoparticles via use of a corrugated surface template. **A:** Thin aluminum film is initially deposited in one direction of the corrugated surface, producing aluminum@aluminum oxide nanoparticles following de-wetting and oxidation. **B:** Atomic force microscopy (AFM) images depicting the aligned nanoparticles produced (a) via deposition of 1.7 nm-thick aluminum film, and (b) 2.5 nm film. Size of images is 200 nm × 200 nm. Reprinted with permission from Maidecchi et al., *ACS Nano.* **7** (2013), 5834–5841, © 2013 American Chemical Society.

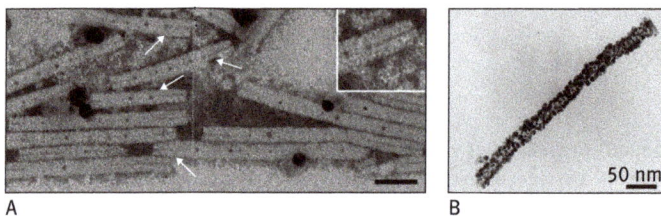

Fig. 8.21: Metal nanoparticle assemblies grown outside and inside a virus template. **A:** Silver nanoparticles, and **B:** Gold nanoparticles grown using tobacco mosaic virus (TMV) as a template. Nanoparticle growth relies on attachment of the nanoparticle precursors on the viral surfaces and subsequent reduction. The positively-charged external surface attracts $AuCl_4^-$ and the negatively-charged inner surface induces Ag^+ docking. The electron microscopy images show the resultant nanoparticle assemblies. Scale bars correspond to 50 nm. Reprinted with permission from Dujardin et al., *Nano Lett.* **3** (2003), 413–417, © 2003 American Chemical Society.

rod morphology, was employed as the biological template by S. Mann and colleagues at the University of Bristol, UK. Intriguingly, the researchers utilized the virus template to grow NPs both inside the hollow rod-shape viral particles as well as on the external surface of the virus. This feat was possible because of the different electrostatic charges at the two surfaces. Specifically, the outer surface of the virus contains high concentration of lysines and arginines which are positively-charged amino acids. The positive surface attracted negative metal NP precursors (such as $AuCl_4^-$) which were consequently reduced, resulting in growth of surface-anchored Au NPs. In comparison, the inner surface of the cylindrical viral rod is rich in negative amino acids such as

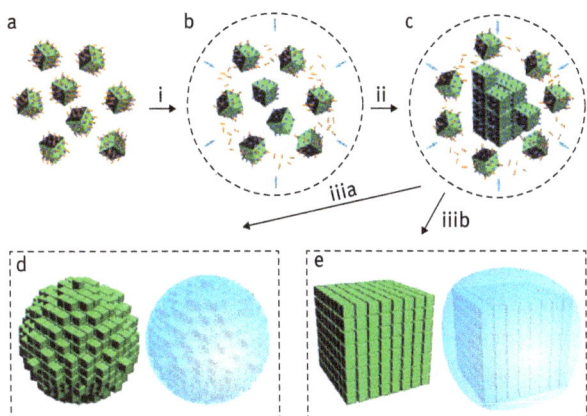

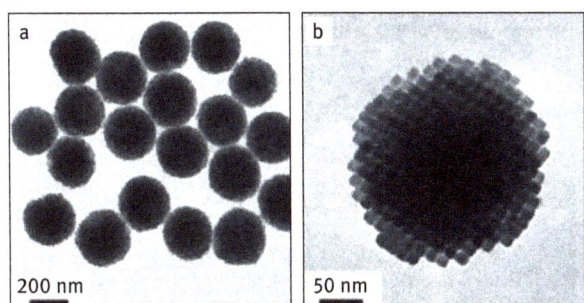

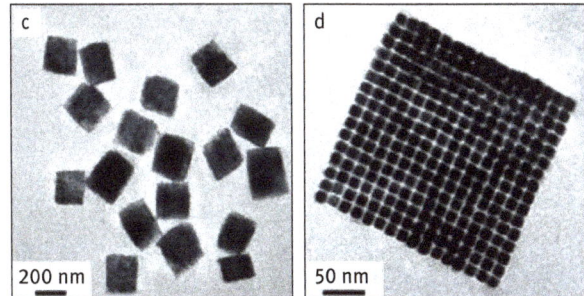

Fig. 8.22: Iron oxide nanocube supercrystals. **A:** Construction of differently-shaped colloidal super-crystals from iron oxide nanocube building blocks. Assembly of the nanocubes is achieved via addition of surfactant molecules acting as linkers. **B:** Electron microscopy images depicting the spherical and cubic supercrystals. Reprinted with permission from Wang et al., *JACS.* **134** (2012), 18225–18228, © 2012 American Chemical Society.

glutamic acid and aspartic acid, which promoted adsorption and consequent reduction of positive metal NP precursors such as Ag^+, giving rise to formation of Ag NPs which were embedded inside the viral particle.

While templates are prominent vehicles for attainment of long-range ordering of metal NPs, NP "superstructures" could even be created without such template agents. Figure 8.22 portrays a system in which *iron oxide nanocubes* were used as building blocks for remarkable ordered "colloidal crystals". Construction of highly symmetrical spherical and cubic supercrystals from the nanocubes was accomplished by Y. C. Cao and colleagues at the University of Florida via addition of surfactant molecules to the nanocube solutions. Affinity between the hydrocarbon chains of the added surfactants and the amphiphilic coating of the nanocubes led to a long-range "crystalline" organization, achieved through careful tuning of the experimental parameters, particularly the type of surfactants and their concentration. The significance of colloidal iron oxide assemblies such as those shown in Figure 8.22 lies not only in the extraordinary feat of complex structures constructed by hydrophobic interactions, but also in the potential applications – for example in modulation of NP superstructures through changing the shapes of the nanoparticle building blocks.

While almost all systems discussed above feature symmetrical organizations of NPs, in some instances nonsymmetrical entities have been sought. Figure 8.23 illustrates an elegant method for preparing "Janus aggregates" comprising magnetic iron oxide NPs (MNPs) in one "pole". The technique, developed by X. Gao and colleagues at the University of Washington, relies on phase separation between two polymeric species exhibiting different hydrophobicity. As shown in Figure 8.23, the two poly-

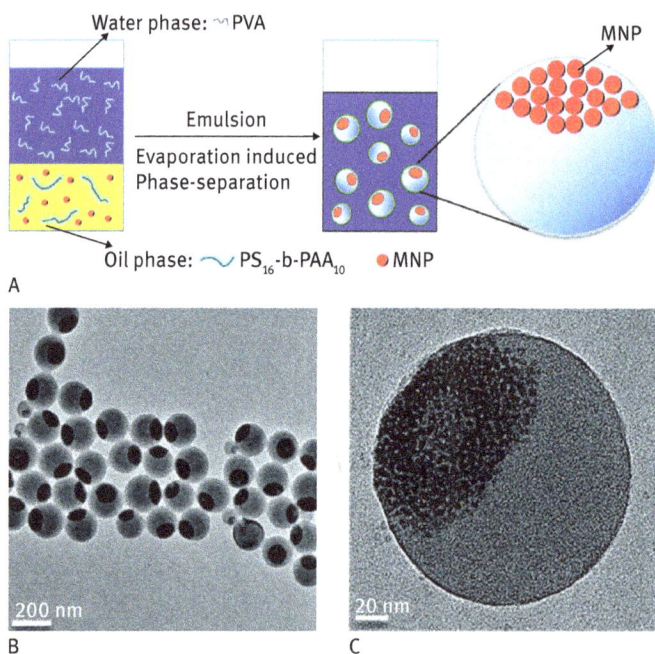

Fig. 8.23: Janus microparticles displaying nonsymmetrical distribution of magnetic nanoparticles. **A:** Microparticle preparation is carried out via initial dissolution of two polymers in immiscible liquids. The iron oxide nanoparticles (e.g. magnetic nanoparticles, MNPs) are also dissolved in the nonaqueous solution. The polymers form phase-separated microparticles following emulsification and evaporation; the MNPs are specifically localized in the hydrophobic polymer domains within the microparticles. **B,C:** Electron microscopy images depicting the Janus microparticles and the MNPs accumulated in one area of the particle. Reprinted with permission from Hu and Gao, *JACS.* **132** (2010), 7234–7237, © 2010 American Chemical Society.

mers and the MNPs were initially present in two phases, aqueous and organic, respectively. Condensed microparticles were subsequently formed following emulsification (dispersion of the oil phase within the water phase) and evaporation of the organic solvent. Crucially, however, the two polymers were still phase-separated within each microparticle. Since the MNPs were hydrophobic (due to the nature of the protective layer on the particle surface), they favored association with the more hydrophobic polymer, and consequently accumulated on one side of the Janus microparticle. The macroscopic magnetic moment of the Janus particles enabled their response to and manipulation by external magnetic fields, particularly useful in biological imaging applications.

8.3 Semiconductor nanoparticle assemblies

In many respects, semiconductor NPs such as quantum dots (QDs) and quantum rods resemble metallic NPs, as they give rise to interesting physical properties when organized in macroscale assemblies. This is particularly reflected in modulation of the optical and electronic properties of semiconductor NPs, which depend on the size of NPs and the coupling between them. Similar to the metal NP assemblies discussed above, one main avenue for creation of ordered semiconductor NP systems is template-directed methods. Such approaches sometimes combine "top-down" lithography to prepare the (surface) templates with "bottom-up" self-assembly of the semiconducting NPs within the templates. Like the metal NP studies discussed above, this field largely focuses on harnessing noncovalent forces between NPs to achieve long range ordering.

An example of a template-directed quantum rod assembly is presented in Figure 8.24. In the experiment, Y. C. Cao and colleagues at the University of Florida first synthesized needle-like structures comprising core-shell CdSe@CdS nanorods. The semiconducting nanoneedles (or nanorods) were deposited within aligned channels via capillary forces on a patterned surface and were subsequently immobilized in a polymer film (made of polydimethylsiloxane, PDMS, a polymer commonly used for creating three-dimensional molds). Subsequent removal of the patterned surface produced "free-standing films" comprising the aligned nanorods. Remarkably, the films exhibited enhanced photoluminescence which was linearly polarized due to the macroscale alignment of the semiconductor core-shell NRs (Fig. 8.24B).

While the use of external templates has been a rather popular strategy for creating semiconductor NP assemblies, methods relying solely on self-assembly phenomena rather than physical scaffolds have also been introduced to produce superstructures. Several studies reported on the use of quantum dots as building blocks for "crystalline" structures, in which the NPs occupied lattice sites – conceptually similar to ions in regular crystals. An elegant demonstration of this concept is provided in Figure 8.25 illustrating self-assembly of cubic PbS nanocrystals into ordered lattices. In-

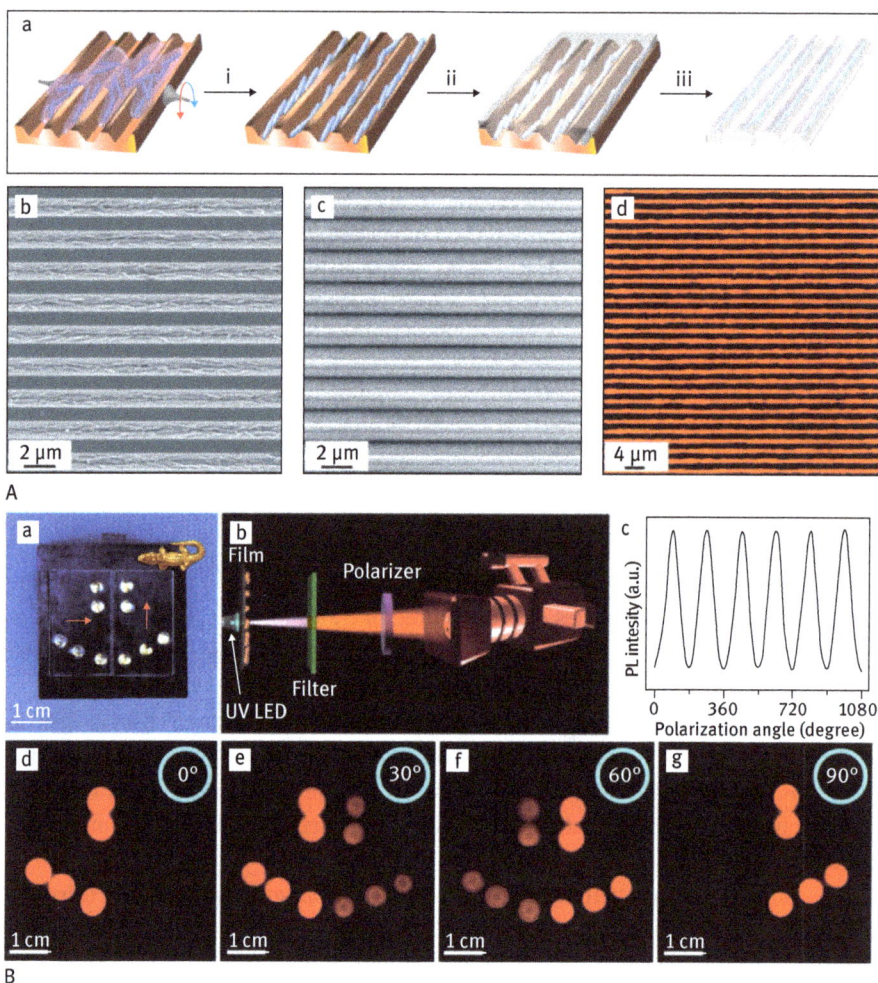

Fig. 8.24: Aligned semiconductor nanorod assemblies. A: Assembly procedure: (a,i) CdSe@CdS core-shell nanorods are deposited on a surface patterned with aligned channels. (ii) A polymer mold is further placed on the surface, and (iii) a "free-standing" film containing the aligned nanorods is obtained following removal of the solid substrate. (b-c) Electron microscopy and (d) optical microscopy images depicting the aligned nanorod film. B: Light polarization accomplished by using linearly-polarized light for photoexcitation of the CdSe@CdS nanorod films. a. Photograph of the experimental setup showing ten ultraviolet light emitting diodes (LEDs) covered by the nanorod films which are placed in perpendicular orientations (indicated by the red arrows); b. Scheme of the experimental apparatus. c. Periodic changes in photoluminescence intensity depending on the angle of incident polarized light. The graph confirms that light emission depends on macroscopic nanorod alignment within the films. d-g: Polarization angle-dependent photoluminescence confirming that nanorod alignment determines the photoluminescent light intensity. Reprinted with permission from Wang et al., *JACS* **135** (2013), 6022–6025, © 2013 American Chemical Society.

triguingly, J. Fang and colleagues at the State University of New York, Binghamton, demonstrated that NP ordering in crystalline lattice could be accomplished spontaneously via careful optimization of nanocube surfactant capping and conditions of the reaction solution. Furthermore, the choice of nanocubes as building blocks (rather than spherical NPs) enabled an unusual tilted packing of the nanoparticles rather than the conventional face-centered cubic (fcc) organization (shown in Fig. 8.22). Notably, the tilted cubic organization likely accounted for the exceptional stability of the self-assembled PbS nanocube superstructures. The experiment outlined in Figure 8.25

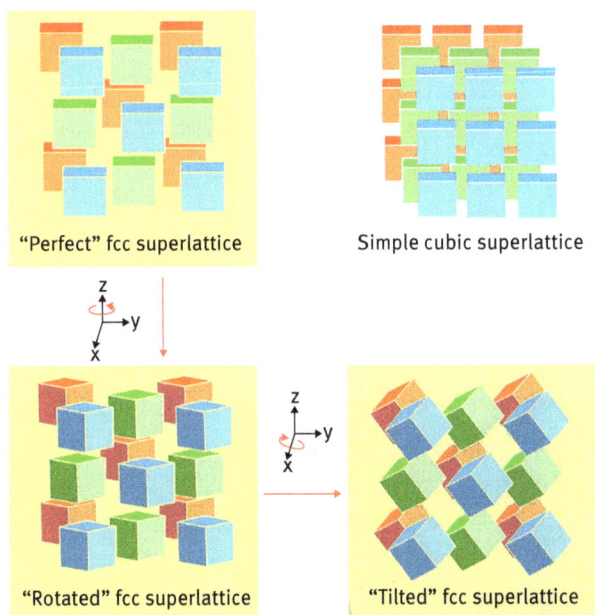

"Perfect" fcc superlattice   Simple cubic superlattice

"Rotated" fcc superlattice   "Tilted" fcc superlattice

A

100 nm

B

Fig. 8.25: PbS nanocube "superlattice". **A:** Diagram showing the formation of a "tilted" cubic lattice of PbS nanocubes. **B:** Electron microscopy image depicting the nanocube superlattice. Reprinted with permission from Quan et al., *Nano Lett.* **12** (2012), 4409–4413, © 2012 American Chemical Society.

is a fine example of the emerging "new chemistry" which uses nanoparticles as the building blocks of new structural materials, in the same way as atoms and molecules have been used to create new materials and functionalities.

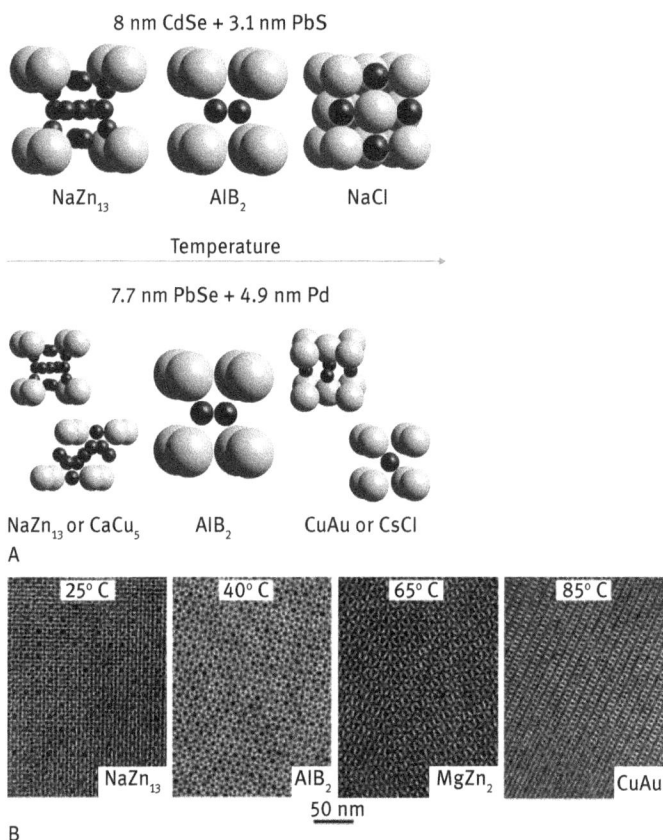

Fig. 8.26: Temperature-controlled superlattice organization of binary semiconductor nanoparticle mixtures. **A:** Different lattice unit cells observed in binary PbS/CdSe and PbSe/Pd nanoparticle mixtures self-assembling in different temperatures. **B:** Electron microscopy images of superlattices formed in a PbS/CdSe nanoparticle mixture at different temperatures; identical crystal organizations of known compounds are indicated in the bottom right of the images. Reprinted with permission from Bodnarchuk et al., *JACS.* **132** (2010), 11967–11977, © 2010 American Chemical Society.

Other strategies have been reported for construction of crystalline phases of semiconductor NP superlattices. Figure 8.26 summarizes experimental data demonstrating that temperature modulation could yield distinct crystalline organizations in a binary mixture of semiconductor NPs. The experimental design by D. V. Talapin and colleagues at the University of Chicago focused on a mixture of CdSe and PbS NPs. The researchers generated highly symmetrical assemblies by drying the NPs from an organic solvent. Analogous to crystallization processes involving ions and molecules, spatial ordering of the NPs was largely determined by the forces exerted between the particles, which in turn affected the system's *enthalpy* and *entropy* – the fundamental thermodynamic parameters responsible for the ultimate structural outcome. In this context, since temperature is an intrinsic physical property associated with enthalpy and entropy, it had a profound effect on the lattice organization of the binary super-lattices.

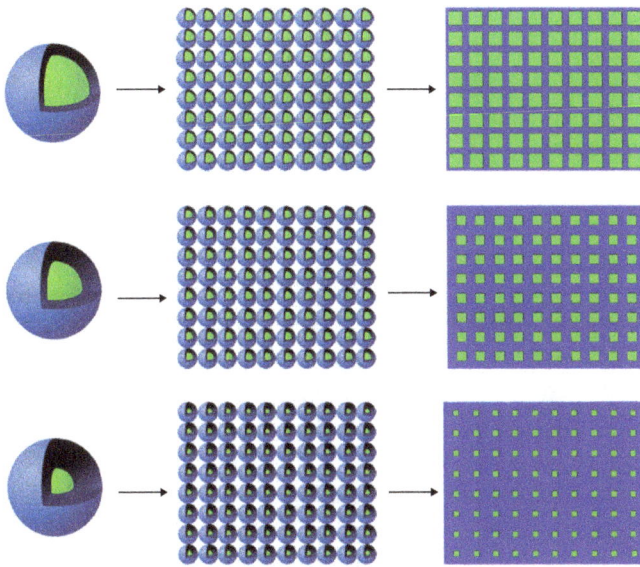

Fig. 8.27: Composite films produced by fusion of core-shell nanoparticles. Diagram showing continuous composite films generated by annealing PbTe@PbS core-shell nanoparticles. Reprinted with permission from Ibanez et al., *ACS Nano.* **7** (2013), 2573–2586, © 2013 American Chemical Society.

While NP assemblies have usually been studied "as prepared" – i.e. the NPs retaining their shapes and chemical identity – in some cases fusion of the NPs has been carried out, yielding composite materials exhibiting interesting functions. The properties of such substances, generally produced via *annealing* of the NP assemblies, depend on the individual NP building blocks. Figure 8.27 provides an example of this approach, showing synthesis of continuous semiconductor films through fusion of ordered core-shell PbTe@PbS NPs. In the experiment, carried out by A. Cabot and colleagues at the Universitat de Barcelona, Spain, the nanocomposite films were prepared by deposition of the core-shell NPs on a solid substrate, drying of the organic solvent, and high-temperature annealing. The films were found to exhibit enhanced electrical conductivity and pronounced *thermoelectricity* (electrical conductance induced by the temperature difference between two contacts), which were ascribed to synergy between the charge carriers (electrons or holes) generated within the two semiconductor domains (the PbTe cores and the fused PbS shells). Notably, the physical properties of the PbTe/PbS nanocomposite films could be tuned by controlling nanoparticle dimensionalities, specifically the diameters of the core and shell.

8.4 Multicomponent nanoparticle assemblies

Nanoparticles comprising more than a single class of materials (for example metal/ semiconductor NPs) have attracted interest in light of their synergistic and complementary properties (see Chap. 6 for a detailed discussion). In analogy, organized assemblies of *different types* of NPs often exhibit unique properties, and such systems have opened new research avenues. Arrays of hybrid NPs have also been prepared, giving rise in some instances to cooperative properties and functionalities different to those of the individual NPs.

Figure 8.28 depicts a mixed NP assembly integrating silicon NPs and quantum dots (QDs), and its application in bioimaging. The composite NP assembly designed by Y. S. Lee and colleagues at Seoul National University, Korea, combined a SiO_2 NP core, an array of QDs attached to the SiO_2 surface, and an additional silica layer coating the entire SiO_2@QD NP construct. Synthesis of the silica@QD@silica NP assembly was carried out in consecutive steps, first preparing the SiO_2 NP core and the semiconductor NPs separately, followed by surface chemical functionalization of the SiO_2 NPs to enable covalent binding of the QDs, and finally coating the SiO_2@QD particle in a thin SiO_2 monolayer. This multicomponent architecture played an important role in shaping the functionalities of the NP assembly. The SiO_2 NP core contributed to immobilization and stabilization of the QDs via surface attachment, thereby facilitating their use in biological applications. The QD array provided the optical reporting mechanism; indeed, the photoluminescence recorded using the SiO_2@QD@ SiO_2 particles was twice as intense compared to the individual QDs, underscoring the intrinsically high sensitivity facilitated by creating the QD array (Fig. 8.28C). Moreover, the external SiO_2 coating endowed biocompatibility to the entire assembly, reducing the risk of cytotoxicity often associated with inorganic QDs.

In most cases, assemblies of NPs belonging to different classes of materials still combine particles of the same shapes (for example the spherical QDs and silica NPs in Fig. 8.28). While such arrangements are conceptually and practically more feasible, there have been reports of organized arrays comprising NPs of different morphologies. These systems, however, are generally more challenging to construct, since different-shaped nanoparticles often assemble in distinct (separate) phases rather than in interspersed mixed assemblies. Accordingly, strategies devised to modulate inter-particle attraction/repulsion have been critical for assembling mixed-shape hybrid NP assemblies. Figure 8.29 illustrates an organized co-assembly of mixed shape/composition NPs comprising semiconductor nanorods and spherical Au NPs. The self-assembly

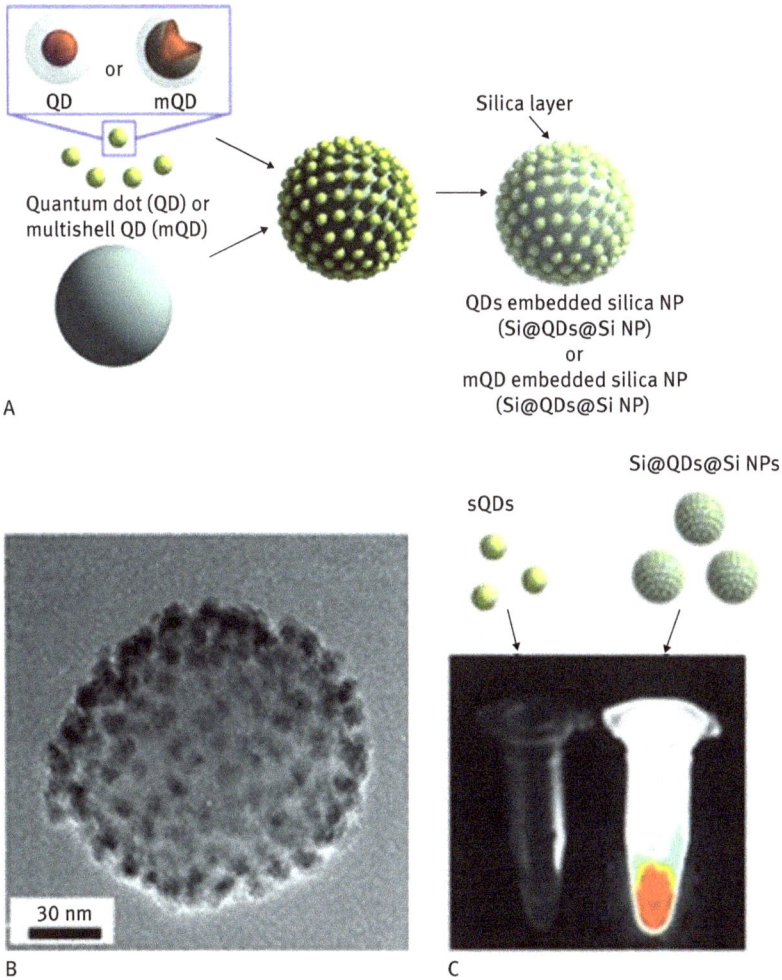

A

B

C

Fig. 8.28: Multicomponent core-shell nanoparticle assembly. **A:** Construction of the silica@quantum-dot (QD)@silica nanoparticles. **B:** Electron microscopy image of the composite nanoparticle. **C:** Photograph showing the enhanced luminescence of the QD/silica assembly as compared to individual quantum dots. Reprinted with permission from Jun et al., *Adv. Funct. Mater.* **22** (2012), 1843–1849, © 2012 John Wiley & Sons.

mechanism, developed by Y. C. Cao and colleagues at the University of Florida, relied on co-addition of a surfactant additive which attracted the coating layers of both the nanorods and the spherical nanoparticles.

A

B

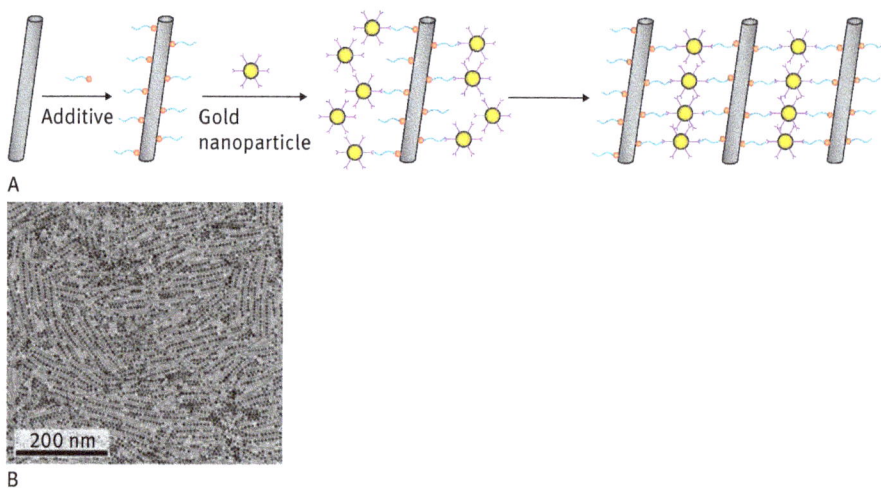

Fig. 8.29: Binary assembly of semiconductor nanorods and spherical gold nanoparticles. **A:** The synthetic route utilizing co-addition of nanorods, nanoparticles and a surfactant additive mediating the interactions between nanoparticles and nanorods y binding the coating layers of both species. **B:** Electron microscopy image showing the nanorod/nanoparticle assembly. Reprinted with permission from Nagaoka et al., *Small.* 2012, 8, 843–846, copyright (2012) John Wiley & Sons.

Further reading

Chapter 2

H. Li et al., One-Dimensional CdS Nanostructures: A Promising Candidate for Optoelectronics, *Adv. Mater.* **25** (2013), 3017–3037.

A. Sitt et al., Band-gap engineering, optoelectronic properties and applications of colloidal heterostructured semiconductor nanorods, *Nano Today*, **8** (2013), 494–513.

C.M. Lieber, Nanoscale Science and Technology: Building a Big Future from Small Things, *MRS Bulletin*, July 2003, 486–491.

I.J. Kramer and E.H. Sargent, The Architecture of Colloidal Quantum Dot Solar Cells: Materials to Devices, *Chem. Rev.* **114**(1) (2014), 863–882.

I.L. Medintz et al., Quantum dot bioconjugates for imaging, labelling and sensing *Nature Materials*. **4** (2005), 435–446.

A.K. Viswanath, From clusters to semiconductor nanostructures. *J Nanosci Nanotechnol.* **14(2)** (2014), 1253–1281.

Chapter 3

K. Saha et al., Gold Nanoparticles in Chemical and Biological Sensing, *Chem. Rev.* **112** (2012), 2739–2779.

K.M. Mayer and J.H. Hafner, Localized Surface Plasmon Resonance Sensors, *Chem. Rev.* **111** (2011), 3828–3857.

R. Bardhan et al., Theranostic nanoshells: from probe design to imaging and treatment of cancer. *Acc Chem Res.* **44(10)** (2011), 936-946.

N. Li et al., Anisotropic gold nanoparticles: synthesis, properties, applications, and toxicity. *Angew Chem Int Ed Engl.* **53** (2014), 1756–1789.

K. Kim and K.S. Shin, Surface-enhanced Raman scattering: a powerful tool for chemical identification. *Anal Sci.* **27**(8) (2011), 775–783.

Y. Xia et al., Gold nanocages: from synthesis to theranostic applications., *Acc Chem Res.* **44** (2011), 914–924.

X. Lu et al., Chemical synthesis of novel plasmonic nanoparticles., *Annu Rev Phys Chem.* **60** (2009), 167–192.

Z. Quan et al., High-Index Faceted Noble Metal Nanocrystals, *Acc. Chem. Res.* **46** (2013), 191–202.

R.J. Chaudhuri and S. Paria, Core/Shell Nanoparticles: Classes, Properties, Synthesis Mechanisms, Characterization, and Applications, *Chem. Rev.* **112** (2012), 2373–2433.

M.T. Jones et al., Templated Techniques for the Synthesis and Assembly of Plasmonic Nanostructures, *Chem. Rev.* **111** (2011), 3736–3827.

Chapter 4

L.H. Reddy, Magnetic Nanoparticles: Design and Characterization, Toxicity and Biocompatibility, Pharmaceutical and Biomedical Applications, *Chem. Rev.* **112** (2012), 5818–5878.

C. Argyo et al., Multifunctional Mesoporous Silica Nanoparticles as a Universal Platform for Drug Delivery, *Chem. Mater.* **26** (2014), 435–451.
X. Chen and S.S. Mao, Titanium Dioxide Nanomaterials: Synthesis, Properties, Modifications, and Applications, *Chem. Rev.* **107** (2007), 2891–2959.
S. Das et al., Cerium oxide nanoparticles: applications and prospects in nanomedicine., *Nanomedicine (Lond).* **8** (2013), 1483–1508.

Chapter 5

G. Rosenman et al., Bioinspired peptide nanotubes: deposition technology, basic physics and nan-
otechnology Applications, *J. Pept. Sci.* **17** (2011), 75–87.
N. Tomczak et al., Polymer-coated quantum dots, *Nanoscale* **5** (2013), 12018-1232.
F. Canfarotta et al., Polymeric nanoparticles for optical sensing, *Biotechnol Adv.* **31** (2013), 1585–1599.

Chapter 6

L. Carbone and P.D. Cozzoli, Colloidal heterostructured nanocrystals: Synthesis and growth mecha-
nisms, *Nano Today* **5** (2010), 449–493.
C. Wang et al., Recent Progress in Syntheses and Applications of Dumbbell-like Nanoparticles, *Adv.
Mater.* **21** (2009), 3045–3052.
R. Costi et al., Colloidal Hybrid Nanostructures: A New Type of Functional Materials, *Angew. Chem. Int.
Ed.* **49** (2010), 4878–4897.

Chapter 7

K.E. Sapsford et al., Functionalizing Nanoparticles with Biological Molecules: Developing Chemistries
that Facilitate Nanotechnology, *Chem. Rev.* **113** (2013), 1904–2074.
F.M. Winnik and D. Maysinger, Quantum Dot Cytotoxicity and Ways To Reduce It, *Acct. Chem. Res.* **46**
(2013), 672–680.
M.A. Zoroddu et al., Toxicity of nanoparticles, Curr Med Chem. **21** (2014), 3837–3853.

Chapter 8

B. Pelaz et al., The State of Nanoparticle-Based Nanoscience and Biotechnology: Progress, Promises,
and Challenges, *ACS Nano.* **6** (2012), 8468–8483.
J.W. Liu et al., Macroscopic-Scale Assembled Nanowire Thin Films and Their Functionalities, *Chem.
Rev.* **112** (2012), 4770–4799.
Y. Yin and D. Yin, The chemistry of functional nanomaterials, *Chem Soc Rev.* **42** (2013), 2484–2487.

Index

www.ingramcontent.com/pod-product-compliance
Lightning Source LLC
Chambersburg PA
CBHW081056220326
41598CB00038B/7116